an Introduction to visualization, modeling, and graphics for Engineering Design

an Introduction to visualization, modeling, and graphics for Engineering Design

Dennis K. Lieu

Professor of Mechanical Engineering
University of California, Berkeley

Sheryl Sorby

Professor of Civil and Environmental Engineering
Michigan Technological University

Australia • Brazil • Japan • Korea • Mexico • Singapore • Spain • United Kingdom • United States

an Introduction to Visualization, Modeling, and Graphics for Engineering Design, 1st Edition
Dennis K. Lieu and Sheryl Sorby

Vice President, Technology and Trades ABU: David Garza

Director of Learning Solutions: Sandy Clark

Managing Editor: Larry Main

Senior Acquisitions Editor: James DeVoe

Development: Ohlinger Publishing Services

Marketing Director: Deborah S. Yarnell

Marketing Manager: Kevin Rivenburg

Marketing Specialist: Mark Pierro

Director of Production: Patty Stephan

Production Manager: Stacy Masucci

Content Project Manager: Michael Tubbert

Art Director: Bethany Casey

Editorial Assistant: Tom Best

Cover Image: Solid model courtesy of Hoyt USA
Background courtesy of Getty Images, Inc.

© 2009 Delmar, Cengage Learning

ALL RIGHTS RESERVED. No part of this work covered by the copyright herein may be reproduced, transmitted, stored or used in any form or by any means graphic, electronic, or mechanical, including but not limited to photocopying, recording, scanning, digitizing, taping, Web distribution, information networks, or information storage and retrieval systems, except as permitted under Section 107 or 108 of the 1976 United States Copyright Act, without the prior written permission of the publisher.

For product information and technology assistance, contact us at
Cengage Learning Customer & Sales Support, 1-800-354-9706

For permission to use material from this text or product, submit all requests online at **www.cengage.com/permissions** Further permissions questions can be emailed to **permissionrequest@cengage.com**

Library of Congress Control Number: 2007942953

ISBN-13: 978-1-4018-4251-2

ISBN-10: 1-4018-4251-8

Delmar
Executive Woods
5 Maxwell Drive, PO Box 8007,
Clifton Park, NY 12065-8007
USA

Cengage Learning is a leading provider of customized learning solutions with office locations around the globe, including Singapore, the United Kingdom, Australia, Mexico, Brazil, and Japan. Locate your local office at:

international.cengage.com/region

Cengage Learning products are represented in Canada by Nelson -Education, Ltd.

For your lifelong learning solutions, visit **delmar.cengage.com**

Visit our corporate website at **cengage.com**

Notice to the Reader
Publisher does not warrant or guarantee any of the products described herein or perform any independent analysis in connection with any of the product information contained herein. Publisher does not assume, and expressly disclaims, any obligation to obtain and include information other than that provided to it by the manufacturer. The reader is expressly warned to consider and adopt all safety precautions that might be indicated by the activities described herein and to avoid all potential hazards. By following the instructions -contained herein, the reader willingly assumes all risks in connection with such instructions. The publisher makes no representations or warranties of any kind, including but not limited to, the warranties of fitness for particular purpose or merchantability, nor are any such representations implied with respect to the material set forth herein, and the publisher takes no responsibility with respect to such material. The publisher shall not be liable for any special, consequential, or exemplary damages resulting, in whole or part, from the readers' use of, or reliance upon, this material.

Printed in the United States of America
1 2 3 4 5 6 7 11 10 09 08

brief contents

Visualization, Modeling, and Graphics for Engineering Design

These optional chapters—as well as all of the textbook chapters—are available electronically and for purchase at www.iChapters.com:

contents

preface

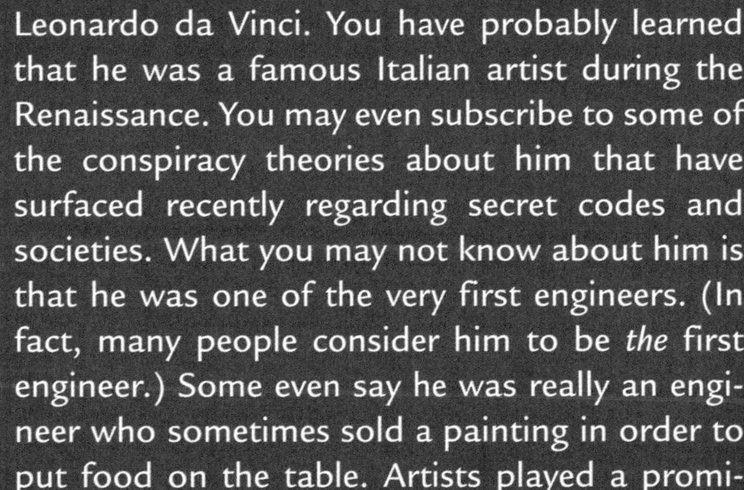

Leonardo da Vinci. You have probably learned that he was a famous Italian artist during the Renaissance. You may even subscribe to some of the conspiracy theories about him that have surfaced recently regarding secret codes and societies. What you may not know about him is that he was one of the very first engineers. (In fact, many people consider him to be *the* first engineer.) Some even say he was really an engineer who sometimes sold a painting in order to put food on the table. Artists played a prominent role at the birth of modern-day engineering, and some of the first artist-engineers included Francesco di Giorgio, Georg Agricola, and Mariano Taccola. These were the individuals who could visualize new devices that advanced the human condition. Their creativity and their willingness to try seemingly "crazy" ideas propelled technology forward at a much faster pace than had occurred in the previous thousand years.

This marriage between art and engineering has diminished somewhat since the early beginnings of the profession; however, creativity in engineering is still of paramount importance. Would the Apollo spaceship have landed on the moon without the creative thinking of hundreds of engineers who designed and tested the various systems necessary for space travel? Would we be able to instantaneously retrieve information and communicate with one another via the World Wide Web without the vision of the engineers and scientists who turned a crazy idea into reality? Would the modern-day devices that enrich and simplify our lives such as washing machines, televisions, telephones, and automobiles exist without the analytical skills of the engineers and technologists who developed and made successive improvements to these devices? The answer to all of these questions is "no." The ability to think of systems that never were and to design devices to meet the changing needs of the human population is the purview of the engineering profession.

Graphical communication has always played a central role in engineering, perhaps due to engineering's genesis within the arts or perhaps because graphical forms of communication convey design ideas more effectively than do written words. Maybe a picture really *is* worth a thousand words. As you might expect, the face of engineering graphics has evolved dramatically since the time of da Vinci. Traditional engineering graphics focused on 2-dimensional graphical mathematics, drawing, and design; knowledge of graphics was considered a key skill for engineers. Early engineering programs included graphics as an integral topic of instruction, and hand-drawn engineering graphics from 50 years ago are works of art in their own right.

However, in the recent past, the ability to create a 2-D engineering drawing by hand has become de-emphasized due to improvements and advances in computer hardware and software. More recently, as computer-based tools have advanced even further, the demand for skills in 3-D geometric modeling, assembly modeling, animation, and data management has defined a new engineering graphics curriculum. Moreover, three-dimensional geometric models have become the foundation for advanced numerical analysis methods, including kinematic analysis, kinetic analysis, and finite-element methods for stress, fluid, magnetic, and thermal systems.

The engineering graphics curriculum has also evolved over time to include a focus on developing 3-D spatial visualization ability since this particular skill has been documented as important to the success of engineers in the classroom and in the field. Spatial visualization is also strongly linked to the creative process. Would da Vinci have imagined his various flying machines without well-honed visualization skills?

We have come full circle in engineering education through the inclusion of topics such as creativity, teamwork, and design in the modern-day graphics curriculum. The strong link between creativity, design, and graphics cannot be overstated. Gone is the need for engineers and technicians who robotically reproduce drawings with little thought involved. With modern-day computational tools, we can devise creative solutions to problems without concern about whether a line should be lightly penciled in or drawn thickly and displayed prominently on the page.

It is for this new, back-to-the-future graphics curriculum that *An Introduction to Visualization, Modeling, and Graphics for Engineering Design* has been designed. This text is a mixture of traditional as well as modern-day topics, a mixture of analytical and creative thinking, a mixture of exacting drawing technique and freeform sketching. Enjoy.

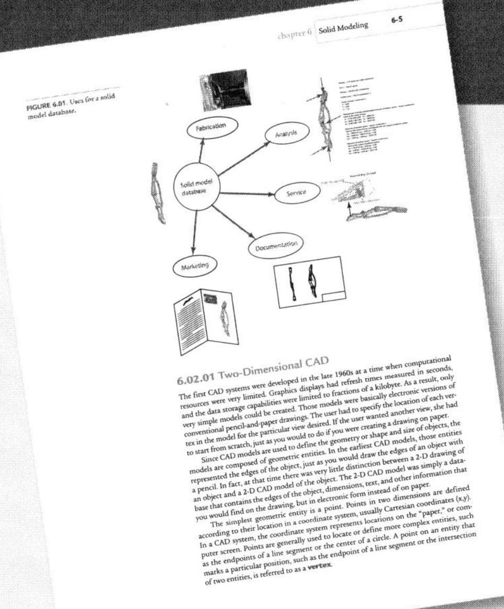

chapter 6 | Solid Modeling 6-5

FIGURE 6.01. Uses for a solid model database.

6.02.01 Two-Dimensional CAD

The first CAD systems were developed in the late 1960s at a time when computational resources were very limited. Graphics displays had refresh times measured in seconds, and the data storage capabilities were limited to fractions of a kilobyte. As a result, only very simple models could be created. Those models were basically electronic versions of conventional pencil-and-paper drawings. The user had to specify the location of each vertex in the model for the particular view desired. If the user wanted another view, she had to start from scratch, just as you would to do if you were creating a drawing on paper.

Since CAD models are used to define the geometry or shape and size of objects, the models are composed of geometric entities. In the earliest CAD models, those entities represented the edges of the object, just as you would draw the edges of an object with a pencil. In fact, at that time there was very little distinction between a 2-D drawing of an object and a 2-D CAD model of the object. The 2-D CAD model was simply a database that contains the edges of the object, dimensions, text, and other information that you would find on the drawing, but in electronic form instead of on paper.

The simplest geometric entity is a point. Points in two dimensions are defined according to their location in a coordinate system, usually Cartesian coordinates (x,y). In a CAD system, the coordinate system represents locations on the "paper," or computer screen. Points are generally used to locate or define more complex entities, such as the endpoints of a line segment or the center of a circle. A point on an entity that marks a particular position, such as the endpoint of a line segment or the intersection of two entities, is referred to as a **vertex**.

Development of the Text

Many of the current graphics textbooks were written several years ago with modern-day topics such as feature-based solid modeling included as a separate add-on to the existing material. In these texts, modern-day computer-based techniques are more of an afterthought: "Oh, by the way, you can also use the computer to help you accomplish some of these common tasks." In fact, some of the more popular texts were written nearly a century ago when computer workstations and CAD software were figments of some forward-looking engineer's imagination. Texts from that era focused on drawing technique and not on graphic communication within the larger context of engineering design and creativity.

Modern engineering graphics curricula—and texts—must follow what is happening in the field. Modern product development techniques allow engineers to use computer hardware and software to examine the proper fit and function of a device. Engineers can "virtually" develop and test a device before producing an actual physical model, which greatly increases the speed and efficiency of the design process. The virtual, computer-based model then facilitates the creation of the engineering drawings used in manufacturing and production—an activity that required many hours of hand drafting just a few short decades ago.

In the real world, modern CAD practices have also allowed us more time to focus on other important aspects of the engineering design cycle, including creative thinking, product ideation, and advanced analysis techniques. Some might argue that these aspects are, in fact, the *most* important aspects of the design process. The engineering graphics curricula at many colleges and universities have evolved to reflect this shift in the design process. However, most engineering graphics textbooks have simply added CAD sections to cover the new topics. Thick textbooks have gotten even thicker. As a wise person has said, "Engineering faculty are really good at addition, but are miserable at subtraction."

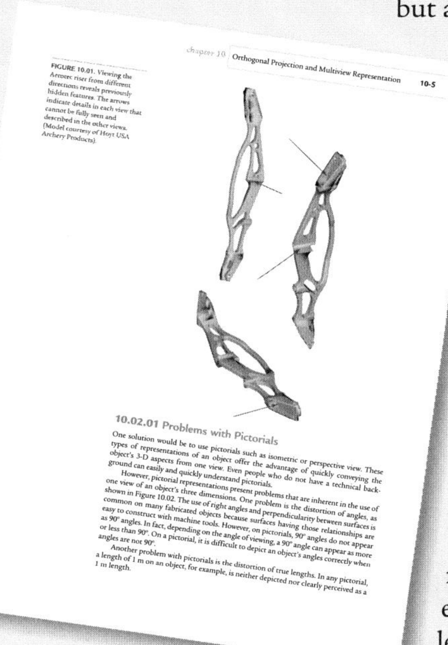

chapter 10 | Orthogonal Projection and Multiview Representation 10-5

FIGURE 10.01. Viewing the Aerotec riser from different directions reveals previously hidden features. The arrows indicate details in each view that cannot be fully seen and described in the other views. (Model courtesy of Hoyt USA Archery Products).

10.02.01 Problems with Pictorials

One solution would be to use pictorials such as isometric or perspective view. These types of representations of an object offer the advantage of quickly conveying the object's 3-D aspects from one view. Even people who do not have a technical background can easily and quickly understand pictorials.

However, pictorial representations present problems that are inherent in the use of one view of an object's three dimensions. One problem is the distortion of angles, as shown in Figure 10.02. The use of right angles and perpendicularity between surfaces is common on many fabricated objects because surfaces having those relationships are easy to construct with machine tools. However, on pictorials, 90° angles do not appear as 90° angles. In fact, depending on the angle of viewing, a 90° angle can appear as more or less than 90°. On a pictorial, it is difficult to depict an object's angles correctly when angles are not 90°.

Another problem with pictorials is the distortion of true lengths. In any pictorial, a length of 1 m on an object, for example, is neither depicted nor clearly perceived as a 1 m length.

When we sat down to plan this text, we wanted to produce the engineering graphics benchmark of the future—an engineering graphics approach that teaches design and design communication rather than a vocational text focused on drafting techniques and standards. We wanted to integrate modern-day design techniques throughout the text, not treat these topics as an afterthought.

A strength of this textbook is its focus not only on "what" to do, but also on "why" you do it (or do not do it) that way—concepts as well as details. This text is intended to be a learning aid as well as a reference book. Step-by-step software-specific tutorials, which are too focused on techniques, are very poor training for students who need to understand the modeling strategies rather than just which buttons to push for a particular task. In fact, we believe that mere *training* should be abandoned in favor of an *education* in the fundamentals. Students need to learn CAD strategy as well as technique. Students need to develop their creative skills and not have these skills stifled through a focus on the minutiae. In order to prepare for a lifelong career in this fast-changing technological world, students will need to understand fundamental concepts. For example, in the current methods-based approach to graphics training, students learn about geometric dimensioning and tolerancing. Many texts describe what the symbols mean, but do not explain how, why, and when they should be used. Yet, these questions of how, why and when are the questions with which most young engineers struggle and the questions that are directly addressed in this text. They are also the *important* questions—if a student knows the answer to these questions, s/he will understand the fundamental concepts in geometric dimensioning and tolerancing. This fundamental understanding will serve far into the future where techniques, and possibly the symbols themselves, are likely to change.

This text is a limited introduction designed to give the reader a flavor for classic, contemporary, and cutting edge topics in engineering design.

Chapter Structure

With a few exceptions, each chapter is organized along similar lines. The material is presented with the following outline:

1. Objectives

   Chapter-opening objectives alert students to the chapter's fundamental concepts.

2. Introduction

   This section provides an overview of the material that will be presented in the chapter, and discusses why it is important.

3. The Problem

   Each chapter directly addresses a certain need or problem in graphical communication. That problem is presented here as if the student had to face such a problem in the field. The presence of a real problem that needs to be solved gives a student added incentive to learn the material in the chapter to solve that problem.

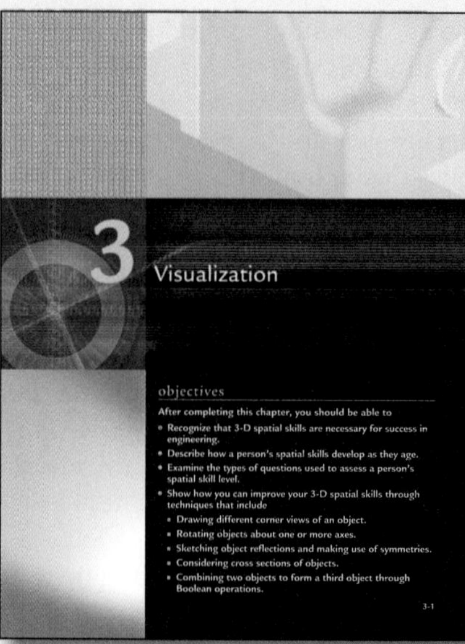

3 Visualization

objectives

After completing this chapter, you should be able to

- Recognize that 3-D spatial skills are necessary for success in engineering.
- Describe how a person's spatial skills develop as they age.
- Examine the types of questions used to assess a person's spatial skill level.
- Show how you can improve your 3-D spatial skills through techniques that include
  - Drawing different corner views of an object.
  - Rotating objects about one or more axes.
  - Sketching object reflections and making use of symmetries.
  - Considering cross sections of objects.
  - Combining two objects to form a third object through Boolean operations.

3-1

6-4 sectiontwo Modern Design Practice and Tools

6.01 introduction

6.02 Tools for Developing Your Idea

Many tools have been developed for creating accurate images of an object as an aid in analyzing its function, recording its history, or visualizing its appearance. One of the simplest tools is a pencil, which is used to make sketches of an object on paper. More formal tools include rulers, protractors, compasses, and various types of manually operated drafting machines. These tools are used to make more accurate, standardized drawings according to precise rules and conventions, as discussed in a previous chapter.

CAD systems are among the most sophisticated graphics and design tools available to engineers and designers. Many types of CAD systems are on the market today. The simplest systems are general purpose drawing or drafting packages that can be used to create 2-D images, similar to the way pencil images are created on paper (except faster and easier). More complex packages allow you to create simulations of 3-D models that can be used not only to generate conventional 2-D drawings of a design but also to create 3-D images for visualization. The core of a CAD model is a geometric **database**. The database includes information about the geometry and other engineering properties of an object. The CAD software uses the database to display the model and to conduct further engineering analysis. A short discussion of CAD history will demonstrate how these systems evolved and provide some insight into the modeling processes used by designers with various CAD systems.

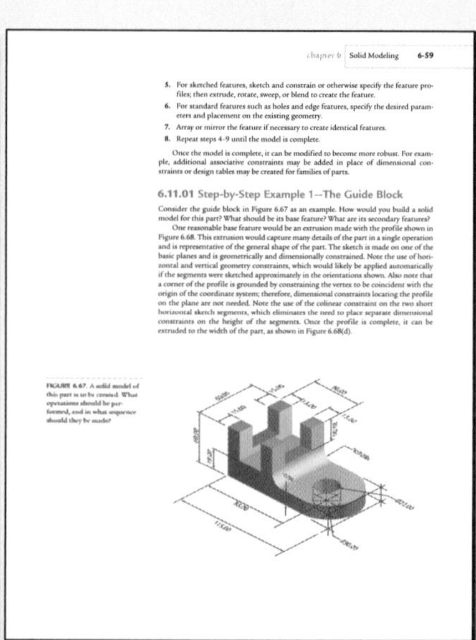

chapter 6 Solid Modeling 6-59

5. For sketched features, sketch and constrain or otherwise specify the feature profiles; then extrude, rotate, sweep, or blend to create the feature.
6. For standard features such as holes and edge features, specify the desired parameters and placement on the existing geometry.
7. Array or mirror the feature if necessary to create identical features.
8. Repeat steps 4-9 until the model is complete.

Once the model is complete, it can be modified to become more robust. For example, additional associative constraints may be added in place of dimensional constraints or design tables may be created for families of parts.

6.11.01 Step-by-Step Example 1—The Guide Block

Consider the guide block in Figure 6.67 as an example. How would you build a solid model for this part? What should be its base feature? What are its secondary features?

One reasonable base feature would be an extrusion made with the profile shown in Figure 6.68. This extrusion would capture many details of the part in a single operation and is representative of the general shape of the part. The sketch is made on one of the basic planes and is geometrically and dimensionally constrained. Note the use of horizontal and vertical geometry constraints, which would likely be applied automatically if the segments were sketched approximately in the orientations shown. Also note that a corner of the profile is grounded by constraining the vertex to be coincident with the origin of the coordinate system; therefore, dimensional constraints locating the profile on the plane are not needed. Note the use of the colinear constraint on the two short horizontal sketch segments, which eliminates the need to place separate dimensional constraints on the height of the segments. Once the profile is complete, it can be extruded to the width of the part, as shown in Figure 6.68(d).

FIGURE 6.67. A solid model of this part is to be created. What operations should be performed, and in what sequence should they be made?

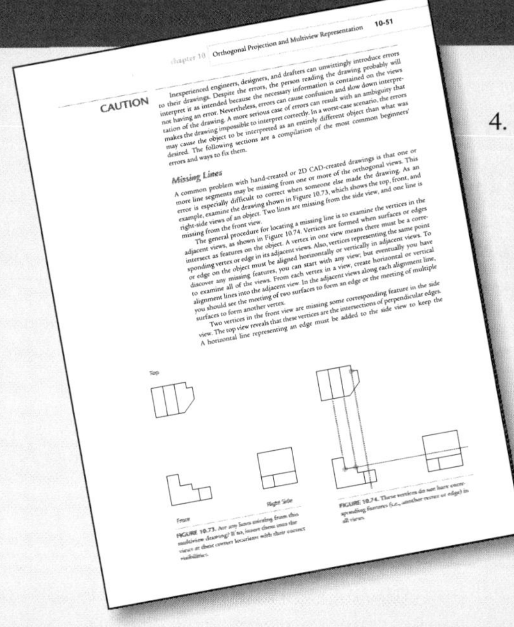

4. Explanation and Justification of Methods

   Engineering graphics has evolved and continues to evolve at an increasingly fast pace due to advances in computer hardware and software. Although new methods associated with new technologies exist, these modern methods must remain compatible with conventional graphics practices. This consistency is required to eliminate possible confusion in the interpretation of drawings, maintain sufficient flexibility to create designs unencumbered by the tools available to document them, and reduce the time and effort required to create the drawings.

5. Guide to Problem Solving

   Sample problems of various types are presented throughout each chapter with detailed, step-by-step solutions.

6. Caution

   Often, people learn more from their mistakes than from their successes. The most common errors—and their potential consequences—are presented as examples in this section.

7. Summary

   This section distills the most important information contained in the chapter.

8. Glossary of Key Terms

   Formal definitions of the most important terms or phrases for the chapter are provided. Each term or phrase is highlighted the first time it is used in the chapter.

9. Questions for Review

   These questions test the student's understanding of the chapter's main concepts.

10. Problems

    A variety of problems and exercises help to develop skill and proficiency of the material covered in the chapter.

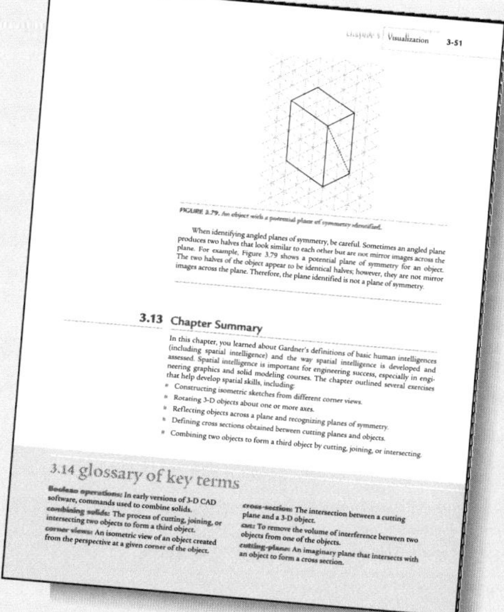

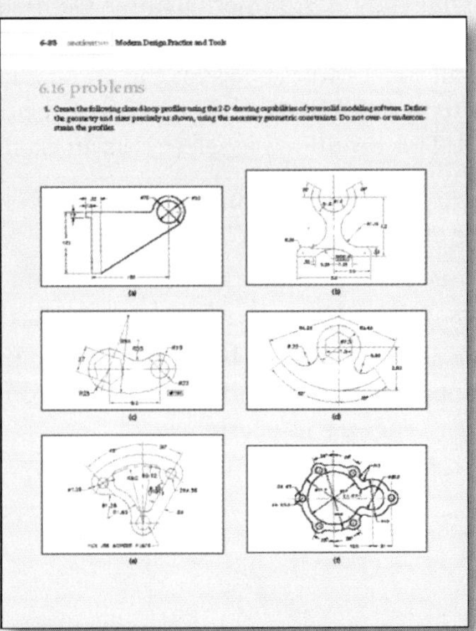

Key Features of the Presentation

We believe that this text will have a broad appeal to engineering graphics students across a wide spectrum of institutions. The following are key features of the text:

- *A focus on learning and fundamental skill development, not only on definitions, tools, and techniques*. This approach prepares students to apply the material to unfamiliar problems and situations rather than simply to regurgitate previously memorized material. In the fast-changing world we live in, an understanding of the fundamentals is a key to further learning and the ability to keep pace with new technologies.
- *Formal development of visualization skills as a key element at an early stage of the curriculum.* Development of these skills is important for students who have not had the opportunity to be exposed to a large number of engineering models and physical devices. Further, the link between visualization and creativity is strong—tools for success over a lifelong career.
- *Use of a problem-based approach.* This approach presents the student with real problems at the beginning of each chapter, shows the graphics solutions, then generalizes the solutions.
- *A casual tone and student-friendly approach.* It is a proven fact that students learn the material better if they are not fast asleep!
- *Examples of poor practices and the potential consequences of these errors.* People seem to learn the most from mistakes. By showing students the mistakes of others, perhaps they can minimize their own.
- *Several common example threads and a common project that are presented in most chapters.* The text shows how the material contained in each chapter was actually applied in the context of product development.

One of the case studies to be presented, for example, is the Hoyt AeroTec™ Olympic style recurve target bow. The unique geometry of this bow was brought to prominence after it was part of the equipment package used to win the Gold Medal in target archery at the summer Olympic Games in Sydney, Australia. The design history of the development of this product is traced starting from its ideation as an improvement to all other target bows on the market at that time. As a student moves through the chapters of this book, the progress of the development of this product will be documented. This product was selected as an example for these reasons:

1. It was a very successful design that accomplished all of the goals set forth by its engineers.
2. It was also a product that is relatively unencumbered by the complexity of mechanisms or electronics, which are not the focus of this book.
3. The design is mature, having made it to the consumer market; this means it offers an opportunity to study some of the non-technical issues that play an important role in engineering design.

Final Remarks

This textbook contains a "core" of material covered in a traditional engineering graphics course and also a number of other chapters on modern graphics techniques. The collected material represents over 50 combined years of personal experience in the learning, application, and teaching of engineering graphics. The result is a text that should appeal to both traditional and contemporary graphics curricula. We, the authors, would like to thank you for considering this text.

Dennis K. Lieu
Professor of Mechanical Engineering
University of California, Berkeley

Sheryl Sorby
Professor of Civil and Environmental Engineering
Michigan Technological University

Contributors
Holly K. Ault, Worcester Polytechnic Institute (Chapter 6 and 17 and Supplemental Chapter 4)
Ron Barr, The University of Texas at Austin (Chapters 5 and 8)
Judy Birchman, Purdue University (Chapter 21)
Ted Branoff, North Carolina State University (Chapters 16 and 17)
Pat Connolly, Purdue University (Supplemental Chapter 6)
Frank Croft, The Ohio State University (Chapter 12)
La Verne Abe Harris, Purdue University (Chapter 21)
Kathy Holliday-Darr, Penn State Erie (Chapter 14 and Supplemental Chapter 5)
Tom Krueger, The University of Texas at Austin (Chapters 5 and 8)
Ann Maclean, Michigan Technological University (Chapter 20)
Kellen Maicher, Purdue University (Supplemental Chapter 6)
Jim Morgan, Texas A&M University (Chapter 4)
Bill Ross, Purdue University (Chapter 19)
Mary Sadowski, Purdue University (Chapter 21)
Tom Singer, Sinclair Community College (Supplemental Chapter 1)
Kevin Standiford, Consultant (Supplemental Chapters 2 and 3)

What is iChapters.com?

iChapters.com, Cengage Learning's online store, is a single destination for more than 10,000 new print textbooks, e-books, single e-chapters, and print, digital and audio supplements.

Choice. Students have the freedom to purchase *a-la-carte* exactly what they need when they need it.

Savings. Students can save 50% for the electronic textbook, and can pay as little as $1.99 for an individual chapter.

Get the best grade in the shortest time possible! Visit www.iChapters.com to receive 25% off over 10,000 print, digital and audio study tools.

acknowledgments

The authors would like to thank the following people for their contributions to the textbook and their support during its development:

George Tekmitchov and Darrin Cooper of Hoyt USA for their assistance and cooperation with the AeroTEC case study material; David Madsen for problem materials; David Goetsch for problem materials; George Kiiskila of U.P. Engineers and Architects for construction drawings; Fritz Meyers for use of his blueprint drawing; Mark Sturges of Autodesk for graphics materials; Rosanne Kramer of Solidworks for graphics materials; Marie Planchard of Solidworks for graphics materials; James DeVoe of Delmar Cengage Learning for his support, confidence, persistence, good humor, and patience in seeing this project through to the end; Monica Ohlinger, editor extraordinaire, of Ohlinger Publishing Services for her organization skills that resulted in the completion of this project.

In addition, the authors are grateful to the following reviewers for their candid comments and criticisms throughout the development of the text:

Tom Bledsaw, ITT
Ted Branoff, North Carolina State University
Patrick Connolly, Purdue University
Richard F. Devon, Penn State University
Kathy Holliday-Darr, Penn State University, Erie
Tamara W. Knott, Virginia Tech
Kellen Maicher, Purdue University
Jim Morgan, Texas A & M
Kellen Maicher, Purdue University
William A. Ross, Purdue University
Mary Sadowski, Purdue University
James Shahan, Iowa State University
Michael Stewart, Georgia Tech

about the authors

Dennis K. Lieu

Prof. Dennis K. Lieu was born in 1957 in San Francisco, where he attended the public schools, including Lowell High School. He pursued his higher education at the University of California at Berkeley, where he received his BSME in 1977, MSME in 1978, and D.Eng. in mechanical engineering in 1982. His major field of study was dynamics and control. His graduate work, under the direction of Prof. C.D. Mote, Jr., involved the study skier/ski mechanics and ski binding function. After graduate studies, Dr. Lieu worked as an advisory engineer with IBM in San Jose CA, where he directed the specification, design, and development of mechanisms and components in the head-disk-assemblies of disk files. In 1988, Dr. Lieu joined the Mechanical Engineering faculty at UC Berkeley. His research laboratory is engaged in research on the mechanics of high-speed electro-mechanical devices and magnetically generated noise and vibration. His laboratory also studies the design of devices to prevent blunt trauma injuries in sports, medical, and law enforcement applications. Prof. Lieu teaches courses in Engineering Graphics and Design of Electro-mechanical Devices. He was the recipient of a National Science Foundation Presidential Young Investigator Award in 1989, the Pi Tau Sigma Award for Excellence in Teaching in 1990, and the Berkeley Distinguished Teaching Award (which is the highest honor for teaching excellence on the U.C. Berkeley campus) in 1992. He is a member of Pi Tau Sigma, Tau Beta Pi, and Phi Beta Kappa. His professional affiliations include ASEE and ASME. Prof. Lieu's hobbies include Taekwondo (in which he holds a 4th degree black belt) and Olympic style archery.

Sheryl Sorby

Professor Sheryl Sorby is not willing to divulge the year in which she was born but will state that she is younger than Dennis Lieu. She pursued her higher education at Michigan Technological University receiving a BS in Civil Engineering in 1982, an MS in Engineering Mechanics in 1985, and a PhD in Mechanical Engineering-Engineering Mechanics in 1991. She was a graduate exchange student to the Eidgenoessiche Technische Hochshule in Zurich, Switzerland studying advanced courses in solid mechanics and civil engineering. She is currently a Professor of Civil and Environmental Engineering at Michigan Technological University. Dr. Sorby is the former Associate Dean for Academic Programs and the former Department Chair of Engineering Fundamentals at Michigan Tech. She has also served as a Program Director in the Division of Undergraduate Education at the National Science Foundation. Her research interests include various topics in engineering education, with emphasis on graphics and visualization. She was the recipient of the Betty Vetter research award through the Women in Engineering Program Advocates Network (WEPAN) for her work in improving the success of women engineering students through the development of a spatial skills course. She has also received the Engineering Design Graphics Distinguished Service Award, the Distinguished Teaching Award, and the Dow Outstanding New Faculty Award, all from ASEE.

Dr. Sorby currently serves as an Associate Editor for ASEE's new online journal, Advances in Engineering Education. She is a member of the Michigan Tech Council of Alumnae. She has been a leader in developing first-year engineering and the Enterprise program at Michigan Tech and is the author of numerous publications and several textbooks. Dr. Sorby's hobbies include golf and knitting.

1

An Introduction to Graphical Communication in Engineering

objectives

After completing this chapter, you should be able to

- Explain and illustrate how engineering graphics is one of the special tools available to an engineer
- Define how engineering visualization, modeling, and graphics are used by engineers in their work
- Provide a short history of how engineering graphics, as a perspective on how it is used today, was used in the past

1.01 introduction

Because engineering graphics is one of the first skills formally taught to most engineering students, you are probably a new student enrolled in an engineering program. Welcome!

You may be wondering why you are studying this subject and what it will do for you as an engineering student and, soon, as a professional engineer. This chapter will explain what engineering is, how it has progressed over the years, and how graphics is a tool for engineers.

What exactly is engineering? What does an engineer do? The term *engineer* comes from the Latin *ingenerare*, which means "to create." You may be better able to appreciate what an engineer does if you consider that *ingenious* also is a derivative of *ingenerare*. The following serves well enough as a formal definition of engineering:

The profession in which knowledge of mathematical and natural sciences, gained by study, experience, and practice is applied with judgment to develop and utilize economically the materials and forces of nature for the benefit of humanity.

A modern and informal definition of engineering is "the art of making things work." An engineered part or an engineering system does not occur naturally. It is something that has required knowledge, planning, and effort to create.

So where and how does graphics fit in? Engineering graphics has played three roles through its history:

1. Communication
2. Record keeping
3. Analysis

First, engineering graphics has served as a means of communication. It has been used to convey concepts and ideas quickly and accurately from one person to another without the use of words. As more people became involved in the development of products, accurate and efficient communication became increasingly necessary. Second, engineering graphics has served as a means of recording the history of an idea and its development over time. As designs became more complex, it became necessary to record the ideas or features that worked well in a design so they could be repeated in future applications. And third, engineering graphics has served as a tool for analysis to determine critical shapes and sizes, as well as other variables needed in an engineered system.

These three roles are still vital today, more so than in the past, because of the technical complexity required to make modern products. Computers, three-dimensional modeling, and graphics software have made it increasingly effective to use engineering graphics as an aid in design, visualization, and optimization.

1.02 A Short History

The way things are done today evolved from the way things were done in the past. You can understand the way engineering graphics is used today by examining how it was used in the past. Graphical communications has supported **engineering** throughout history. The nature of engineering graphics has changed with the development of new graphics tools and techniques.

1.02.01 Ancient History

The earliest documented forms of graphical communication are cave paintings, such as the one shown in Figure 1.01, which showed human beings depicting organized social

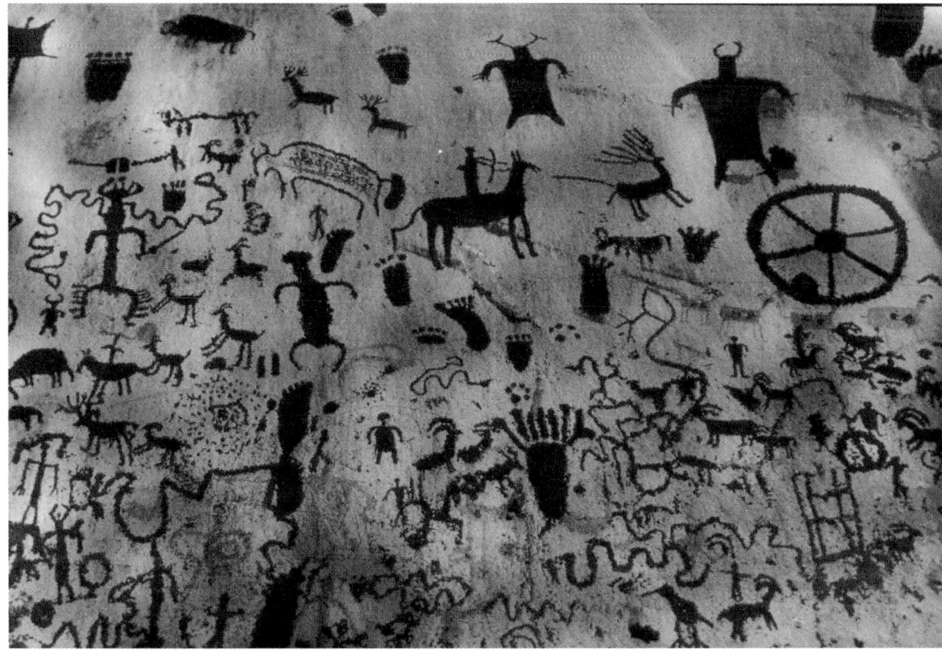

FIGURE 1.01. Undated cave painting showing hunting and the use of tools.
Source: ACE STOCK LIMITED/ Alamy

behavior, such as living and hunting in groups. The use of tools and other fabricated items for living comfort and convenience were also communicated in cave paintings. However, these paintings typically depicted a lifestyle, rather than any instructions for the fabrication of tools, products, or structures. How the items were made is still left to conjecture.

The earliest large structures of significance were the Egyptian pyramids and Native American pyramids. Some surviving examples are shown in Figure 1.02. The Egyptian pyramids were constructed as tombs for the Pharaohs. The Native American pyramids were built for religious ceremonies or scientific use, such as observatories. Making these large structures, with precision in the fitting of their parts and with the tools that were available at the time, required much time, effort, and planning. Even with modern tools and construction techniques, these structures would be difficult to re-create today. The method of construction for the pyramids is largely unknown—records of the construction have never been found—although there have been several theories over the years.

FIGURE 1.02. Mayan pyramid, Yucatan, Mexico (left), and Pharaoh Knufu and Pharaoh Khafre Pyramids, Giza, Egypt (right).
Sources: Brand X Pictures/Alamy, above; DIOMEDIA/Alamy, below.

FIGURE 1.03. Ancient Egyptian hieroglyphics describing a life story.
Source: © Bettmann/CORBIS

Egyptian hieroglyphics, which were a form of written record, included the documentation of a few occupational skills, such as papermaking and farming, although, for the most part, they documented lifestyle. An example of a surviving record is shown in Figure 1.03. As a result of those records, papermaking and farming skills could be maintained and improved over time. Even people who were not formally trained in those skills could develop them by consulting the written records.

Two engineering construction methods helped the Roman Empire expand to include much of the civilized European world. These methods were used to create the Roman arch and the Roman road.

The Roman arch, shown in Figure 1.04, was composed of stone that was precut to prescribed dimensions and assembled into an archway. The installation of the keystone at the top of the arch transferred the weight of the arch and the load it carried into the

FIGURE 1.04. Pont-du-Gard Roman aqueduct (left) built in 19 BC to carry water across the Gardon Valley to Nimes. Spans of the first- and second-level arches are 53–80 feet. The Ponte Fabricio bridge in Rome (right) built in 64 BC spans the bank of the River Tiber and Tiber Island.
Photos by William G. Godden. Reprinted with permission from EERC Library, Univ. of California, Berkeley.

FIGURE 1.05. An engraving showing the operation of an Archimedes screw to lift water. *Courtesy of Time Life Pictures/Getty Images*

remaining stones that were locked together with friction. This structure took advantage of the compressive strength of stone, leading to the creation of large structures that used much less material. The Roman arch architecture was used to create many large buildings and bridges. Roman era aqueducts, which still exist today in Spain and other countries in Europe, are evidence of the robustness of this **design**.

The method used to construct Roman roads prescribed successive layers of sand, gravel, and stone (instead of a single layer of the native earth), forming paths wide enough for commercial and military use. In addition to the layered construction methods, these roads were also crowned to shed rain and had gutters to carry away water. This construction method increased the probability that the roads would not become overgrown with vegetation and would remain passable even in adverse weather. As a result, Roman armies had reliable access to all corners of the empire.

The Roman Empire is long gone, but the techniques used for the construction of the Roman arch and the Roman road are still in use today. The reason for the pervasiveness of those designs was probably due to Marcus Vitruvius, who, during the Roman Empire, took the trouble to carefully document how the structures were made.

The Archimedes screw, used to raise water, is an example of a mechanical invention developed during the time of the Greek Empire. Variations of the device were used for many centuries because diagrams depicting its use were (and still are) widely available. One of those diagrams is shown in Figure 1.05. These early documents were precursors to modern engineering **drawings**. Because the documents graphically communicated how to build special devices and structures, neither language nor language translation was necessary.

1.02.02 The Medieval Period

Large building construction helped define the medieval period in Europe. Its architecture was more complicated than the basic architecture used for the designs of ancient buildings. The flying buttress, a modification of the Roman arch, made it possible to construct larger and taller buildings with cavernous interiors. This type of structure was especially popular in Europe for building cathedrals, such as the one shown in Figure 1.06. The walls of fortresses and castles became higher and thicker. Towers were included as an integral part of the walls, as shown in Figure 1.07, to defend the inhabitants from many directions, even when attackers had reached the base of a wall.

FIGURE 1.06. Flying buttress construction used to support the exterior walls of Notre Dame Cathedral in Paris.
Courtesy of Getty Images

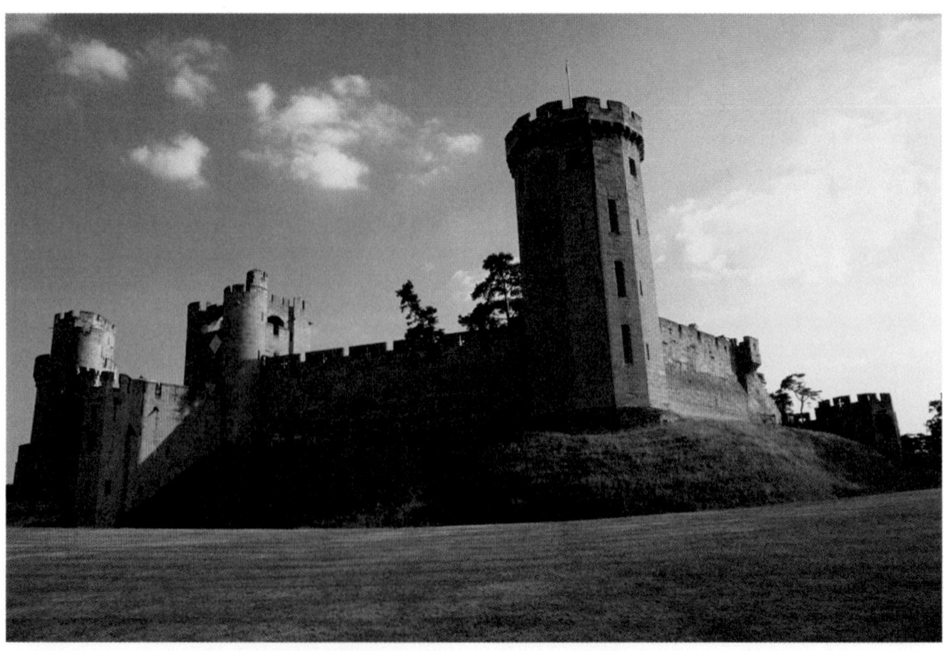

FIGURE 1.07. Warwick Castle, England, circa 1350, is an example of a medieval style fortification.
Source: © Royalty-Free/CORBIS

FIGURE 1.08. The Great Wall of China, built during the medieval period, used simple engineering principles despite the large scale of the project. *Source: Nigel Hicks/Alamy*

In Asia, large fortifications, shrines, and temples, as shown in Figure 1.08, were built to last hundreds of years. The complexity of techniques to build those structures required planning and documentation, especially when raw materials had to be transported from long distances. Building structures of such sizes required an understanding of the transmission of forces among the supporting members and the amount of force those members could withstand. That knowledge was especially important when wood was the primary building material.

Large-scale civil engineering **projects** were begun during the medieval era. Those projects were designated by a civilian government to benefit large groups or the general population, as opposed to projects constructed for private or military use. The windmills of Holland, shown in Figure 1.09, are an example of a civil engineering project. The windmills harvested natural wind energy to pump large amounts of water out of vast swampland, making the land suitable for farming and habitation.

Windmills and waterwheels were used for a variety of tasks, such as milling grain and pumping water for irrigation. Both inventions were popular throughout Europe and Asia; a fact that is known because diagrams showing their construction and use have been widely available.

1.02.03 The Renaissance

The beginning of the Renaissance in the 1400s saw the rise of physical scientific thinking, which was used to predict the behavior physical **systems** based on empirical observation and mathematical relationships. The most prominent person among the scientific physical thinkers at that time was Leonardo da Vinci, who documented his ideas in drawings. Some of those drawings, which are well known today, are shown in Figure 1.10. Many of his proposed devices would not have worked in their original form, but his drawings conveyed new ideas and proposals as well as known facts.

Prior to the Renaissance, nearly all art and diagrams of structures and devices were records of something already in existence or were easily extrapolated from something already tried and known to work. When inventors applied physical science to

FIGURE 1.09. The network of windmills in Holland, used to drain water from flooded land, is an example of an early large-scale civil engineering project.
Source: © PaulAlmasy/CORBIS

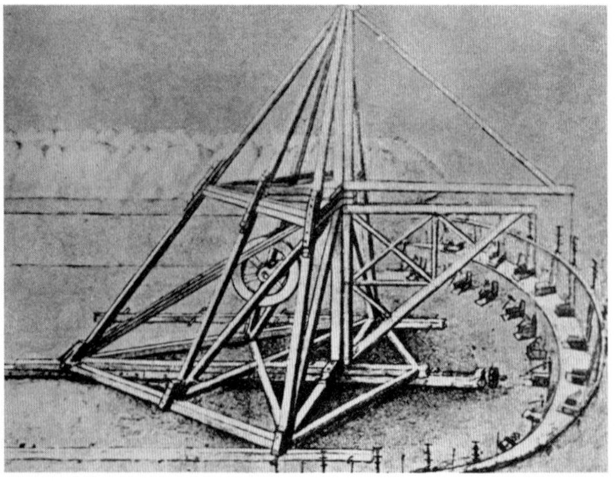

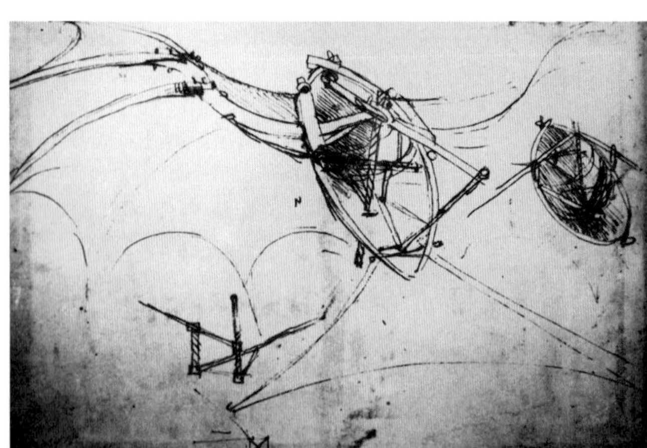

FIGURE 1.10. Images of original da Vinci drawings: a machine used for canal excavation (left) and a flying ship (right). Codex Atlanticus, folio 860; drawing from Il Codice Atlantico di Leonardo da Vinci nella biblioteca Ambrosiana di Milano, Editore Milano Hoepli 1894–1904; the original drawing is kept in the Biblioteca Ambrosiana in Milan.
Sources: © Bettmann/CORBIS

FIGURE 1.11. A perpetual motion machine by medieval inventors: an Archimedes screw driven by a waterwheel is used to mill grain.
Source: Timewatch Images/Alamy

engineering, they could conceive things that theoretically should have worked without having been previously built. When inventors did not understand the science behind the proposed devices, the devices usually did not work. The many perpetual motion machines proposed at that time, as shown in Figure 1.11, are evidence of inventors' lack of understanding of the physical science and their resultant failed attempts to build the machines.

Engineers began to realize that accurate sizing was an element of the function of a structure or device. Diagrams made during the Renaissance paid more attention to accurate depth and perspective than in earlier times. As a result, drawings of proposed and existing devices looked more realistic than earlier drawings.

Gunpowder was introduced during the Renaissance, as was the cannon. The cannon made obsolete most of the fortresses built during the medieval era. The walls could not withstand impact from cannon projectiles. Consequently, fortresses needed to be redesigned to survive cannon fire. In France, a new, stronger style of fortification was designed. The fortification was constructed with angled walls that helped to deflect cannon fire and did not crumble as flat vertical walls did when struck head on. The new fortresses were geometrically more complicated to build than their predecessors with vertical walls. Further, the perimeter of the fortress had evolved from a simple rectangular shape to a pentagonal shape with a prominent extension at each apex. That perimeter shape, coupled with the angled walls, resulted in walls that intersected at odd angles that could not be seen and measured easily or directly. Following is a list of questions that builders of earlier fortresses could easily answer but that builders of the angled wall fortresses could not:

- What is the surface area of a wall?
- What is the fill volume?

- What are the specific lengths of timbers and beams needed to construct and brace the walls?
- What are the true angles of intersection between certain surfaces?
- What are the distances between lines and other lines, between points and lines, and between points and surfaces?

Fortunately, the French had Gaspard Monge, who developed a graphical analysis technique called **descriptive geometry**. Analytical techniques using mathematics were not very sophisticated at that time, nor were machines available to do mathematical calculations. But mechanical **instruments**, such as compasses, protractors, and rulers, together with the graphical method, were used to analyze problems without the need to do burdensome math. Descriptive geometry techniques enabled engineers to create any view of a geometric object from two existing views. By creating the proper view, engineers could see and measure an object's attributes, such as the true length of its lines, the true shape of planes, and true angles of intersection. Such skills were necessary, especially for the construction of fortifications, as shown in Figure 1.12. The complex geometry, odd angles of intersection, and height of walls were intended to maximize the cross fire on an approaching enemy, while not revealing the interior of the fortress. Another objective was to construct the ramparts and walls by moving the minimum amount of material for maximum economy.

The astuteness of the French at building fortifications kept France the prime military power in Europe until the 1700s. At that time, descriptive geometry was considered a French state secret; divulging it was a crime punishable by death. As a result of the alliance between France and the newly constituted United States, many U.S. fortifications used French designs. An example is Fort McHenry (shown in Figure 1.13), which was built in 1806 and is exquisitely preserved in Baltimore, Maryland. Fort McHenry survived bombardment by the British during the War of 1812 and is significant because it inspired Francis Scott Key to write "The Star Spangled Banner."

By the 1800s, most engineering was either civil engineering or military engineering. Civil engineering specialized in the construction of buildings, bridges, roads, commerce ships, and other structures, primarily for civilian and trade use. Military engineering specialized in the construction of fortifications, warships, cannons, and other items for military use. In both fields of engineering, as projects became more complicated, more people skilled in various subspecialties were needed. Clear, simple,

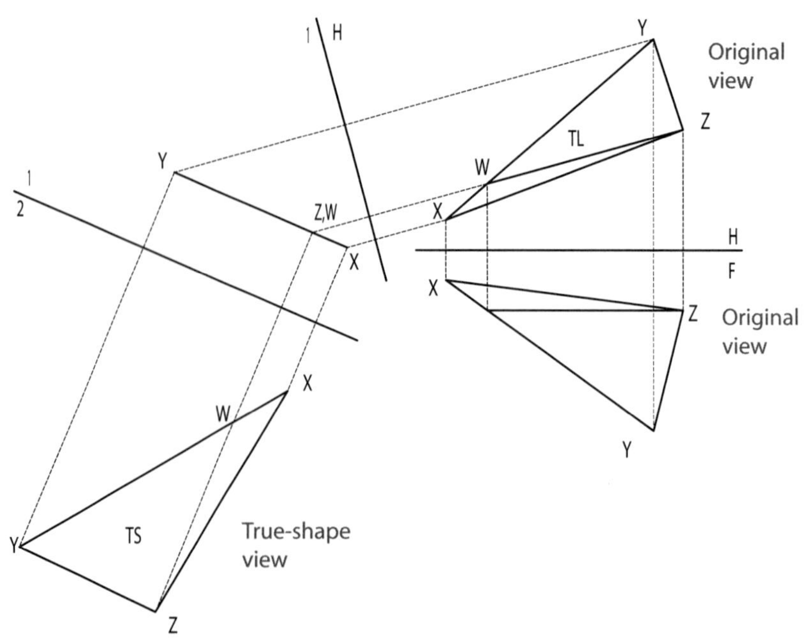

FIGURE 1.12. Using descriptive geometry to find the area of a plane.
Courtesy of D. K. Lieu

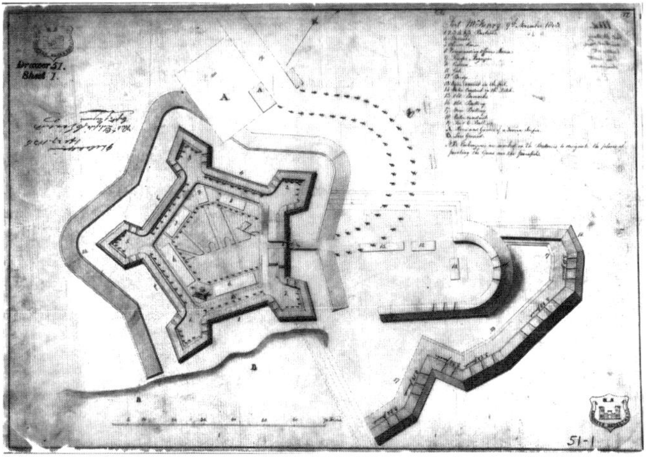

FIGURE 1.13. French fortification design principles (left) and Fort McHenry (right) in Baltimore, Maryland, whose design was based on those principles.
Courtesy of The National Archives, College Park, Maryland, left; Courtesy of The National Park Service, Fort McHenry NMHS, right.

and universal communication was necessary to coordinate and control the efforts of specialists interacting on the same project. Different people needed to know what other people were doing in order for various **parts** and **assemblies** to fit together and function properly. To fill that need, early forms of scaled drawings began to be used as the medium for communications in constructing a building or device.

1.02.04 The Industrial Revolution

The industrial revolution began in the early 1800s with the new field of mechanical engineering. This revolution was, in part, a result of the need for new military weapons. Before the 1800s, ships and guns were fabricated one at a time by skilled craftsmen. No original plans of any ships from the Age of Discovery exist, because shipwrights did not use plans drawn on paper or parchment. The only plans were in the master shipwright's mind, and ships were built by eye. As the demand for ships grew, production methods changed. It was far more economical to build many ships using a single design of common parts than to use a custom design for each ship. Constructing from a common design required accurate specifications of the parts that went into the design.

The hardware products that general and military consumers needed then were no longer produced by skilled craftsmen but were mass-produced according the techniques and machines specified by engineers. Mass production meant that each product had to be identical to all other products, had to be fabricated within predictable and short production times, had to be made from parts that were interchangeable, and had to be produced economically in volumes much larger than in the past. The consistent and repetitive motions of machines required efficient, large-scale production which replaced manufacturing operations that had needed the skilled motions of craftsmen. Also, engines, boilers, and pressure vessels were required to provide power to machines. An early manufacturing facility with machine tools and an early steam engine are shown in Figure 1.14 and Figure 1.15, respectively.

Creating not only a product but also the machines to produce it was beyond the abilities of individual craftsmen—each likely to have a different set of skills needed for the production of a single product. The high demand for creating machines as well as products meant that the existing master-apprentice relationship could no longer supply the demand for these skills. To meet the growing demand, engineering schools had to teach courses in basic physics, machine-tool design, physical motion, and energy transfer.

FIGURE 1.14. A photo showing early factory conditions during the industrial revolution.
Courtesy of Time Life Pictures/Getty Images

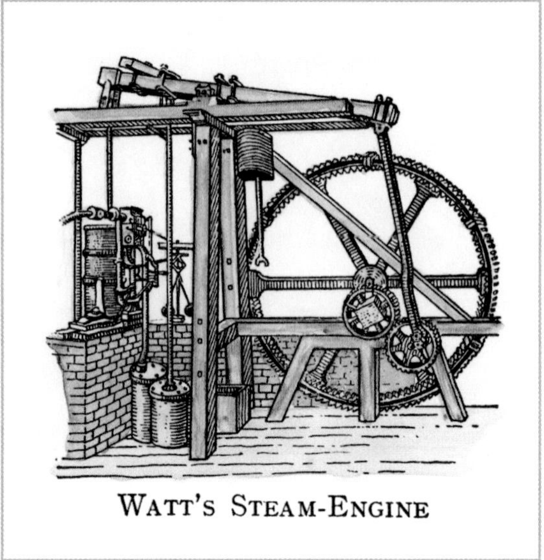

FIGURE 1.15. A schematic drawing of a James Watt steam engine; the type commonly used to power production machinery during the early years of the industrial revolution.
Source: North Wind Picture Archives/Alamy

Communication was necessary to coordinate and control the efforts of different people with different skills. Each craftsman, as well as each worker, on a project needed to know what others were doing so the various pieces, devices, structures, and/or systems would fit together and function properly. The ideas of the master designer had to be transferred without misinterpretation to those who worked at all levels of supporting roles. In the design stage, before things were actually built, the pictorial diagrams once used were soon found to be insufficient and inaccurate when new structures with new techniques were being built. More accurate representations, which would provide exact sizes, were needed. That need eventually led to the modern engineering drawing, with its multiple-view presentation, identification of sizes, and specification of allowable errors.

Around the time of the industrial revolution, patents started to become important. As a method of stimulating innovation in an industrialized society, many governments offered patents to inventors. The owners of patents were guaranteed exclusive manufacturing rights for the device represented in the patent for a prescribed number of years in exchange for full disclosure of how the device operated. Since a single successful patented invention could make its owner rich, many people were inspired to create new products. From the start, the difference between patent drawings and engineering has been that engineering drawings are made to be viewed by those formally trained in engineering skills and to show precise sizes and locations. Patent drawings, on the other hand, are made to teach others how and why a device operates. Consequently, patent drawings often do not show the actual or scaled sizes of the parts. In fact, sizes are commonly distorted to make the device more difficult for potential competitors to copy. An example of a patent drawing is shown in Figure 1.16.

1.02.05 More Recent History

As technology advanced over time, additional engineering specialties were born. In the late 1800s, as electric power became more popular and more available, electrical engineering was born. Electrical engineering at that time was concerned with the production, distribution, and use of electrical energy. The information derived from the study of

FIGURE 1.16. A U.S. patent drawing showing function but not necessarily the true sizes of the parts.
Courtesy of D. K. Lieu

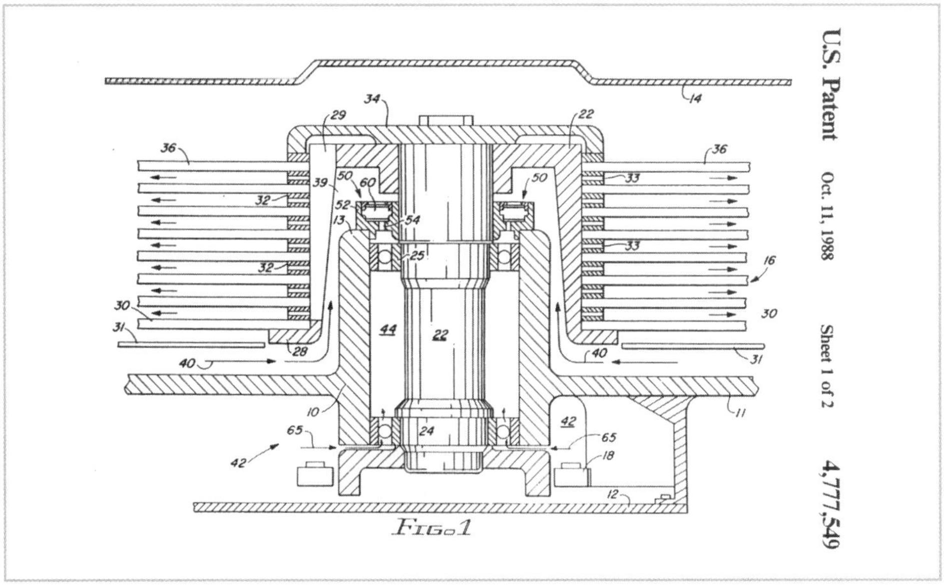

electric motors, generators (shown in Figure 1.17), power conversion, and transmission lines needed for their design was more than other engineers—not specifically electrical engineers—could be expected to know and use. Chemical engineering, as a special engineering discipline, emerged at the beginning of the twentieth century with the need for large-scale production of petroleum products in refineries, as shown in Figure 1.18, and the production of synthetic chemicals.

During the 1950s, industrial engineering and manufacturing engineering emerged from the necessity to improve production quality, control, and efficiency. Nuclear engineering emerged as a result of the nuclear energy and nuclear weapons programs.

Some of the more recent engineering disciplines include bioengineering, information and computational sciences, micro-electro-mechanical systems (MEMS), and nano-engineering. The design of a MEMS device (for example, the valve shown in

FIGURE 1.17. Later during the industrial revolution, steam engines were replaced by electric power supplied, for example, by these generators at the Long Island Railway shown (circa 1907). Electrical engineering was born.
Courtesy of SMITHSONIAN INSTITUTION Neg.#44191D

FIGURE 1.18. The demand for chemical and petroleum products led to the construction of sophisticated plants and refineries and the disciplines of chemical and petroleum engineering.
Courtesy of DOE/NREL, photo by David Parsons.

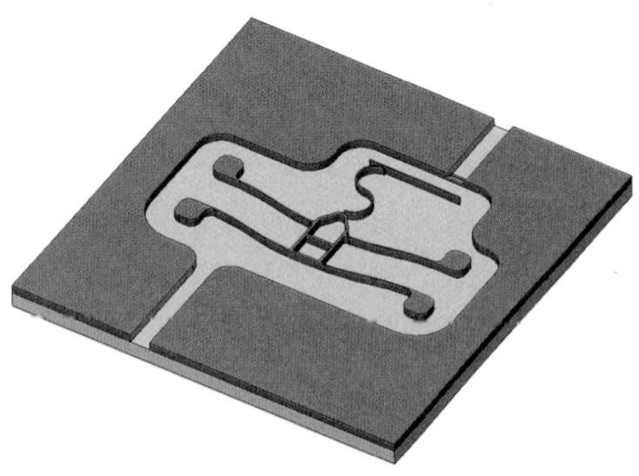

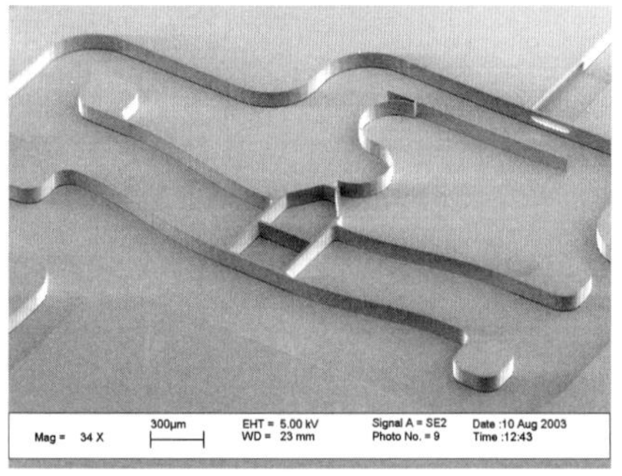

FIGURE 1.19. This MEMS valve was designed with a solid modeler and was fabricated using semiconductor processing techniques.
Courtesy of the Berkeley Sensor and Actuator Center, University of California.

FIGURE 1.20. This nano-device for sorting molecules does not actually appear as shown, but the use of graphics aids in understanding its operating principles.
Courtesy of Kenneth Hsu.

Figure 1.19) requires skills from both electrical and mechanical engineering. A nano-engineered device cannot be seen with conventional optics. Its presumed appearance, such as that shown in Figure 1.20, and function are based on conjecture using engineering graphics tools.

With the emergence of a new discipline comes formal intensive training, specifically in the specific discipline, as opposed to subspecialty training within an existing discipline. Most complex engineering projects today require the combined skills of engineers from a variety of disciplines. Engineers from any single discipline cannot accomplish landing an astronaut on the moon or putting a robotic rover, shown in Figure 1.21, on Mars.

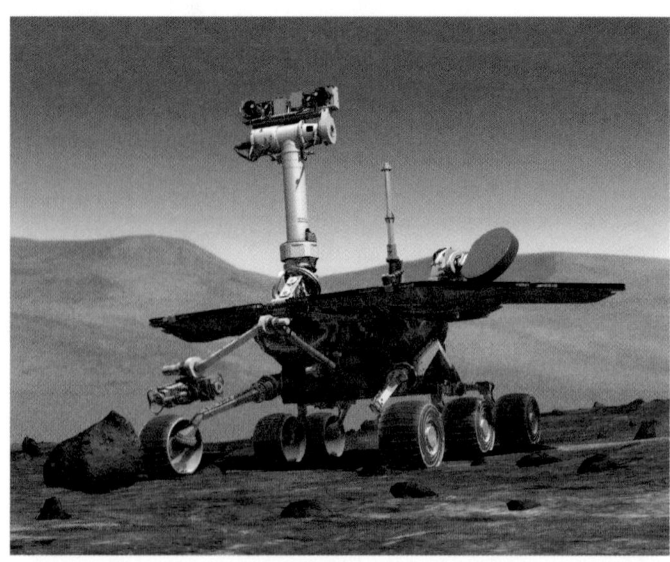

FIGURE 1.21. Complex engineering projects, such as interplanetary space missions, require interdisciplinary engineering skills.
Source: NASA/JPL/Cornell University

1.03 The People and Their Skills

Today few engineering projects exist where a single person or a small group of people is responsible for all aspects of the project from beginning to end. Many people with many different types of technical and nontechnical skills participate in the development and production phases of a project. Whether that person is the engineer who conceives the overall idea or the fabricator who makes the individual pieces or the technician who assembles the parts to make the system operate, they all have common questions:

- What is this part, device, or structure supposed to do?
- What is it supposed to look like?
- What are the precise geometries and sizes of its features?
- What is it made of?
- How is it made?
- How does it fit into other parts, devices, or structures?
- How do I know if everything is made the way it was supposed to be made?

To answer these questions, a clear, unbroken, and unambiguous flow of communication must take place, as depicted in each diagram in Figure 1.22.

The object envisioned by the engineer must be the same object produced by the fabricator and the same object assembled into the working system by the technician. Graphical communication that follows universally accepted standards for representing shapes and sizes makes that happen.

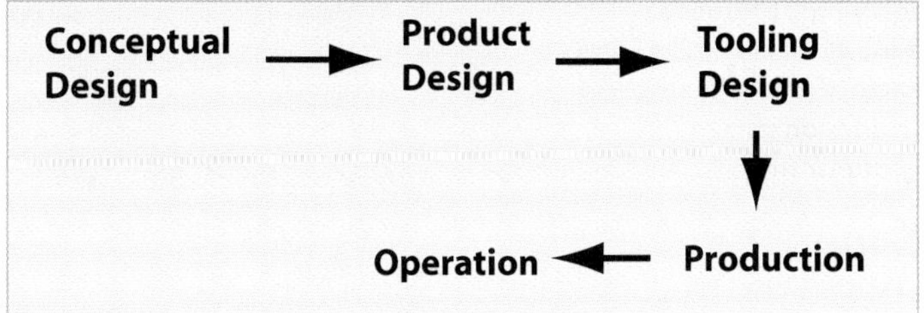

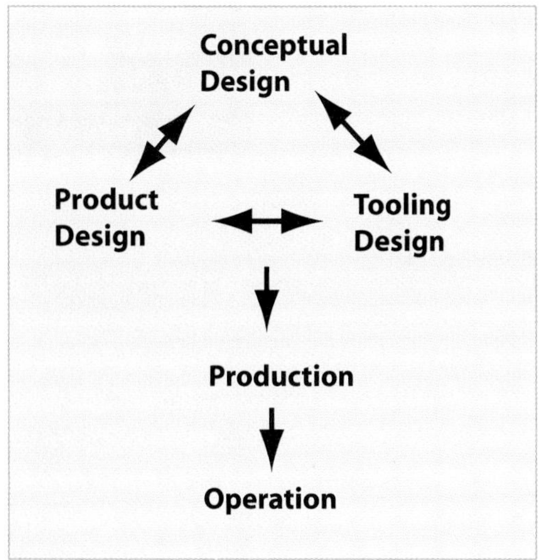

FIGURE 1.22. In conventional product design (above), phases of the development cycle occur sequentially. Concurrent engineering (below) combines two or more phases to accelerate the cycle.
Courtesy of D. K. Lieu

1.03.01 Organization of Project Life Phases

An engineering project may be as simple as a one-piece can opener or as complex as an interplanetary space mission—or anything in between. The number of people involved can be as few as one to as many as several thousand. Regardless of the complexity of the project or the number of people involved, any project can be broken into several phases over its lifetime. These life phases are as follows:

- Concept
- Design
- Fabrication
- Installation
- Operation
- Disposal

For example, consider the wind-powered electric generation facility located at California's Altamont Pass approximately 100 km east of San Francisco. This facility, composed of about 7,200 large wind turbines covering an area of several hundred square kilometers, is one of the largest wind-energy-producing facilities in the world. Many of the lessons learned in the construction and operation of the Altamont facility were incorporated into the plans to build a new wind-powered generation facility for the Solano County area, which is near the Sacramento River delta in California. A small portion of the Solano facility is shown in Figure 1.23.

FIGURE 1.23. A few of the 100-meter-high wind turbines at the Solano Wind Power Generation Facility (above) and one of the turbines on the ground before mounting (below).
Courtesy of D. K. Lieu

Before the facility could be built, it had to be decided during the Concept Phase whether the project would be economically viable and socially and environmentally acceptable. Other decisions involved the size and density of turbines.

During the Design Phase, the types of turbines were selected. The shapes and sizes of the various parts for the turbines and their supporting structures were developed, as was a scheme for collecting, controlling, and distributing the electric power that would be produced.

Parts that could not be purchased as finished units were custom-built during the Fabrication Phase. During the fabrication of custom parts for any project, the appropriate manufacturing processes are selected to reduce costs as much as possible. For large projects, it is just as common to use foreign and domestic suppliers for prefabricated as well as custom-fabricated parts.

The Installation Phase involves taking the individual parts and putting them together to create individually operating turbines. The individual turbines needed to be networked to supply power as an entire system.

The Operation Phase includes not only the control of the system but also the maintenance of the individual turbines and the networked power grid. When parts wear out or are damaged, they must be repaired or replaced.

Finally, when the entire facility has reached the end of its useful life, there must be a plan in the Disposal Phase for removal of the facility, disposal or recycling of its components, and return of the land to other uses.

1.03.02 Organization of Functional Groups

The larger the number people within each phase and between phases, the greater the need for effective communication. In a complex engineering project, the work needs to be divided into many subspecialties that are usually performed by different organizations. The personnel involved in each phase of a project generally can be organized into the following functions:

- Research and Development
- Design
- Manufacturing
- Sales and/or Buying
- Service
- Subcontractors

Depending on the complexity of the project, the same person can handle each function as a specialty, or an entire group of people may be responsible for each function. In the case of the Solano wind-powered generation facility, let's examine one of the project phases. During the Design Phase, certain people were responsible for seeking and evaluating new materials, devices, and technologies that would be of immediate use in designing and building the turbines. Those people filled the Research and Development function. Other people in the Design function were responsible for specifying the shapes and sizes of premade and custom parts so the parts would fit and function as intended. The people who actually made the parts or assembled them into operating prototypes filled the Manufacturing function. Any premade items or raw materials that had to be purchased were done by people in the Buying function. People in the Service function were responsible for operating the prototypes and, if necessary, to collect data needed to evaluate the design for improvement. Subcontractors supplied items or services that could be produced more quickly and efficiently by third parties. For the project life phases, similar responsibilities could be identified for each of the functions just mentioned.

1.03.03 Organization of Skills

Within each function of each project phase, the responsibility of the engineering personnel can be further subdivided as follows:

- Engineers
- Designers
- Drafters
- Fabricators
- Inspectors
- Technicians

Engineers are responsible for ensuring that systems and devices are specified to operate within their theoretical limits, specifying the materials and sizes of parts and assemblies so that failures do not occur, specifying the methods in which the devices are maintained and operated, and evaluating and preparing the environment in which large projects are to be placed.

Designers are responsible for the project's fit and finish, that is, specifying the geometry and sizes of components so they properly mate with each other and are ergonomically and aesthetically acceptable within the operating environment.

Drafters are responsible for documentation—the formal graphical records of parts and assemblies that are required not only for record keeping but also for unambiguous communication between people working on the project.

Fabricators are responsible for making the parts according to the specifications of the engineers and designers, using the documentation provided by the drafter as a guide.

Inspectors are responsible for checking; they take parts made by the fabricators and compare the actual sizes of the parts' features to the desired sizes. This is done to ensure that the parts are properly made and will fit and function as intended. Some projects are installed over large pieces of land. In those cases, inspectors ensure that the land has been properly prepared and that the various elements that compose the project have been made and installed according to the specifications of the engineers.

Technicians are responsible for operation and maintenance; they typically assemble various components to create working devices or structures, operate them, and maintain them.

Depending on the particular phase of the project and the particular function group within that phase, a group will have different combinations of engineering personnel. For example, the Design Phase of the wind-powered generation facility had many engineers and designers but few technicians. However, during the Operation Phase, the facility had mostly technicians, with only a few engineers. One interesting problem that engineers faced during the Operation Phase of the Altamont Pass wind power facility was how to reduce the number of birds, including large raptors, that were killed every year by the spinning turbine blades. No one foresaw this problem during the earlier phases of the project. Special avian experts had to be consulted during the Operation Phase to assist the engineers with possible solutions. Those same avian experts were consulted during the Design Phase of the Solano wind power facility. The new turbines at Solano were designed to have slower blade rotation speeds, and their towers were designed to make bird nesting difficult.

Regardless of the makeup of the engineering group, whenever a number of people participate in any aspect of a project, such as designing and constructing a part, everyone must know what that part is supposed to look like, what the part is made of, and what the part is supposed to do.

1.03.04 Concurrent Engineering

You should not finish the preceding section thinking that the only way engineering projects are done, or even the preferred way to get them done, is through formally

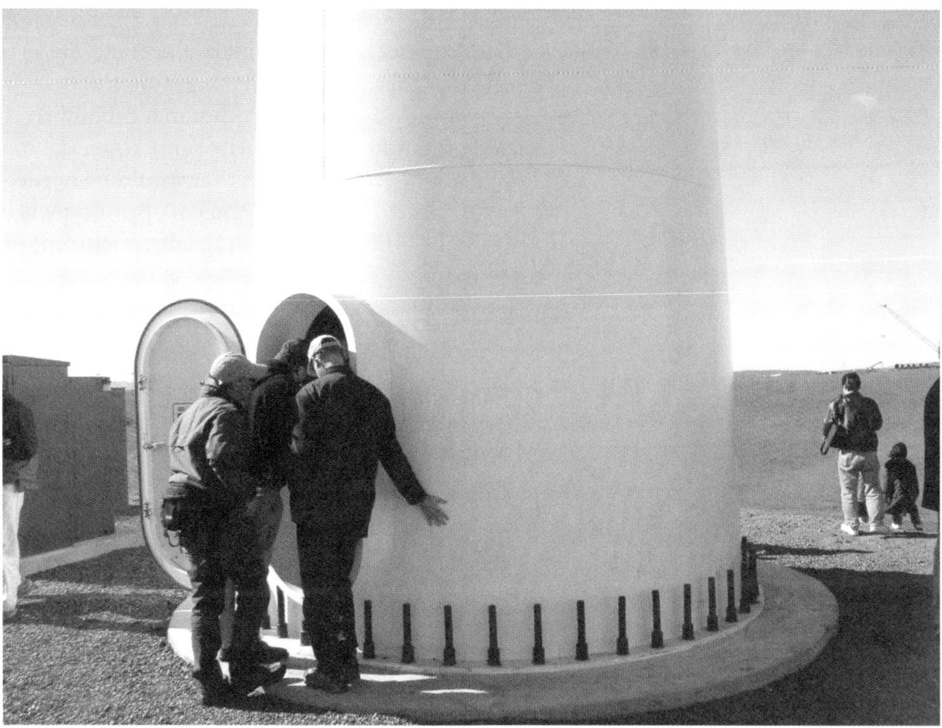

FIGURE 1.24. Entering the maintenance hatch in one of the Solano wind turbine towers. *Courtesy of D. K. Lieu*

organizing functional groups and separating skills. Separate functions and skills may be the classic way to do things, but many modern products use concurrent engineering to reduce the time needed for the product design and production cycles. Concurrent engineering is a process where the design and certain aspects of the fabrication phases are combined. The engineers responsible for the design of a product and the engineers responsible for manufacturing or construction work together closely. Thus, as a part is being conceived, its method of fabrication and assembly into other parts is being given careful consideration. The design of the part is then altered to facilitate its fabrication and, when economically feasible, its assembly into a larger system. Concurrent engineering also considers the method of disposal once the part has reached the end of its useful life.

As an example of concurrent engineering, consider the support tower for one of the wind turbines at the Solano facility. This structure supports the turbine, transmission gearbox, and generator; it also provides access (via a very long stairway) to these devices for maintenance, shown in Figure 1.24. Assume you are the designer of this structure and you want it made from steel.

Since several thousands of these structures may be needed for the many wind power installations around the world, you need to consider the economics of fabricating them. Using a conventional engineering timeline approach, you need to determine the required material and geometry, then make the drawings for the structure. You would have a prototype fabricated, installed on a prototype wind turbine, and tested to prove that it will do what you designed it to do and, especially, that it will not fail. Then you would turn the part and its drawings over to a manufacturing engineer to figure out the best and most economical way to **fabricate** large numbers of the structures to satisfy the worldwide needs of wind-power generation installations. For example, one way of fabricating the tower would be to make it from many small curved plates of steel, with all of the plates welded together. But it may be more economical to produce large sections of complete tubes that are the diameter of the tower and then connect those sections. However, the cost of any special tooling required to make and transport the large tubes would need to be included in the final cost of the structure. Different fabrication processes are possible for making the sections. Each process has different advantages and disadvantages in terms of cost and efficiency.

With concurrent engineering, engineers from all phases of the project work together. Engineers who ordinarily become involved later in the process get together with the designer at the early stages of the design. For example, the manufacturing engineer would advise the part designer on how to change a part so it would be easier to fabricate or handle. At nearly the same time, the manufacturing engineer would begin to design any specialized tools that would be needed to fabricate, handle, and assemble the part in large numbers. As the part prototype is being fabricated, these special tools also would be built. The advantage of concurrent engineering is that product development is reduced. The disadvantage is that large errors in design are expensive because any changes in design also require changes to the production tooling.

1.04 Engineering Graphics Technology

Mechanical drawing instruments have been a tremendous aid for the creation of engineering graphics. These instruments greatly improve the precision with which graphics can be produced and reproduced, reducing any distortion and making analyses easier and more accurate. The improvement of engineering graphics technology over the years has been a major factor in the improvement of engineering design and communication.

1.04.01 Early Years

Up until the time of the Renaissance, most drawing was done by hand without mechanical devices, because none were available. As a result, many of the drawings that were made to depict some sort of engineering device were distorted. The amount of distortion depended on the skill of the person making the drawing. **Two-dimensional (2-D) drawings** were common because they were easy to make. Attempts at drawing objects showing depth had mixed results. Leonardo da Vinci was one of few people who was good at it, but he was also a skillful artist. In general, though, handmade drawings were good for conveying ideas and some rough sizing. They were poor when precision was necessary, mostly because it was not possible to determine exact sizes from them. In fact, the inch and foot as units of measurement in Europe were not standardized until the twelfth century, and the meter was not defined until the eighteenth century. As a result, when different craftsmen built the same item, the sizes of the parts would be slightly different. Those differences made part interchangeability, and thus mass production, extremely difficult.

1.04.02 Instrument Drawing

Early instruments used to make drawings included straightedges with graduated scales, compasses and dividers, and protractors. They were generally custom-made items for the convenience of those who could afford them. Mechanical instruments for drawing did not become widely available until the industrial revolution, when, for a reasonable cost, machines could produce accurate instruments for both drawing and measuring. Both standardized units and accurate drawings made it possible for different fabricators to make the same part. With careful specifications, those parts would be interchangeable between the devices in which they functioned. Now that engineering drawing made it possible to fabricate the same part at different manufacturers, engineering drawing became a valuable means of communication.

From the industrial revolution to the late twentieth century, drawing instruments slowly improved in quality and became less expensive. Drawing instrument technology reached its most effective and highest level of use during the 1970s. Some companies and individuals today still retain, and even prefer, to use mechanical instruments for making engineering drawings. Classic drawing instruments, some of which are shown

FIGURE 1.25. Tools for instrument drawing (left) and a drafting machine (right).
Sources: © Peter Harholdt/SuperStock, left; © Françoise Gervais/CORBIS, right.

in Figure 1.25, are available from architecture, art, and engineering supply shops; these instruments include the following:

- Drafting board—a large, flat table with straight, square edges for alignment of drawing instruments
- Drawing vellum—a tough, dimensionally stable, and age-resistant paper on which drawings are made when placed on the drafting board
- T square—an instrument used to make horizontal and vertical lines by using the edges of the drafting board for reference
- Triangle—an instrument used to make lines at common angles
- Protractor—an instrument used to measure angles or make lines at arbitrary angles
- Scale—an instrument used to measure linear distances
- Drafting machine—a special machine used to hold scales at arbitrary angles while the scales are allowed to translate across the drawing, thus replacing many of the previously listed instruments
- Compass—an instrument used to make circles and arcs
- French curve—an instrument used to make curves
- Template—an instrument used to make common shapes

Using pencil or ink, engineers use instruments to draw directly on the desired sized vellum sheet. Large drawings are reproduced on special copy machines. Up until the 1980s, engineering students often were burdened with having to learn how to use the drawing instruments.

1.04.03 The Computer Revolution

During the 1970s, many large companies, particularly those in the automotive and aerospace industries, recognized the advantages of computer-based drawing and graphics: ease of storage and transmission of data, precise drawing data, and ease of data manipulation when drawings needed to be changed. Several large companies began developing computer-aided drawing (**CAD**) tools for their own use. Mainframe computers were just reaching the point where their cost, computation power, and storage capability would support computer-based drawing. The CAD systems consisted of computer terminals connected to a mainframe computer. However, the conversion to computer-based drawing was slow. Mainframe computers were expensive, the user had to have some computer skills, computer hardware and software were not very reliable, and special input and output devices were necessary. Thus, the average engineer or drafter still had a difficult time making the transition from mechanical tools to computer-based tools.

In the late 1970s and early 1980s, several companies specializing in CAD developed freestanding computer-drawing stations based on small independent computers called workstations. Those companies marketed the computer hardware and software as a complete, ready-to-operate unit known as a turnkey system. The workstation approach to CAD made the software more affordable for smaller companies. Also, CAD software became more sophisticated and easier to use. It began to grow in popularity. As personal computers (PCs) began to proliferate in the 1980s, CAD software made specifically to run on PCs became popular. One company that became a leader in this application was Autodesk, with its AutoCAD software. Companies that formerly supplied mainframe computer-based or turnkey CAD systems either quickly adapted their products for PC use or went out of business. As PCs became more powerful, cheaper, easier to use, and more prolific, CAD software did the same. Drafting boards were quickly replaced by PCs. An example of a PC-based CAD system is shown in Figure 1.26.

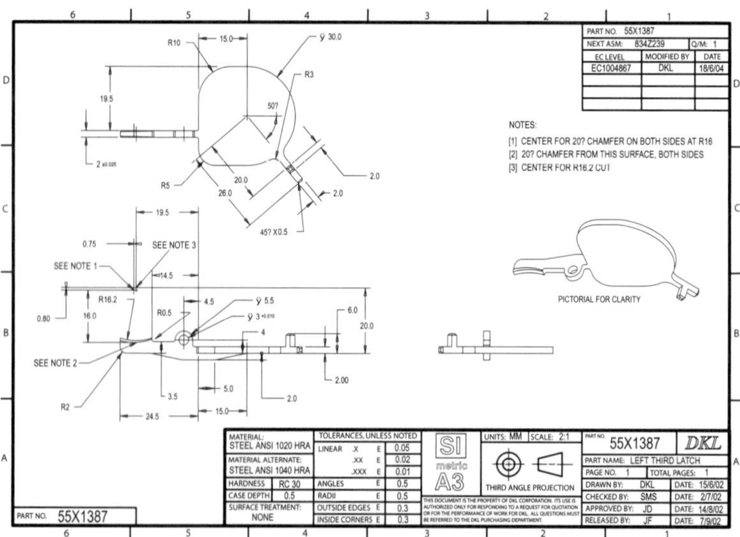

FIGURE 1.26. Computer graphics stations have replaced mechanical drawing instruments in most applications. A CAD drawing can be created by itself or extracted from a solid model.
Courtesy of D. K. Lieu

1.04.04 Graphics as a Design Tool

Computer-based **three-dimensional (3-D) modeling** as an engineering design tool began in the 1980s. CAD was a great convenience, but it produced only drawings. In this sense, CAD was just a very accurate instrument for making drawings. A drawing's representation of an object in three dimensions had to be visualized by the person reading the drawing. It was the same for any fit or function of an assembly—the person reading the drawing had to visualize it. One problem was that not all readers visualized a drawing the same way. Three-dimensional modeling addressed those problems directly. Unlike a 2-D CAD drawing, which was a collection of 2-D objects used to represent specified views of an object, computer-based solid models had 3-D properties.

The field of mechanical engineering quickly adopted 3-D modeling, calling it **solid modeling**, for the design and analysis of mechanical parts and assemblies. Extrusion or revolution of 2-D shapes created simple 3-D geometries. More complex geometries were created by Boolean operation with simple geometries. The computer calculated a 3-D pictorial **image** of the part, which the engineer could see on a computer monitor. The biggest advantage of solid modeling over CAD was that it permitted viewing a 3-D object from different perspectives, greatly easing the **visualization** of a proposed object. Multiple parts could be viewed together as an assembly and examined for proper fitting. With solid modeling, graphics became more of an engineering design tool, rather than merely a drawing tool. An example of a solid **model** for a single part is shown in Figure 1.27. An assembly model is shown in Figure 1.28.

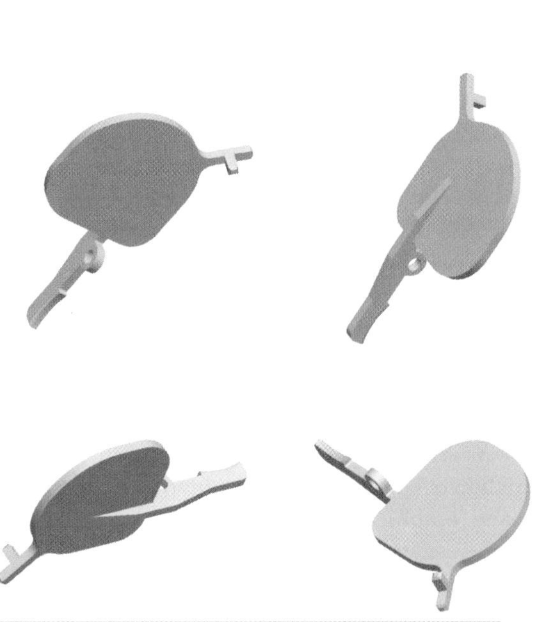

FIGURE 1.27. Solid modeling allows a proposed part to be easily visualized in a variety of orientations.
Courtesy of D. K. Lieu

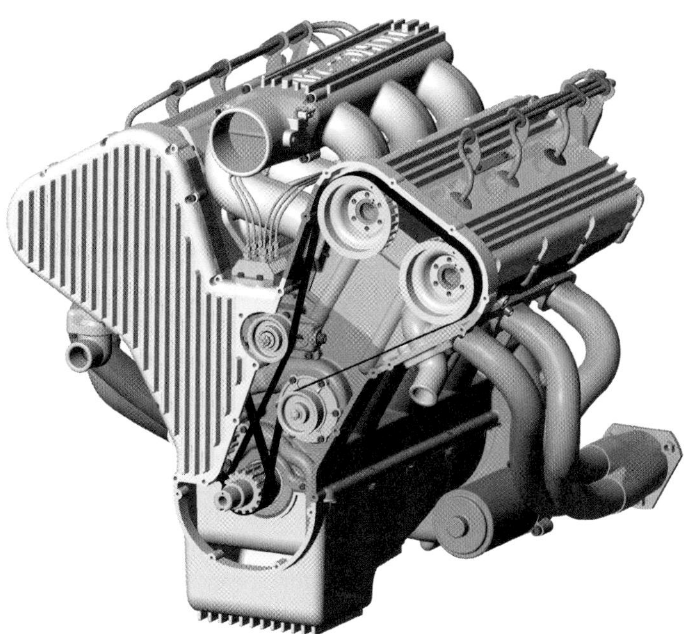

FIGURE 1.28. An assembly model of an Omnica 3.2-liter V-6 engine made from a collection of solid model parts.
Courtesy of SolidWorks Corporation

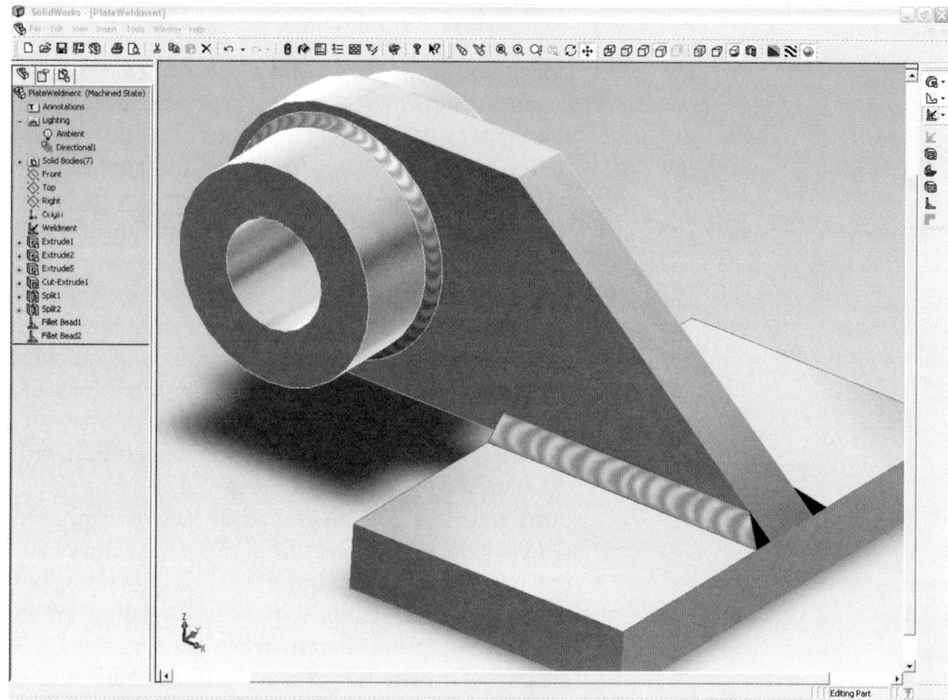

FIGURE 1.29. The graphical user interface of a solid modeling software program.
Courtesy of SolidWorks Corporation

As you may have realized, solid modeling required more computation power and memory to process files than CAD did. That is why solid modeling was originally introduced on computer workstations using UNIX operating systems, which were relatively costly at the time. In the late 1980s, a new software algorithm increased the utility of solid modeling by making it possible to link the sizes and locations of features on an object to variables that could be input and changed easily. The process was known as parametric design. Those products made it easy for an engineer to add, delete, or change the geometry and sizes of features on a part and see the results almost immediately. Dynamic viewing, which enabled the engineer to twist and turn the part image in real time, was also a powerful software feature. A particular facility of that software—the quick and easy extraction of engineering drawings from the 3-D model—made the total software package a valuable drawing tool as well as a modeling tool.

As PCs continued to become more powerful, in the 1990s solid modeling was introduced as a PC software product. The migration of solid modeling from expensive workstations to less expensive PCs made the software popular among small companies and individuals. The later development of new graphical user interfaces, such as the one shown in Figure 1.29, as opposed to the text menus prevalent at the time, made solid modeling easy to use, even for casual users. PC-based solid modeling with graphical user interfaces soon became a standard.

1.04.05 Graphics as an Analysis Tool

Prior to the 1970s, before the days of inexpensive digital computers and handheld calculators, many types of mathematical problems were solved using graphical techniques. Those types of problems included graphical vector analysis, roots and intersections of nonlinear functions, and graphical calculus. Numerical techniques now solve these problems more quickly and easily than graphical techniques, so graphical techniques are not used much anymore. Although solid modeling has decreased the usefulness of descriptive geometry as an analytical tool in many mechanical engineering applications, descriptive geometry still has useful applications in some large-scale civil, architectural, and mining projects. For the most part, drafting boards have been replaced with computers and CAD software, considerably improving accuracy as

FIGURE 1.30. Design of many large structures, such as the Forth Road Bridge, Scotland, shown here, still requires the use of classical two-dimensional drawing and analysis techniques.
Sources: Photo by William G. Godden. Reprinted with permission from EERC Library, Univ. of California, Berkeley, above. Photo by Fredrick T. Godden. Reprinted with permission from EERC Library, Univ. of California, Berkeley, below.

well as ease of use. However, the classical methods of finding distances, areas, inclines, and intersections used for land characterization and modifications are still used. Many recent large-scale construction and landscaping projects, such as the one shown in Figure 1.30, used classical 2-D graphical analysis and presentation methods.

Using solid modeling, the calculation of important mechanical properties of parts and assemblies can be done easily. The volume that a part or assembly occupies usually can be calculated with a single command after the computer model has been built. Properties of volume, such as mass, center of mass, moments of inertia, products of inertia, and principal axes, can also be calculated. Without a solid modeler, the calculation of these properties would be laborious, especially for complex geometries.

The analysis capability of 3-D modeling also has made it popular for certain types of analyses in civil engineering applications. Two-dimensional topographic maps, such

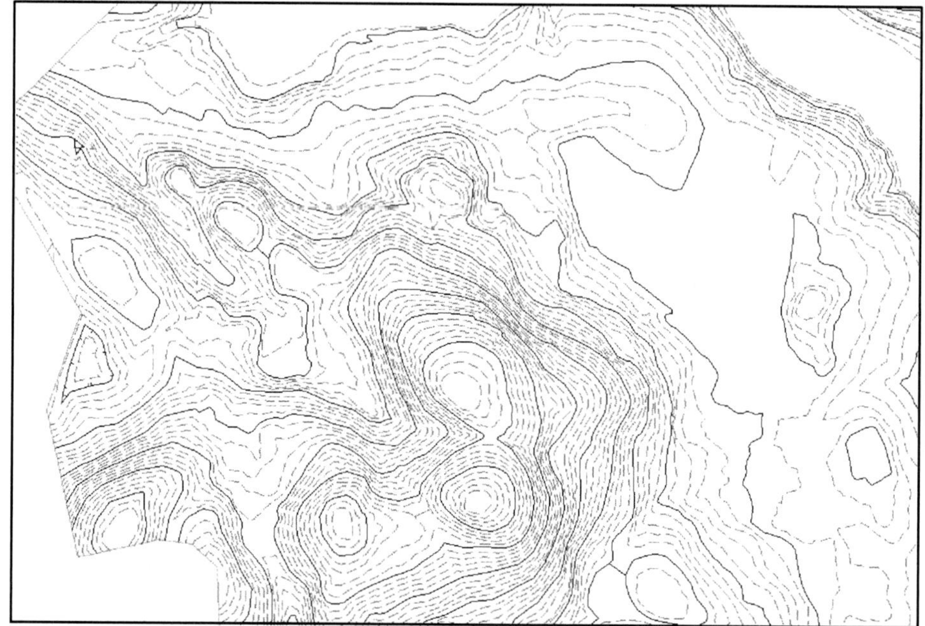

FIGURE 1.31. Classical two-dimensional presentation of land-height contours, natural landscape (top) and development for roads and housing (bottom).
Images courtesy of Autodesk Corporation.

as the one shown in Figure 1.31, shows land elevations at development sites for proposed residential areas before and after the addition of roads and building pads. The elevation contours of the land change, because certain locations are excavated while other locations are filled with earth to accommodate the roads and pads.

The use of 3-D land models, shown in Figure 1.32, generated from surveying data has made it easier for both engineers and nonengineers to visualize the appearance of a landscape before and after a proposed development. Further, the analytical capability of 3-D modeling in civil engineering applications has made it possible to quickly calculate the volumes of earth that must be removed or added to accommodate the development. It is even possible to match the total addition to the total removal of earth to minimize the volume changed from the site.

FIGURE 1.32. Three-dimensional images of the contours shown in the previous figure, original land (top) and modifications to accommodate roads and buildings (bottom). *Courtesy of Autodesk ® Civil 3D © 2005 software.*

1.04.06 Graphics as a Presentation Tool

An engineer must be able to communicate not only ideas and designs but also precise engineering data. Whether this data is empirical, as collected from experiments, or analytical, as calculated from mathematical models, they must be presented so other people can understand them quickly and easily. Traditional methods of data presentation are in the form of charts and graphs. Charts include familiar items such as pie charts and bar charts commonly used for presenting data to the general public. Graphs, which are usually more technical, show data trends when the relationship between two or more variables is plotted on orthogonal axes. Examples of these types of data presentation are shown in Figure 1.33.

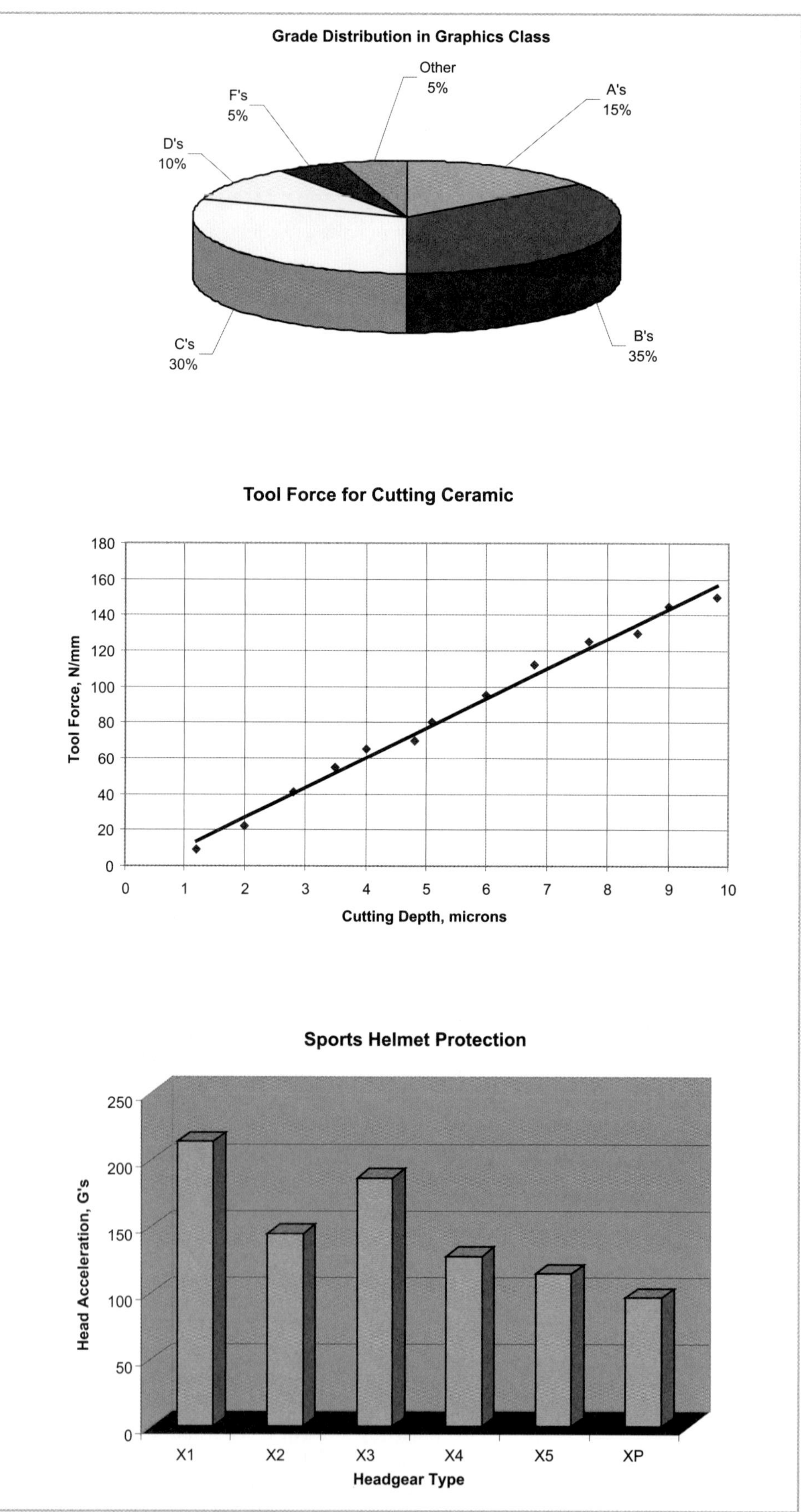

FIGURE 1.33. Data presentation and analysis is a vital part of engineering.
Courtesy of D. K. Lieu

Three-dimensional modeling software is also used to build geometric models that can be exported for finite element analysis (FEA). FEA is a numerical analysis method used to calculate results such as stress distribution, temperature distribution, or deformation in a part. Although FEA usually is not considered a formal part of engineering graphics, one of the most efficient and effective methods of presenting FEA results is to show the predicted contours of variables such as stress, deflection, or temperature atop a pictorial of the object. Different colors are used to represent different magnitudes of a variable. For example, Figure 1.34 shows how a solid model of the teeth, steel, and magnets of a small electric motor are created for geometry analysis. The same model is then used to generate a FEA mesh in preparation for an analysis of the magnetic-flux density distribution in the structure. The flux densities are calculated and their contours are plotted directly atop the original solid model image to show the location and magnitude of the flux densities in the motor.

A popular and effective data presentation method is to show the stress distribution in a part by plotting stress contours directly on the part image, as shown in Figure 1.35. In this way, the location and level of the highest stress in the part can be located easily. The same technique can be used for plotting the temperature distribution and magnetic flux densities in a part.

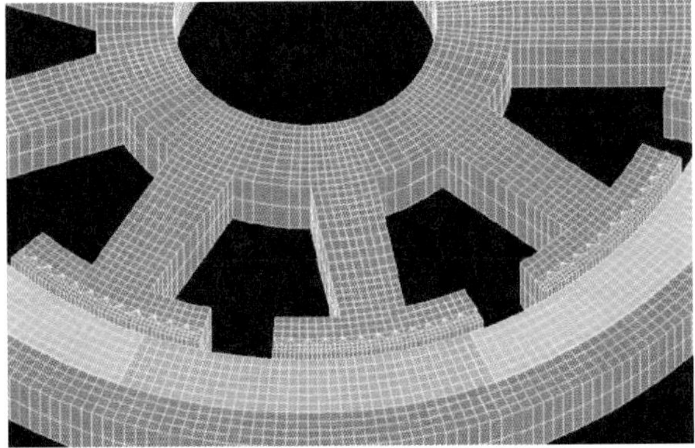

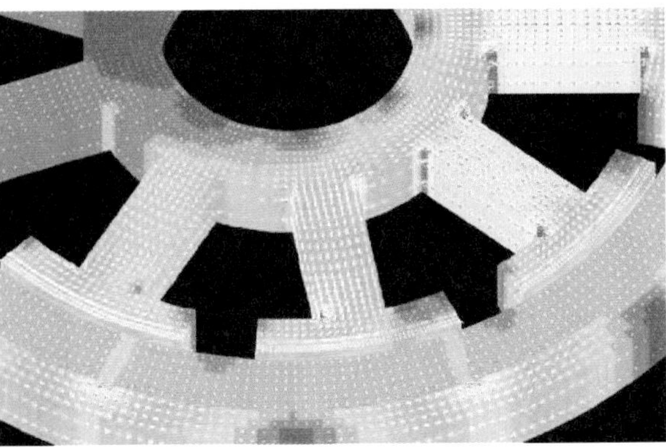

FIGURE 1.34. A three-dimensional model (top) of an electric motor is used to create a FEM mesh (center) from which magnetic flux densities can be calculated and presented (bottom).
Courtesy of D. K. Lieu

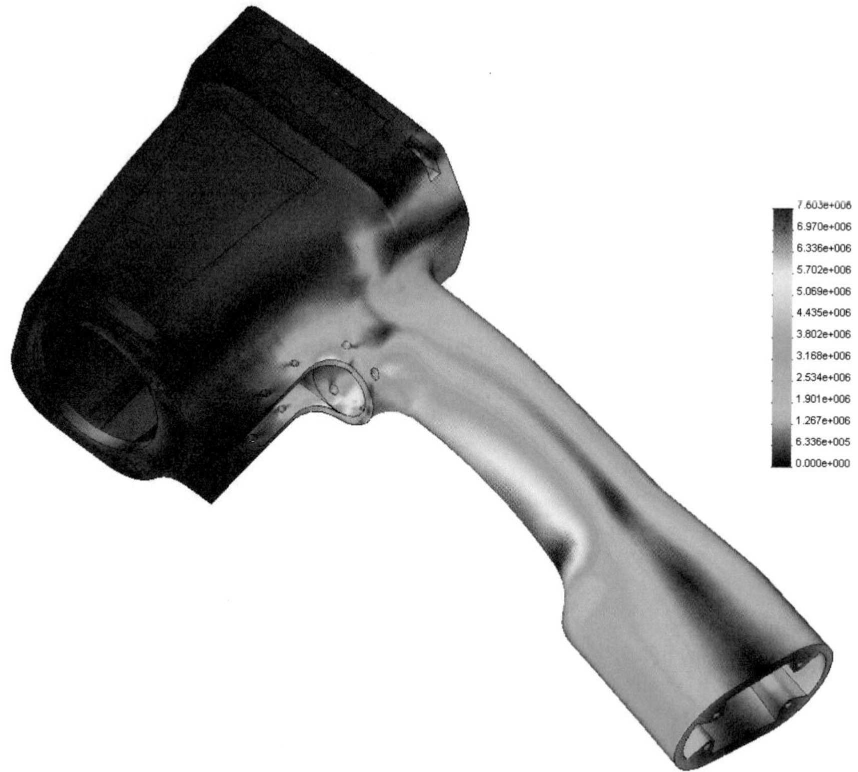

FIGURE 1.35. Graphical representation of stress, such as that produced by forces applied to the part shown, is an important part of presenting the results of a finite element analysis. *Courtesy of SolidWorks Corporation*

1.05 The Modern Role of Engineering Graphics

Although the role of engineering graphics has evolved over the years, many aspects remain the same. Graphics remains the medium for communicating ideas and technical information. The best way to communicate an idea for a part or device is to show a picture of it. In the past, the pictures were crude handmade drawings, which required time and skill to create. Now pictures are computer-generated images of 3-D models that can be turned and rotated so they are viewable from any direction, providing more accurate depictions. Because the models are easy to create, many variations can be created and viewed in a short time. This advantage makes 3-D modeling useful not only as a means of communication but also as a means of design.

Recording the history of a design also remains an important role of engineering graphics. In the past, recording the history of designs usually meant saving master hard-copy drawings in cabinets in some sort of vault. The smallest change in a design meant changing the master copy and then sending updated copies of the master to whoever needed them. Copies commonly suffered distortion or reduced resolution due to the machines that made the copies. While hard-copy drawings are still necessary, most drawing data are now stored as electronic files. There are enormous advantages in the cataloging, retrieval, and transmission of data stored in this manner. Today model and drawing data and their updates can be sent across the world in a fraction of a second with no loss in resolution.

Engineering graphics remains an analysis tool, but the type of analysis has changed. Graphical means are no longer used to solve vector algebra, mathematics, or calculus problems. Instead, graphical models are now used to do things like examine the proper fit and function of parts within assemblies. Using 3-D models, engineers can examine parts in their final assembled state for proper motion and location. Engineers can extract the volumetric and inertial properties of the parts and assemblies, to ensure that they fit as specified. Based on externally applied forces, the stresses and deflections in the material also can be examined to ensure that failure of the device does not occur.

Formal engineering drawing remains a part of the overall design process. The traditional role of formal engineering drawings was to ensure that parts would be fabricated to specified sizes, that they would appear as specified, and that various parts would fit together properly. Prior to the 1990s, most engineering graphics classes concentrated on drawing technique and accuracy and on proper use of mechanical drawing instruments. Since engineering drawings can now be created easily and accurately with computers and software tools, the effort required by the formal drawing process is greatly reduced from what it was in the past. Since most computer graphics tools are easy to master, modern graphics classes concentrate mainly on visualization, analysis, function, and **optimization** of designs.

The development of visualization skills is a particular goal of modern engineering graphics courses. Developing visualization skill is necessary for envisioning, specifying, and creating complex designs with functional features in the three-spatial dimensions. Traditionally, these skills are developed through hand-eye coordination involving physical parts. Hands-on experience, such as repairing an automobile or a bicycle, constructing models, or playing with building toys, is helpful for developing visualization skills. In an engineering graphics curriculum, these skills can be developed by doing special visualization exercises and by building and working with solid models. Another method of developing visualization skills is to disassemble and reassemble engineered devices in a process known as mechanical dissection. During this process, students examine the operating concepts and their practical implementation, as shown in Figure 1.36.

Sketching also has proven to be a valuable technique for developing visualization skills, as shown in Figure 1.37. Sketches, which can be prepared quickly, provide a simple graphical representation of an idea with a great deal of information on concepts and appearance, without the need for formal drawing tools. For this reason, even though powerful computers and software are available, sketching remains a part of engineering graphics, both as a learning tool and as a practical skill, as you will see throughout this textbook.

FIGURE 1.36. The construction and function of a device can be learned from its disassembly, examination, and reassembly in a process known as mechanical dissection.
Courtesy of D. K. Lieu

FIGURE 1.37. Sketching is not only a useful skill but also an excellent exercise for developing spatial reasoning abilities. *Courtesy of D. K. Lieu*

1.06 Chapter Summary

The history of graphical communication has shown it to be vital in nearly all aspects of engineering. The development of technology, tools, and techniques used for engineering graphics has advanced, with all of the developments supporting each other. Technological tools have made the tasks associated with classical engineering graphics much easier. The technical sophistication and simple human interface of new tools have enabled engineers to concentrate on learning and developing the techniques offered by the tools, instead of merely operating the tools. Advances in computing, modeling, and display tools have increased the speed and accuracy with which communication, visualization, and analytical problems are performed. More complex designs can be produced more quickly with better functionality and fewer errors than in the past. Engineering drawing has become quicker and simpler; making it possible for engineers to concentrate on what they do best, which is to examine the functionality of a design and to optimize it for its intended environment. Engineers have new responsibilities associated with the new tools, including following protocols for the construction of proper computer models, the electronic transmission of data, and data management.

1.07 glossary of key terms

assembly: A collection of parts that mate together to perform a specified function or functions.

CAD: Computer-aided drawing. The use of computer hardware and software for the purpose of creating, modifying, and storing engineering drawings in an electronic format.

descriptive geometry: A two-dimensional graphical construction technique used for geometric analysis of three-dimensional objects.

design (noun): An original manifestation of a device or method created for performing one or more useful functions.

design (verb): The process of creating a design (noun).

drawing: A collection of images and other detailed graphical specifications intended to represent physical objects or processes for the purpose of accurately re-creating those objects or processes.

1.07 glossary of key terms (continued)

engineer (verb): To plan and build a device that does not occur naturally within the environment.

engineer (noun): A person who engages in the art of engineering.

engineering: The profession in which knowledge of mathematical and natural sciences gained by study, experience, and practice is applied with judgment to develop and utilize economically the materials and forces of nature for the benefit of humanity.

fabricate: To make something from existing materials.

image: A collection of printed, displayed, or imagined patterns intended to represent real objects, data, or processes.

instruments: In engineering drawing, mechanical devices used to aid in creating accurate and precise images.

model: A mathematical representation of an object or a device from which information about its function, appearance, or physical properties can be extracted.

optimization: Modification of shapes, sizes, and other variables to achieve the best performance based on predefined criteria.

part: A single object fabricated to perform one or more functions.

project: In engineering, a collection of tasks that must be performed to create, operate, or retire a system or device.

solid modeling: Three-dimensional modeling of parts and assemblies originally developed for mechanical engineering use but presently used in all engineering disciplines.

system: A collection of parts, assemblies, structures, and processes that work together to perform one or more prescribed functions.

three-dimensional (3-D) modeling: Mathematical modeling where the appearance, volumetric, and inertial properties of parts, assemblies, or structures are created with the assistance of computers and display devices.

two-dimensional (2-D) drawing: Mathematical modeling or drawing where the appearance of parts, assemblies, or structures are represented by a collection of two-dimensional geometric shapes.

visualization: The ability to create and manipulate mental images of devices or processes.

1.08 questions for review

1. Why are most cave drawings and hieroglyphics not considered to be engineering drawings?
2. In what ways did the design of military fortifications change after the discovery of gunpowder and the invention of the cannon?
3. Why did engineering drawings need to become more precise during the industrial revolution?
4. What were the three traditional roles of engineering graphics?
5. What are some of the new roles of engineering graphics created by computer graphics?
6. What are some of the advantages and disadvantages of using mechanical drawing instruments, as opposed to mathematical tools, for problem solving?
7. What are some of the advantages and disadvantages of using mechanical drawing instruments, as opposed to computational tools, for problem solving?
8. How is solid modeling different from CAD?
9. What is visualization?
10. In what ways can visualization skills be developed?

1.09 problems

Graphical communications makes the lives of engineers easier in many ways. The following exercises are intended to give you a feeling of what communication and analysis would be like without the tools and techniques used in engineering graphics. Do not become discouraged if you find these exercises to be difficult or cumbersome or if you find that the results are not accurate, which is the point of these exercises. In the chapters that follow, you will be introduced to methods of addressing the difficulties you encounter here.

1a. Do this exercise with one of your classmates. Select one or more of the objects shown in Figure P1.1, but

1.09 problems (continued)

do not show the object(s) to your partner. Using only words, give your partner a complete description of the objects you selected. Then have your partner sketch a picture of the objects based on your verbal descriptions. Reverse roles using different objects. What errors occurred between the objects that were being described and the objects that were envisioned? What can be done to reduce these errors?

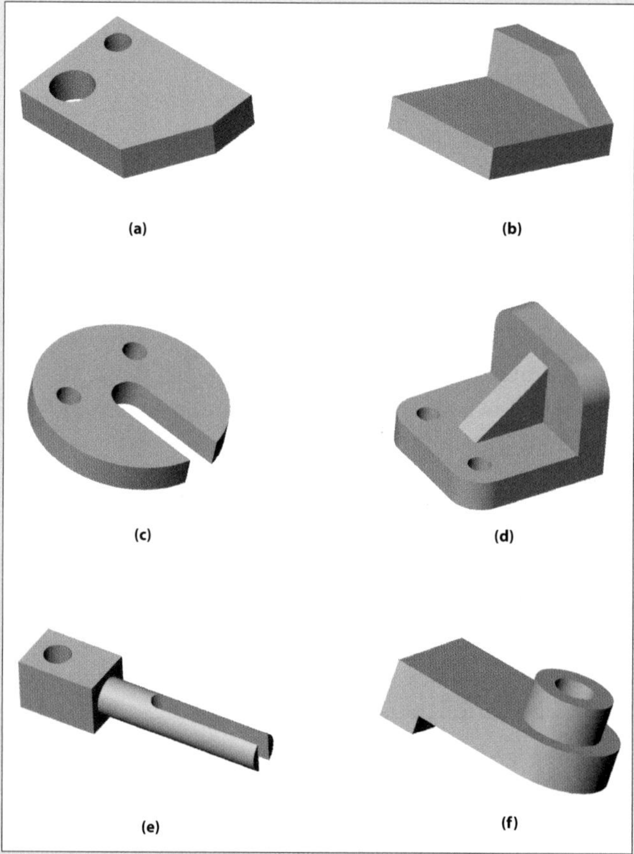

FIGURE P1.1. (a)–(f) Verbally describe these objects to your partner; then have your partner sketch a picture of the object.
Courtesy of D. K. Lieu

1b. Give a third classmate the sketches made in Part A of this exercise. Without revealing what the original objects in the figure look like, give a complete description of the errors in the sketches and have this person make corrections to the sketches. Reverse roles using different objects. How much closer are the sketches to representing the objects shown in the figure? What additional problems occur when a third person is involved?

2. Do this exercise with a group of classmates. Select one or more of the objects shown in Figure P1.2 but do not show the figure(s) to the rest of the group. Make sketches of the object(s) you have selected, give them to the first person in the group, have that person examine them carefully, and then retrieve your sketches. Have that person use the memory of your sketches to make new sketches. Then give the new person's sketches to the second person in the group. Do not show the previous sketches to the new person. Repeat for all of the classmates in the group. When the last person is done, compare the final set of sketches to the objects selected by the first person in the original figure. What errors occurred between the final sketches and the objects that were selected? What happens to the sketches with each revision? What can be done to reduce these errors?

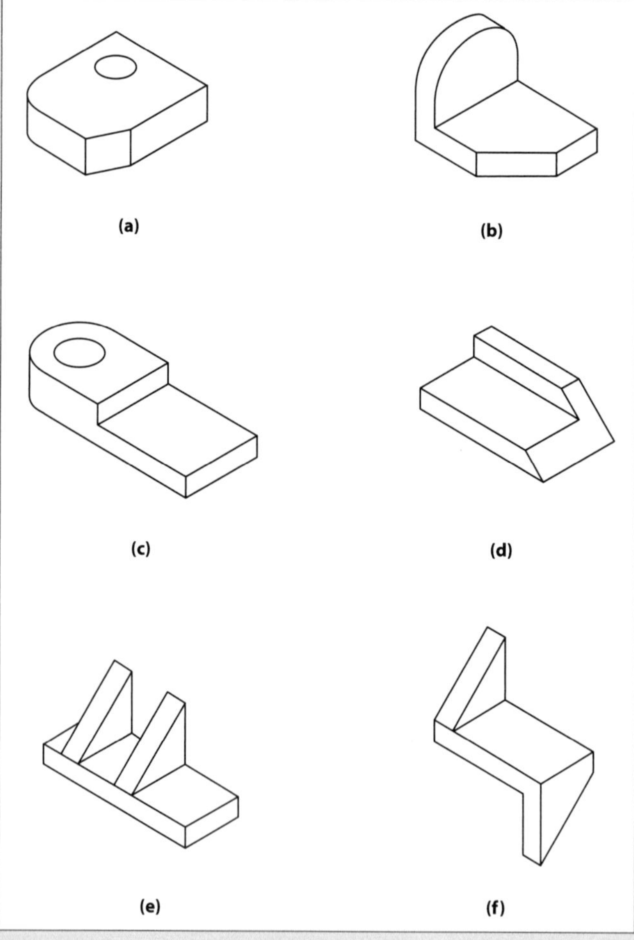

FIGURE P1.2. (a)–(f) Show one or more of these objects to your partner and have your partner sketch the object(s) from memory. Repeat the process with the newly created sketch. Compare the sketch to the original object.
Courtesy of D. K. Lieu

1.09 problems (continued)

3. For the geometric elements shown in each of the three panels of Figure P1.3, develop formulas for finding the length, angle, area, or volume, whichever is required in each panel, using analytical methods. Generalize the solution in terms of x-, y-, and z-coordinates of the points given. What problems do you envision if the person making the calculations has no access to computers, calculators, or any other computational aids? What happens to the solution formulas as the geometries become more complicated or are rotated and translated in space?

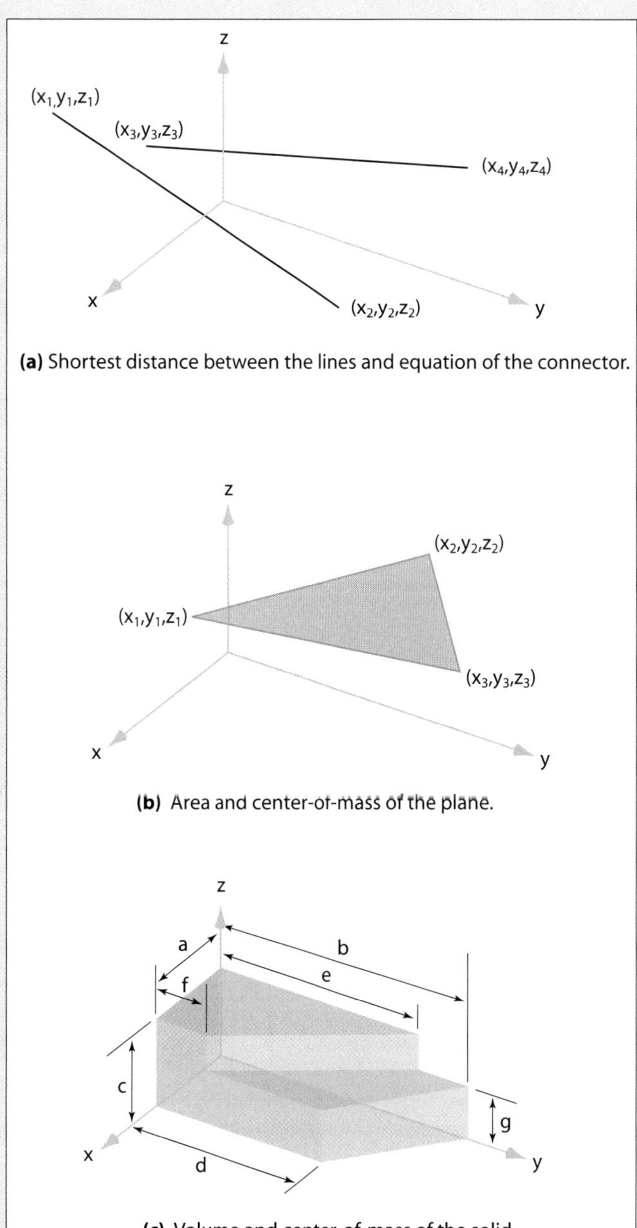

FIGURE P1.3. (a)–(c) Find the specified geometric properties of the objects.
Courtesy of D. K. Lieu

2 Sketching

objectives

After completing this chapter, you should be able to

- Explain the importance of sketching in the engineering design process
- Make simple sketches of basic shapes such as lines, circles, and ellipses
- Use 3-D coordinate systems, particularly right-handed systems
- Draw simple isometric sketches from coded plans
- Make simple oblique pictorial sketches
- Use advanced sketching skills for complex objects

2.01 introduction

Sketching is one of the primary modes of communication in the initial stages of the design process. Sketching also is a means to creative thinking. It has been shown that your mind works more creatively when your hand is sketching as you are engaged in thinking about a problem.

This chapter focuses on one of the fundamental skills required of engineers and technologists—freehand sketching. The importance of sketching in the initial phases of the design process is presented, as are some techniques to help you create sketches that correctly convey your design ideas. The definition of 3-D coordinate systems and the way they are portrayed on a 2-D sheet of paper will be covered, along with the difference between right-handed and left-handed coordinate systems. The chapter will investigate how to create simple pictorial sketches. Finally, the advanced sketching techniques of shading and cartooning will be presented with a framework for creating sketches of complex objects. You will begin to explore these topics in this chapter and will further refine your sketching abilities as you progress through your graphics course.

2.02 Sketching in the Engineering Design Process

As you may remember from Chapter 1, engineers communicate with one another primarily through graphical means. Those graphical communications take several forms, ranging from precise, complex drawings to simple sketches on the back of an envelope. Most of this text is focused on complex drawings; however, this chapter focuses on simple sketches.

Technically speaking, a sketch is any drawing made without the use of drawing instruments such as triangles and T squares. Some computer graphics packages allow you to create sketches; however, you will probably be more creative (and thus more effective) if you stick to hand sketching, particularly in the initial stages of the design process. In fact, carefully constructed, exact drawings often serve as a hindrance to creativity when they are employed in the initial stages of the design process. Typically, all you need for sketching are a pencil, paper, an eraser, and your imagination.

Your initial sketches may be based on rough ideas. But as you refine your ideas, you will want to refine your sketches, including details that you left out of the originals. For example, suppose you were remodeling the bathroom in your house. Figure 2.01 shows two sketches that define the layout of the bathroom, with details added as ideas evolve. Once you have completed the layout to your satisfaction, you can create an official engineering drawing showing exact dimensions and features that you can give to the contractor who will perform the remodeling work for you.

When engineers sit down to brainstorm solutions to problems, before long, one of them usually takes out a sheet of paper and sketches an idea on it. The others in the

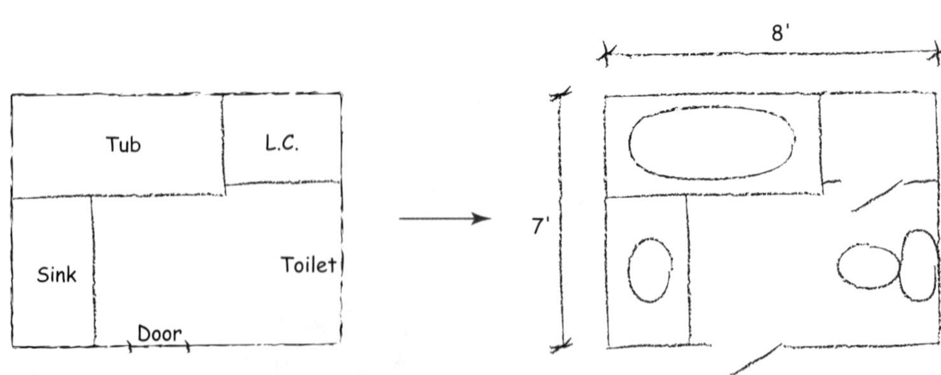

FIGURE 2.01. Sketches for a bathroom remodel.

discussion may add to the original sketch, or they may create sketches of their own. The paper-and-pencil sketches then become media for the effective exchange of ideas. Although few "rules" regulate the creation of sketches, you should follow some general guidelines to ensure clarity.

2.03 Sketching Lines

Most of your sketches will involve basic shapes made from lines and circles. Although you are not expected to make perfect sketches, a few simple techniques will enable you to create understandable sketches.

When drawing **lines**, the key is to make them as straight as possible. If you are right-handed, you should sketch your vertical lines from top to bottom and your horizontal lines from left to right. If you are sketching an angled line, choose a direction that matches the general inclination of the line—for angled lines that are mostly vertical, sketch them from top to bottom; for angled lines that are mostly horizontal, sketch them from left to right. If you are left-handed, you should sketch your vertical lines from top to bottom, but your horizontal lines from right to left. For angled lines, left-handed people should sketch from either right to left or top to bottom, again depending on the inclination of the line. To keep your lines straight, focus on the endpoint as you sketch. The best practices for sketching straight lines are illustrated in Figure 2.02.

FIGURE 2.02. Techniques for sketching straight lines.

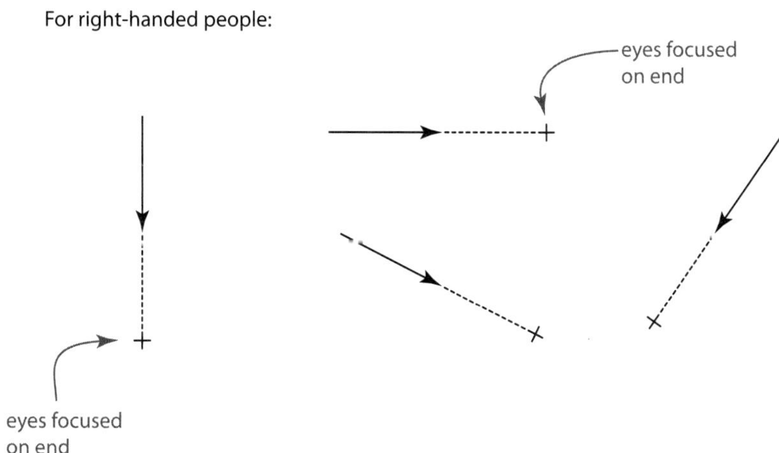

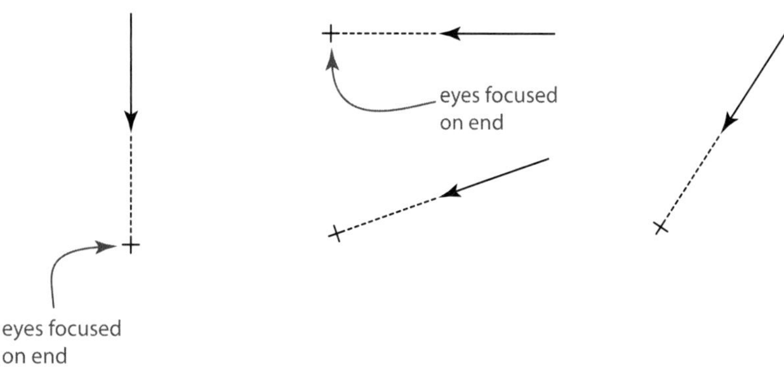

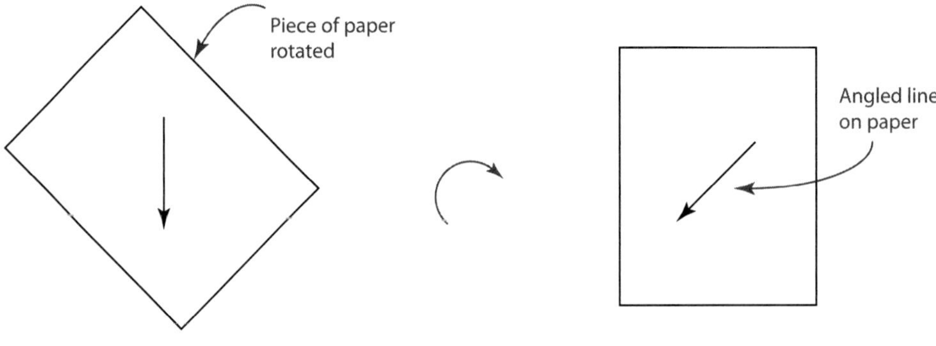

FIGURE 2.03. Rotating the paper to draw an angled line.

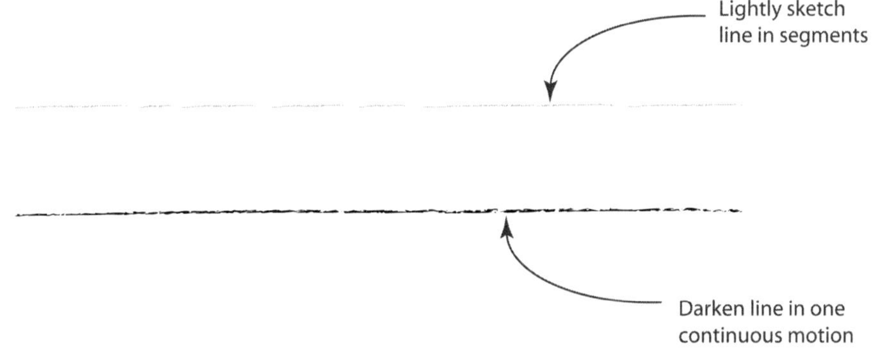

FIGURE 2.04. Sketching long lines in segments.

You also can try rotating the paper on the desk to suit your preferences. For example, if you find that drawing vertical lines is easiest for you and you are confronted with an angled line to sketch, rotate the paper on the desk so you can sketch a "vertical" line. Or you can rotate the paper 90 degrees to sketch a horizontal line. Figure 2.03 illustrates rotation of the paper to create an angled line.

One last point to consider when sketching lines is that you initially may have to create "long" lines as a series of connected segments. Then you can sketch over the segments in a continuous motion to make sure the line appears to be one entity and not several joined end to end. Using segments to define long lines is illustrated in Figure 2.04.

2.04 Sketching Curved Entities

Arcs and **circles** are other types of geometric entities you often will be required to sketch. When sketching arcs and circles, use lightly sketched square **bounding boxes** to define the limits of the curved entities and then construct the curved entities as tangent to the edges of the bounding box. For example, to sketch a circle, you first lightly sketch a square (with straight lines). Note that the length of the sides of the bounding box is equal to the diameter of the circle you are attempting to sketch. At the centers of each edge of the box, you can make a short **tick mark** to establish the point of tangency for the circle, then draw the four arcs that make up the circle. Initially, you may find it easier to sketch one arc at a time to complete the circle; but as you gain experience, you may be able to sketch the entire circle all at once. Figure 2.05 shows the procedure used to sketch a circle by creating a bounding box first.

One problem you may have when using a bounding box to sketch a circle occurs when the radius of the circle is relatively large. In that case, the arcs you create may be too flat or too curved, as shown in Figure 2.06. To avoid this type of error, you might try marking the radius at points halfway between the tick marks included on the bounding box. Using simple geometry, when you draw a line between the center of the circle and the corner of the bounding box, the radius is about two-thirds of the distance (technically, the radius is 0.707, but that number is close enough to two-thirds for your purposes). Then you can include some additional tick marks around the circle to guide your sketching and to improve the appearance of your circles. This technique is illustrated in Figure 2.07.

Sketching an arc follows the same general procedure as sketching a circle, except that your curved entity is only a portion of a circle. Sketching an **ellipse** follows the same general rules as sketching a circle, except that your bounding box is a rectangle and not a square. Sketching arcs and ellipses is illustrated in Figure 2.08.

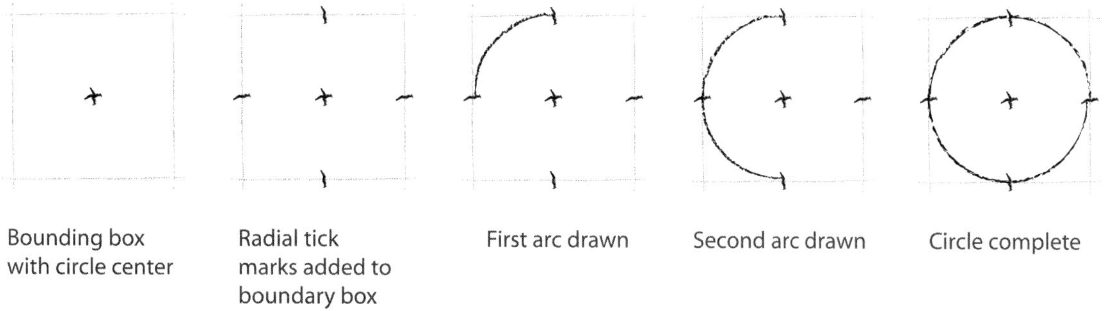

FIGURE 2.05. Sketching a circle using a bounding box.

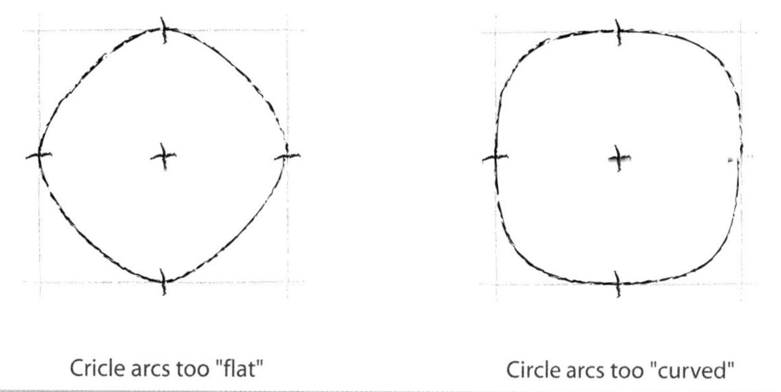

Cricle arcs too "flat"
Circle arcs too "curved"

FIGURE 2.06. Circles sketched either too flat or too curved.

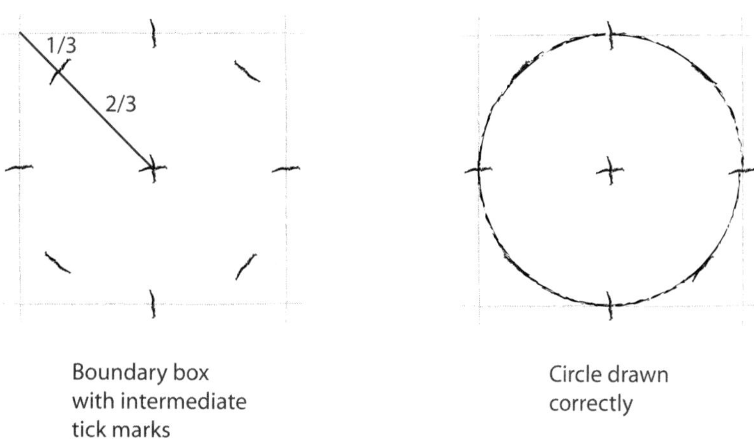

FIGURE 2.07. Using intermediate radial tick marks for large circles.

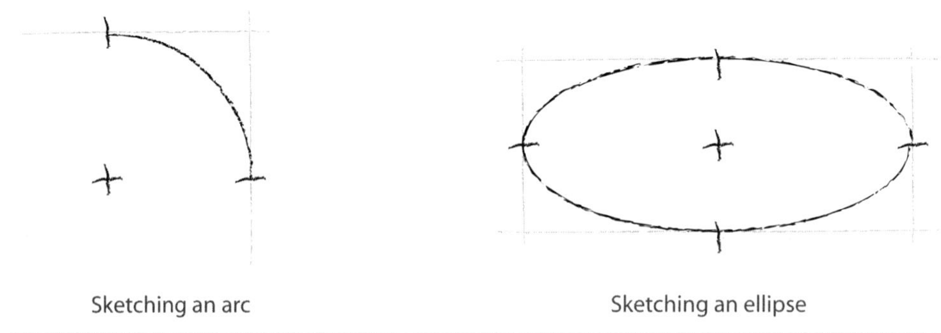

FIGURE 2.08. Using boundary boxes to sketch arcs and ellipses.

2.05 Construction Lines

Similar to the way you used bounding boxes to create circles and ellipses, other construction lines help with your sketching. Using construction lines, you outline the shape of the object you are trying to sketch. Then you fill in the details of the sketch using the construction lines as a guide. Figure 2.09 shows the front view of an object you need to sketch. To create the sketch, you lightly draw the construction lines that outline the main body of the object and then create the construction lines that define the prominent features of it. One rule of thumb is that construction lines should be drawn so lightly on the page that when it is held at arm's length, the lines are nearly impossible to see. The creation of the relevant construction lines is illustrated in Figure 2.10.

Using construction lines as a guide, you can fill in the details of the front view of the object until it is complete. The final result is shown in Figure 2.11.

Another way you can use construction lines is to locate the center of a square or rectangle. Recall from your geometry class that the diagonals of a box (either a rectangle or a square) intersect at its center. After you create construction lines for the edges of the box, you sketch the two diagonals that intersect at the center. Once you find the center of the box, you can use it to create a new centered box of smaller dimensions—a kind of concentric box. Locating the center of a box and creating construction lines for a newly centered box within the original box are illustrated in Figure 2.12.

Once you have created your centered box within a box, you can sketch a circle using the smaller box as a bounding box, resulting in a circle that is centered within the larger box as shown in Figure 2.13. Or you can use these techniques to create a square with four holes located in the corners of the box as illustrated in Figure 2.14.

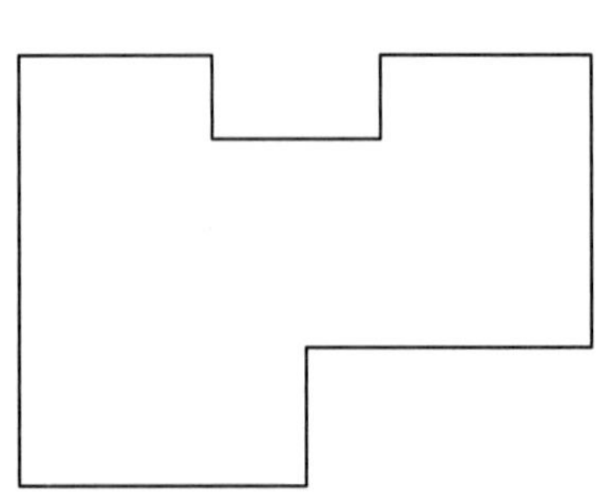

FIGURE 2.09. The front view of an object to sketch.

FIGURE 2.10. Construction lines used to create a sketch.

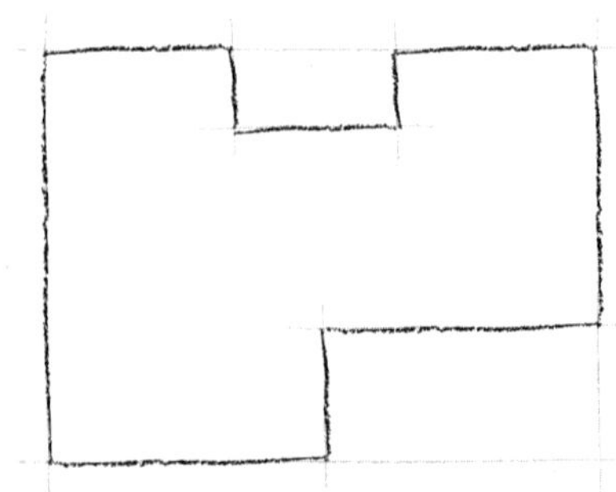

FIGURE 2.11. Completed sketch using construction lines as a guide.

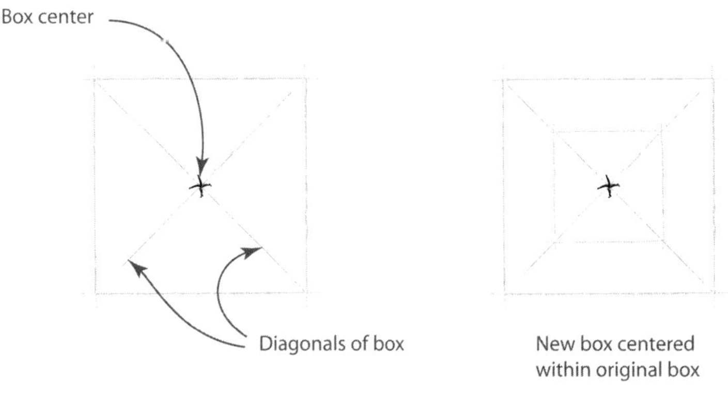

FIGURE 2.12. Creating concentric bounding boxes.

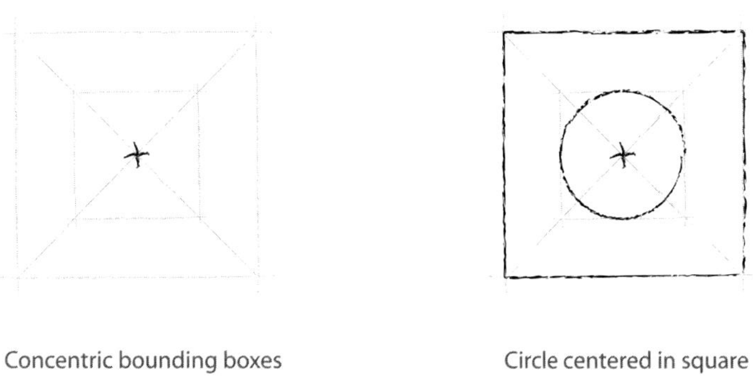

FIGURE 2.13. Sketching a circle in a box.

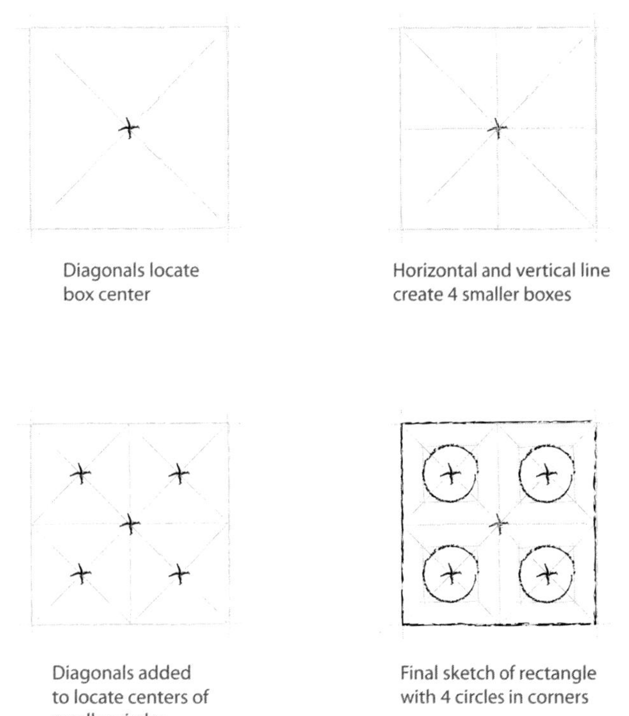

FIGURE 2.14. Using diagonal construction lines to locate centers.

2.06 Coordinate Systems

When sketching, you often have to portray 3-D objects on a flat 2-D sheet of paper. As is usually the case with graphical communication, a few conventions have evolved over time for representing 3-D space on a 2-D sheet of paper. One convention, called the **3-D coordinate system**, is that space can be represented by three mutually perpendicular coordinate axes, typically the x-, y-, and z-axes. To visualize those three axes, look at the bottom corner of the room. Notice the lines that are formed by the intersection of each of the two walls with the floor and the line that is formed where the two walls intersect. You can think of these lines of intersection as the x-, y-, and z-coordinate axes. You can define all locations in the room with respect to this corner, just as all points in 3-D space can be defined from an origin where the three axes intersect.

You are probably familiar with the concept of the three coordinate axes from your math classes. In Figure 2.15, a set of coordinate axes, notice the positive and negative directions for each of the axes. Typically, arrows at the ends of the axes denote the positive direction along the axes.

For engineering, the axes usually define a right-handed coordinate system. Since most engineering analysis techniques are defined by a right-handed system, you should learn what this means and how to recognize such a system when you see it. A **right-handed system** means that if you point the fingers of your right hand down the positive x-axis and curl them in the direction of the positive y-axis, your thumb will point in the direction of the positive z-axis, as illustrated in Figure 2.16. This procedure is sometimes referred to as the **right-hand rule**.

Another way to think about the right-hand rule is to point your thumb down the positive x-axis and your index finger down the positive y-axis; your middle finger will then automatically point down the positive z-axis. This technique is illustrated in Figure 2.17. Either method for illustrating the right-hand rule results in the same set of coordinate axes; choose the method that is easiest for you to use.

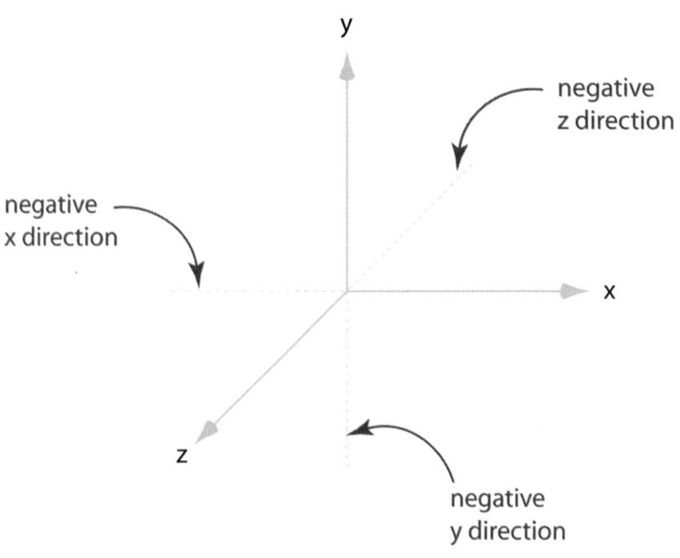

FIGURE 2.15. The x-, y-, and z- coordinate axes.

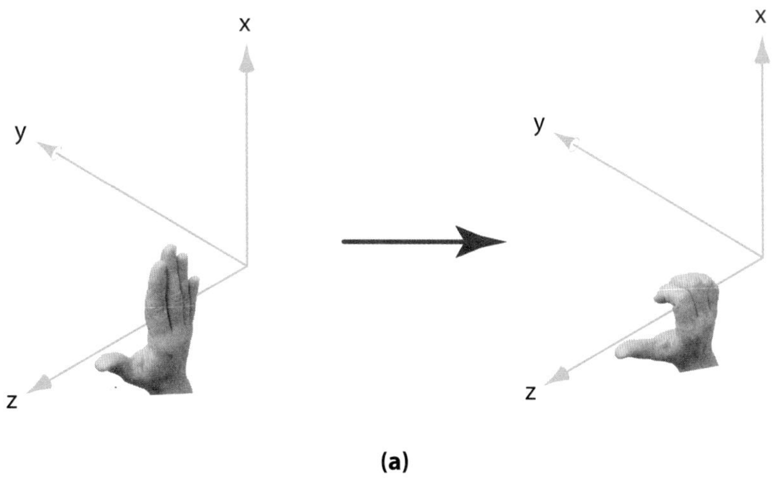

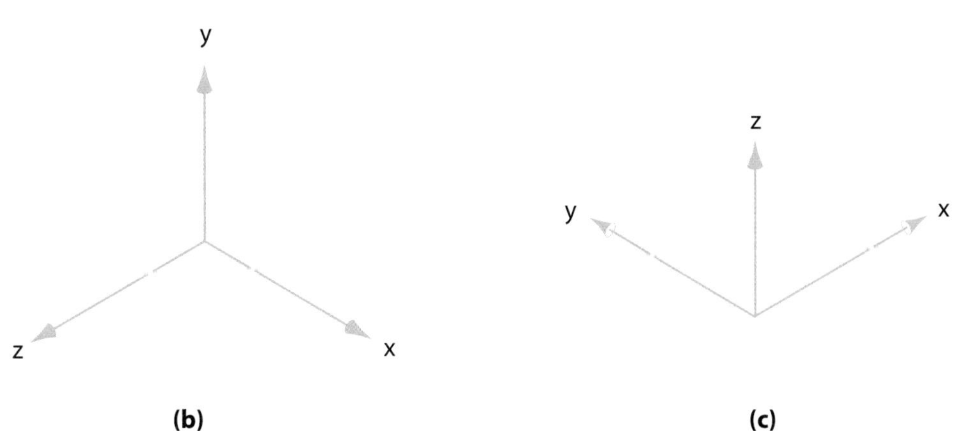

FIGURE 2.16. Curling the fingers to check for a right-handed coordinate system in (a) and alternative presentations of right-handed coordinate systems in (b) and (c).

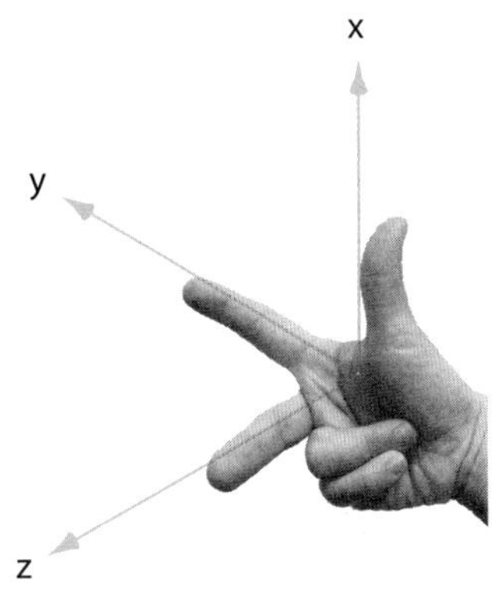

FIGURE 2.17. An alternative method to check for a right-handed coordinate system.

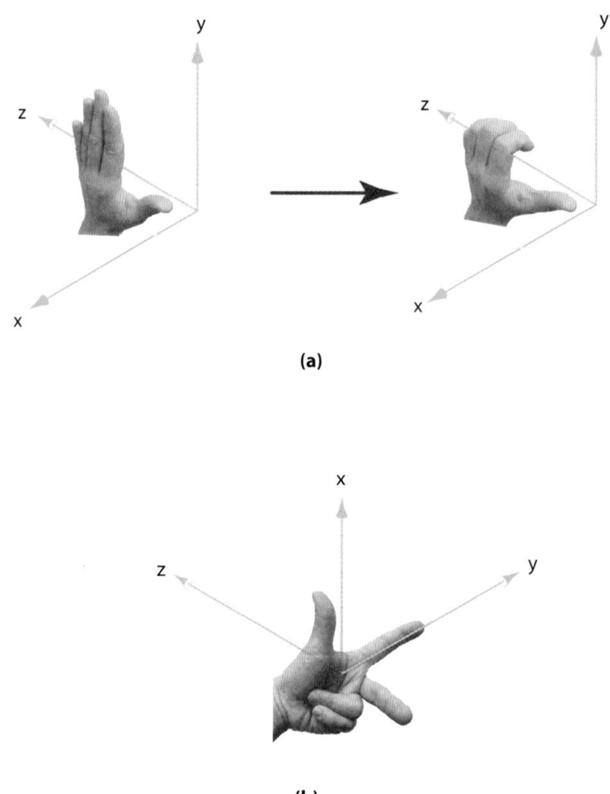

FIGURE 2.18. The result of using the left hand to test for a right-handed coordinate system.

Notice that if you try either technique with your left hand, your thumb (or middle finger) will point down the negative z-axis, as illustrated in Figure 2.18.

A **left-handed system** is defined similarly to a right-handed system, except that you use your left hand to show the positive directions of the coordinate axes. Left-handed systems are typically used in engineering applications that are geologically based—positive z is defined as going down into the earth. Figure 2.19 illustrates left-handed coordinate systems. (Use the left-hand rule to verify that these are left-handed coordinate systems.)

The question remains about how to represent 3-D space on a 2-D sheet of paper when sketching. The answer is that the three coordinate axes are typically represented as oblique or isometric, depending on the preferences of the person making the sketch. You are probably most familiar with oblique representation of the coordinate axes, which seems to be the preferred method of many individuals. With this method, two axes are sketched perpendicular to each other and the third is drawn at an angle, usually 45 degrees to both axes. The angle of the inclined line does not have to be 45 degrees, but

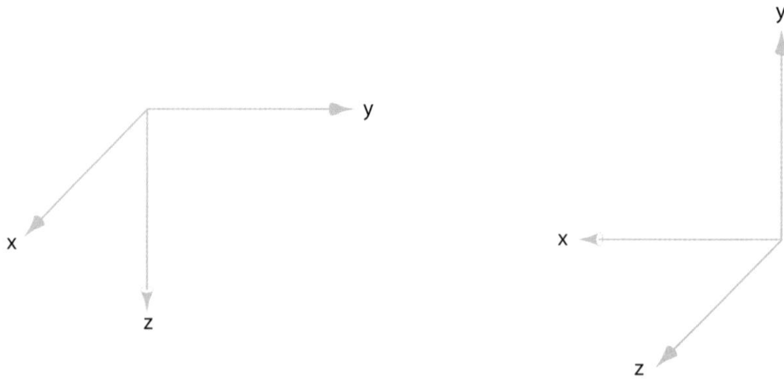

FIGURE 2.19. Left-handed coordinate systems.

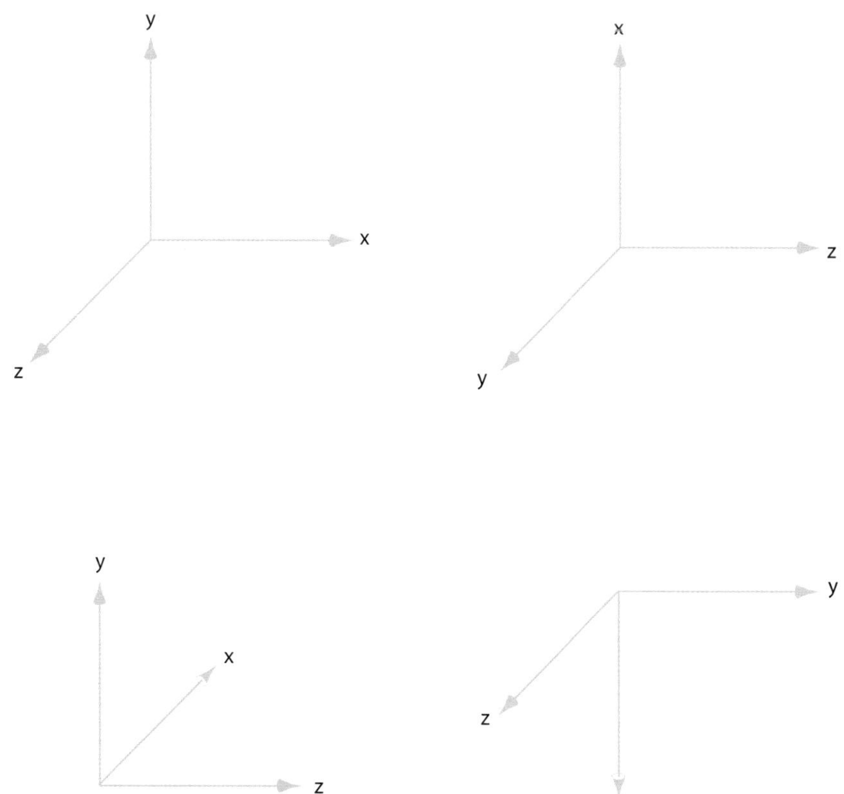

FIGURE 2.20. An oblique representation of right-handed coordinate systems.

it is usually sketched that way. Your math teachers probably sketched the three coordinate axes that way in their classes. Figure 2.20 shows multiple sets of coordinate axes drawn as oblique axes. Notice that all of the coordinate systems are right-handed systems. (Verify this for yourself by using the right-hand rule.)

Another way of portraying the 3-D coordinate axes on a 2-D sheet of paper is through isometric representation. With this method, the axes are projected onto the paper as if you were looking down the diagonal of a cube. When you do this, the axes appear to be 120 degrees apart, as shown in Figure 2.21. In fact, the term *isometric* comes from the Greek *iso* (meaning "the same") and *metric* (meaning "measure"). Notice that for **isometric axes** representations, the right-hand rule still applies.

Isometric axes also can be sketched with one of the axes extending in the "opposite" direction. This results in angles other than 120 degrees, depending on the orientation of the axes with respect to the paper, as shown in Figure 2.22.

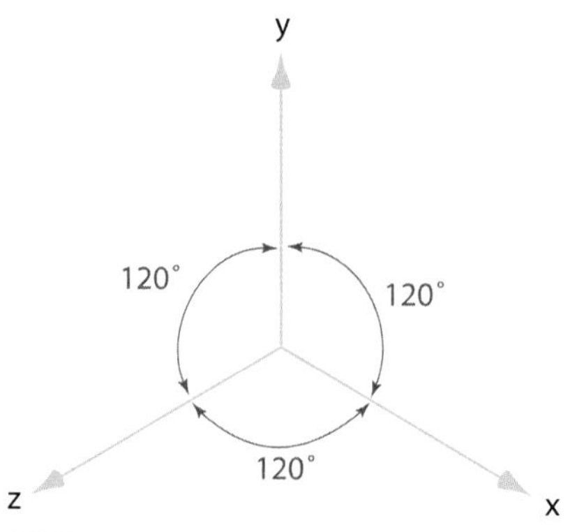

FIGURE 2.21. An isometric representation of a right-handed coordinate system.

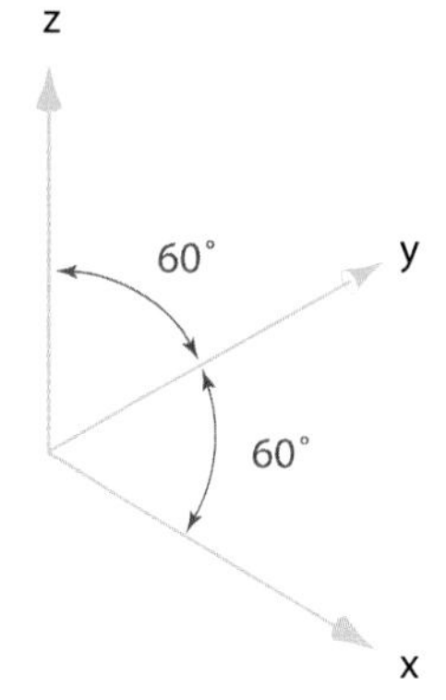

FIGURE 2.22. An isometric representation of axes with angles less than 120 degrees.

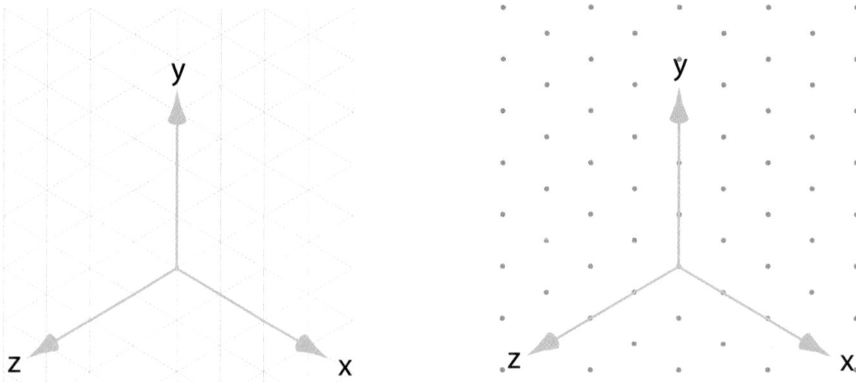

FIGURE 2.23. Isometric grid and dot paper.

Grid or dot paper can help you make isometric sketches. With **isometric dot paper**, the dots are oriented such that when you sketch lines through the dots, you end up with standard 120 degree axes. With grid paper, the lines are already drawn at an angle of 120 degrees with respect to one another. Isometric grid paper and isometric dot paper are illustrated in Figure 2.23.

2.07 Isometric Sketches of Simple Objects

Creating isometric drawings and sketches of complex objects will be covered in more detail in a later chapter; however, this section serves as an introduction to the topic for simple objects. Mastering the techniques used to create isometric sketches of simple objects may help as you branch out to tackle increasingly complex objects. Figure 2.24 shows how **isometric grid paper** is used to sketch a 3 × 3 × 3 block. Notice that there is more than one orientation from which the block can be sketched on the same sheet of grid paper. Ultimately, the orientation you choose depends on your needs or preferences.

Coded plans can be used to define simple objects that are constructed entirely out of blocks. The numerical values in the coded plan represent the height of the stack of blocks at that location. The object then "grows" up from the plan according to the numbers specified. Figure 2.25 shows a coded plan on isometric grid paper and the object that results from it.

The object shown in Figure 2.25 clearly outlines all of the blocks used to create it. When isometric sketches of an object are made, however, standard practice dictates that lines appear only where two surfaces intersect—lines between blocks on the same surface are not shown. Figure 2.26 shows the object from Figure 2.25 after the unwanted lines have been removed. Notice that the only lines on the sketch are those formed from the intersection of two surfaces. Also notice that object edges hidden from view on the back side are not shown in the sketch. Not showing hidden edges on an **isometric pictorial** also is standard practice in technical sketching.

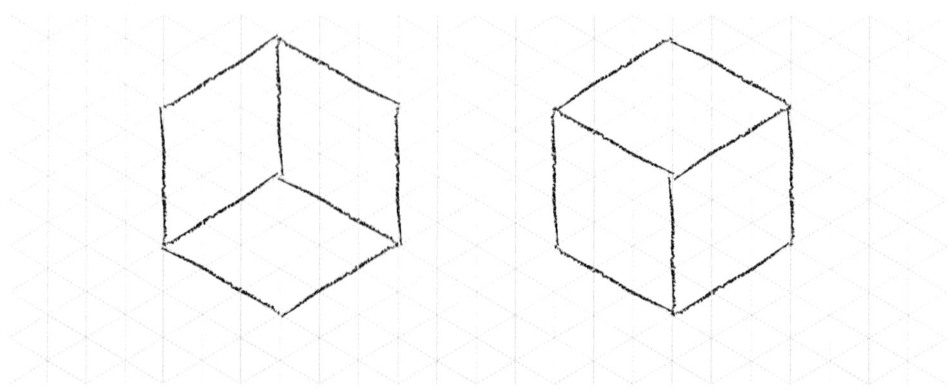

FIGURE 2.24. Using isometric grid paper to sketch a block.

Sometimes when you are creating an isometric sketch of a simple object, part of one surface is obscured by one of the more prominent features of the object. When creating the sketch, make sure you show only the visible part of the surface in question, as illustrated in Figure 2.27.

Figure 2.28 shows several coded plans and the corresponding isometric sketches. Look at each isometric sketch carefully to verify that it matches the defining coded plan: those lines are shown only at the edges between surfaces (not to define each block), that no hidden edges are shown, and that only the visible portions of partially obscured surfaces are shown.

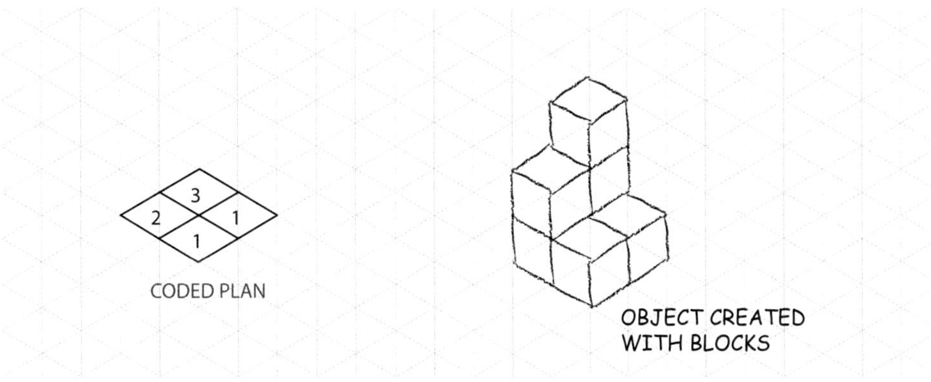

FIGURE 2.25. A coded plan and the resulting object.

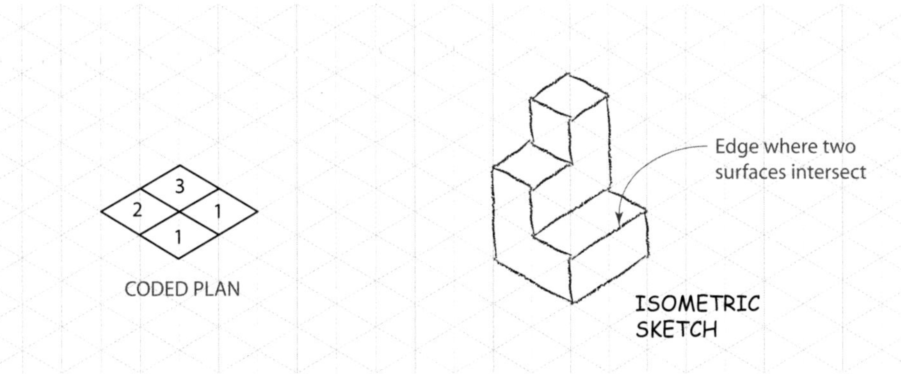

FIGURE 2.26. A properly drawn isometric sketch of the object from the coded plan.

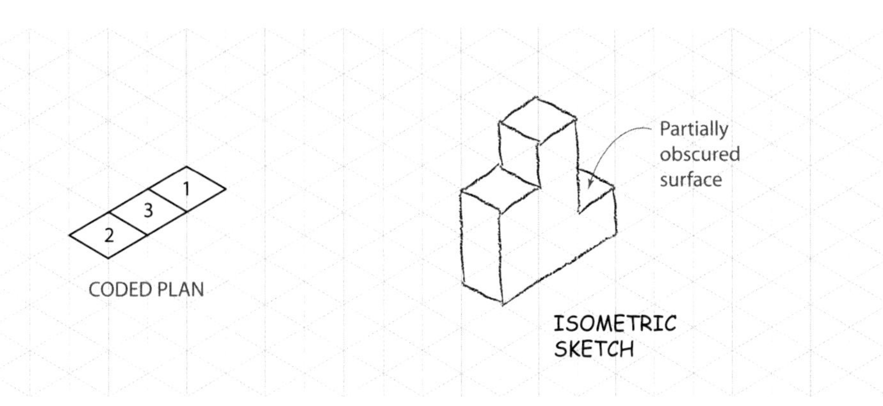

FIGURE 2.27. The partially obscured surface on an isometric sketch.

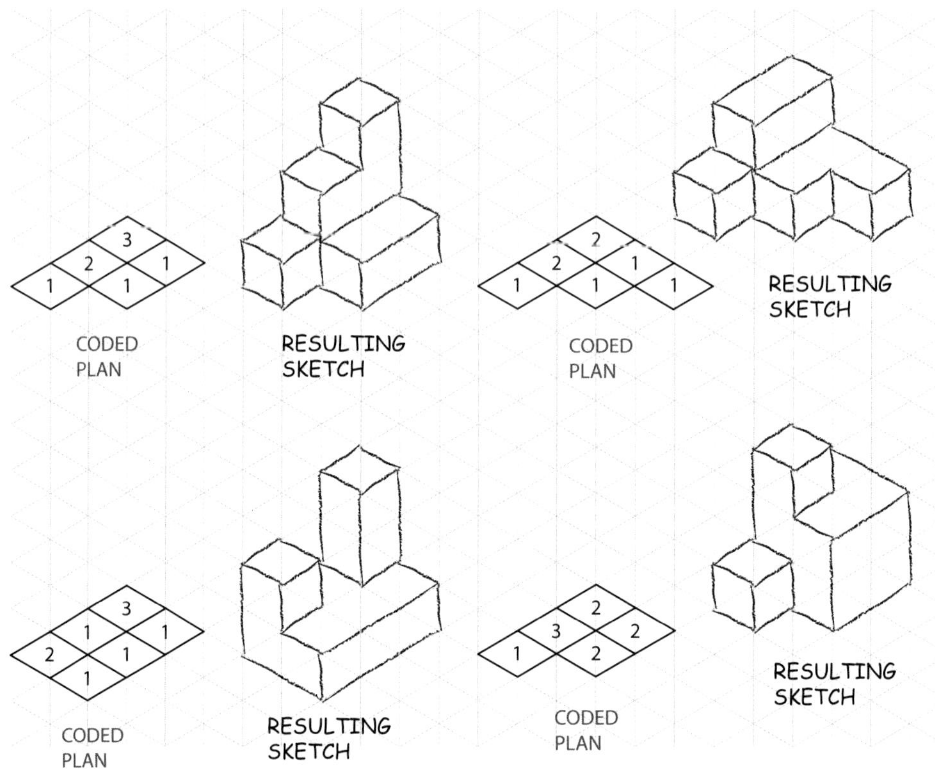

FIGURE 2.28. Four coded plans and the resulting isometric sketches.

2.07.01 Circles in Isometric Sketches

Look back at the 3 × 3 × 3 block shown in Figure 2.24. In reality, you know that all of the surfaces of the block are 3 × 3 squares; yet in the isometric sketch, each surface is shown as a parallelogram. The distortion of planar surfaces is one disadvantage of creating isometric sketches. The isometric portrayal of circles and arcs is particularly difficult. Circles appear as ellipses in isometric sketches; however, you will not be able to create a rectangular bounding box to sketch the ellipse in isometric as described earlier in this chapter. To create an ellipse that represents a circle in an isometric sketch, you first create a square bounding box as before; however, the bounding box will appear as a parallelogram in the isometric sketch. To create your bounding box, locate the center of the circle first. From the center, locate the four radial points. The direction you move on the grid corresponds to the lines that define the surface. If you are sketching the circle on a rectangular surface, look at the sides of the rectangle as they appear in isometric and move that same direction on the grid. Figure 2.29 shows a 4 × 4 × 4 cube with a circle center and four radial points located on one of the sides.

Once you have located the center of the circle and the four radial points, the next step is to create the bounding box through the radial points. The edges of the bounding box should correspond to the lines that define this particular surface. The edges will be parallel to the edges of the parallelogram that define the surface if that surface is square or rectangular. Figure 2.30 shows the cube with the circle center and the bounding box located on its side.

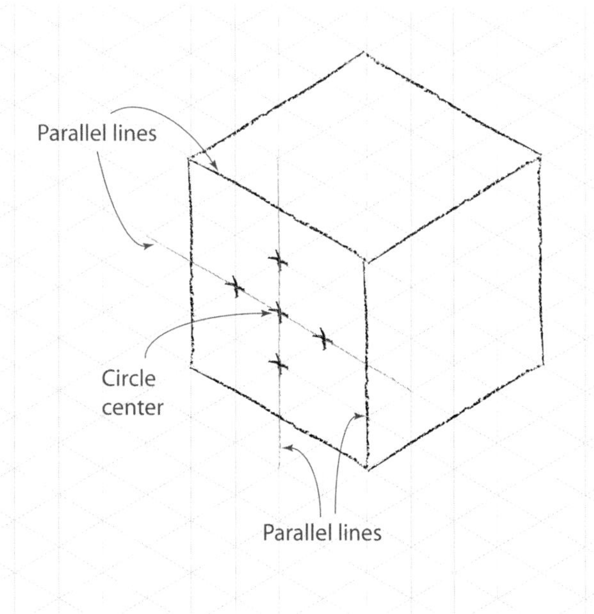

FIGURE 2.29. A cube with a circle center and radial points located.

Four arcs that go through the radial points define the ellipse, just like an ellipse drawn in a regular rectangular bounding box. The difference is that for the isometric ellipse, the arcs are of varying curvatures—two long arcs and two short arcs in this case. The arcs are tangent to the bounding box at the radial points, as before. It is usually

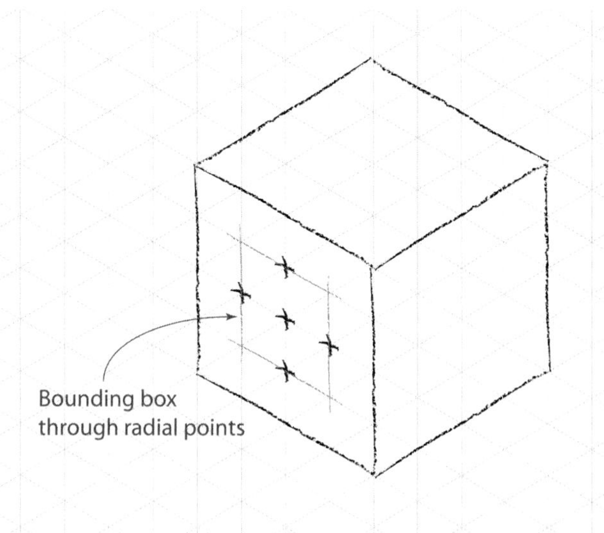

FIGURE 2.30. A cube with the circle center and bounding box on the side.

best if you start by sketching the long arcs, and then add the short arcs to complete the ellipse. Sketching the arcs that form the ellipse is illustrated in Figure 2.31.

Creating ellipses that represent circles on the other faces of the cube is accomplished in a similar manner, as illustrated in Figure 2.32 and Figure 2.33.

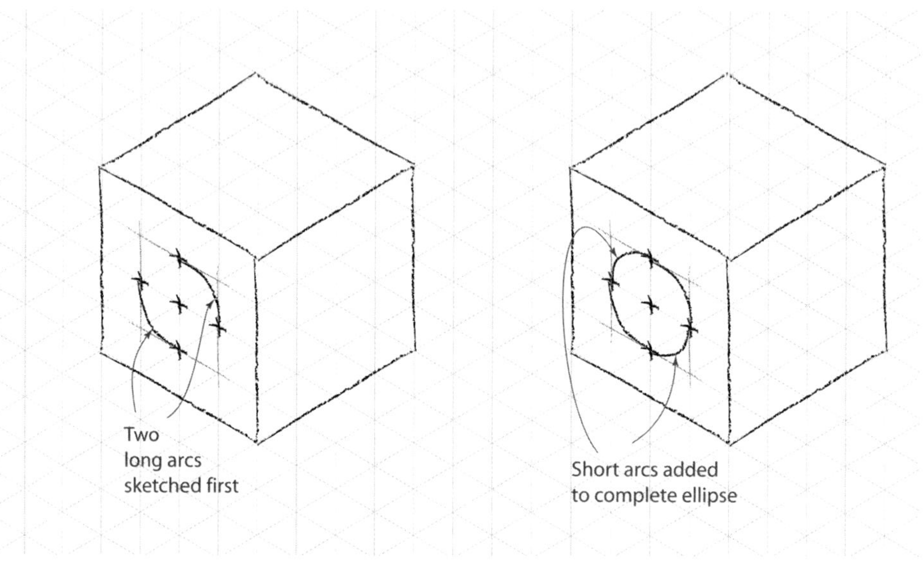

FIGURE 2.31. Sketching arcs to form an ellipse.

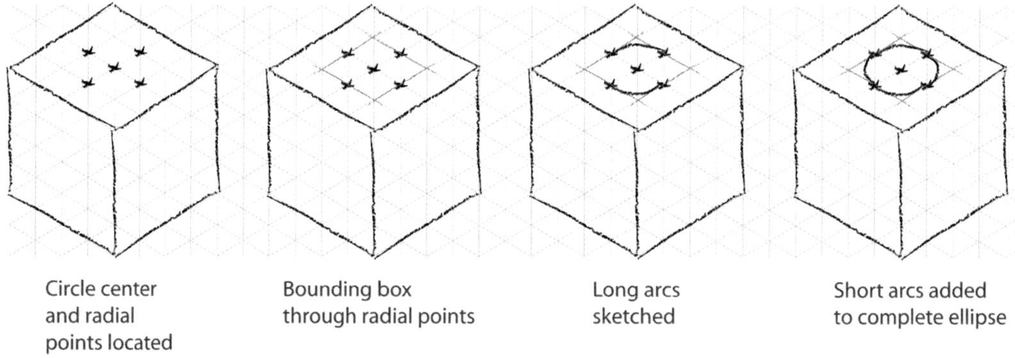

FIGURE 2.32. Sketching an ellipse on the top surface of a cube.

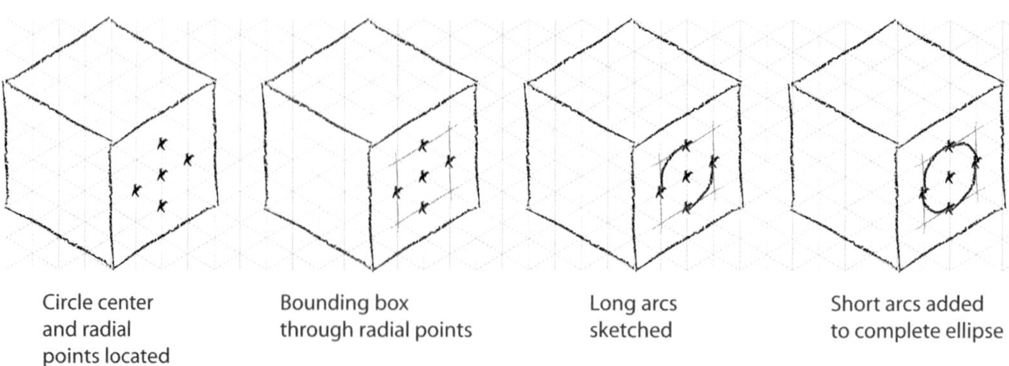

FIGURE 2.33. Sketching an ellipse on the side face of a cube.

2.07.02 Circular Holes in Isometric Sketches

One of the most common occurrences that produces a circular feature in an isometric sketch is a hole in the object. You will learn more about circular holes and object features in a later chapter, but a short introduction follows here. A circular hole usually extends all the way through an object. In an isometric pictorial, a portion of the "back" edge of a circular hole is often visible through the hole and should be included in your sketch. As a rule of thumb, the back edge of a hole is partially visible when the object is relatively thin or the hole is relatively large; when the object is thick or the diameter of the hole is small, the back edge of the hole is not visible. Figure 2.34 shows two blocks with circular holes going through them. Notice in the "thin" block that you can see a portion of the back edge of the hole; in the thicker block, though, the back edge is not visible.

To determine whether a part of the back edge of a hole is visible in an isometric sketch, you first need to locate the center of the back hole. To locate the back center, start from the center of the hole on the front surface and move in a direction perpendicular to the front surface toward the back of the object a distance equal to the object's dimension in that direction. Figure 2.35 shows the location of the center of the two back circles for the objects in Figure 2.34.

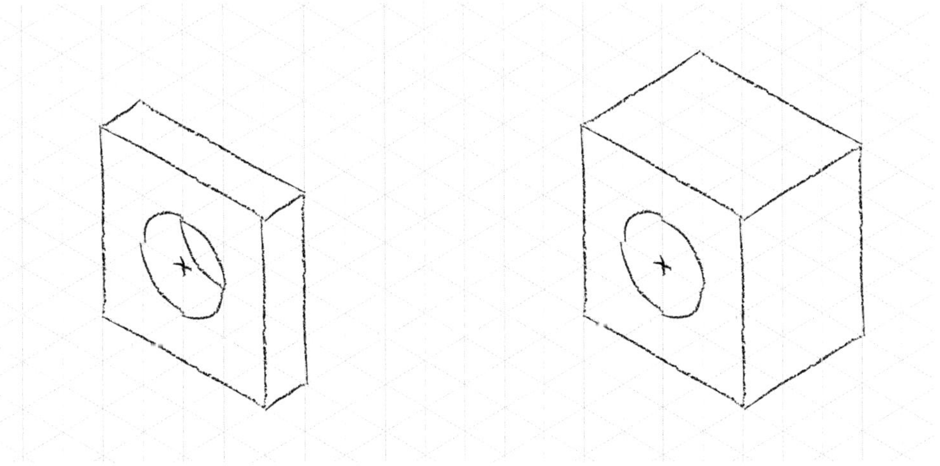

FIGURE 2.34. Blocks with circular holes in them.

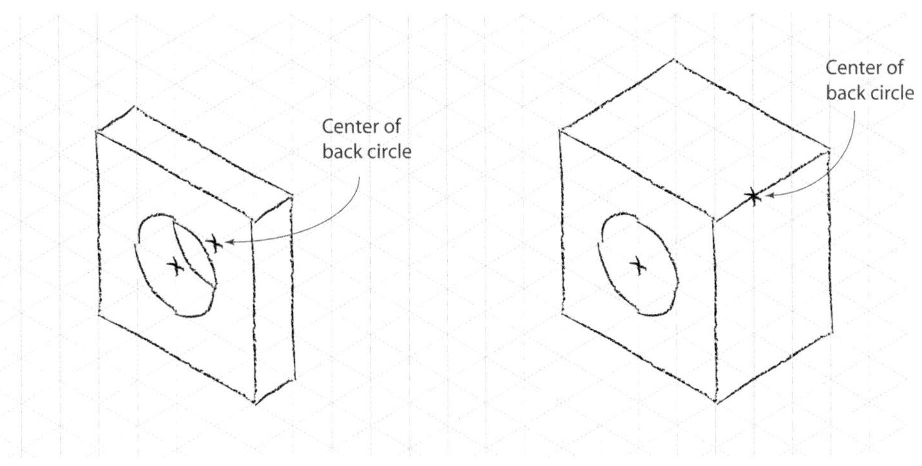

FIGURE 2.35. Centers of back circles located.

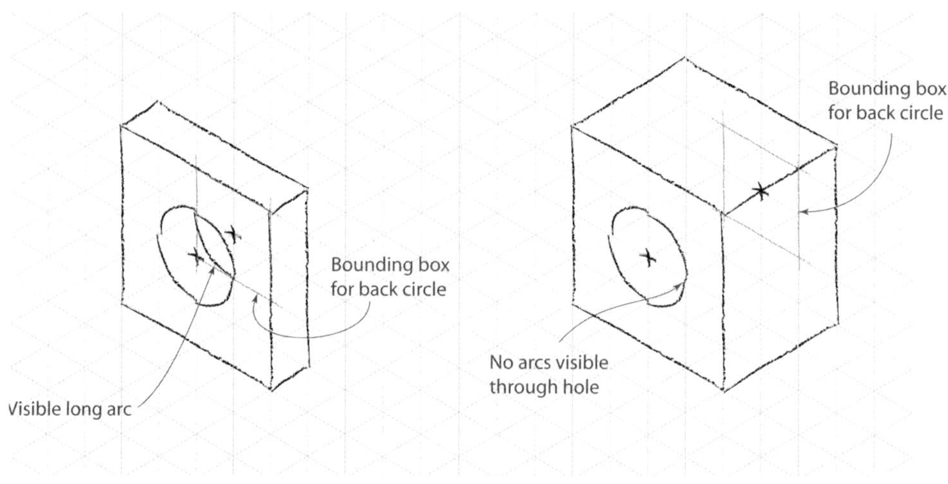

FIGURE 2.36. Determining visibility of back circles.

Starting from the back center point, lightly sketch the radial points and the bounding box for the back circle similar to the way you did for the front circle. Then add the long arc that is visible through the hole. (Note that only *one* of the long arcs is typically visible through the hole.) Add segments of the short arcs as needed to complete the visible portion of the back edge of the hole. Conversely, if after you sketch the back bounding box you notice that no portion of the ellipse will be visible on the sketch, do not include any arcs within the hole on the sketch and erase any lines associated with the bounding box. Figure 2.36 illustrates the inclusion and noninclusion of segments of the back edges of holes for the objects in Figure 2.34 and Figure 2.35.

2.08 Oblique Pictorials

Oblique pictorials are another type of sketch you can create to show a 3-D object. Oblique pictorials are usually preferred for freehand sketching because a specialized grid is not required. With oblique pictorials, as with **oblique axes**, the three dimensions of the object are shown with the height and width of the object in the plane of the paper and the third dimension (the depth) receding off at an angle from the others. Although the angle is usually 45 degrees, it can be any value.

The advantage that oblique pictorials have over isometric pictorials is that when one face of the object is placed in the plane of the paper, the object will appear in its true shape and size in that plane—it will be undistorted. This means that squares remain squares, rectangles remain rectangles, and circles remain circles. Figure 2.37 shows two pictorial representations of simple objects—one in isometric and one in

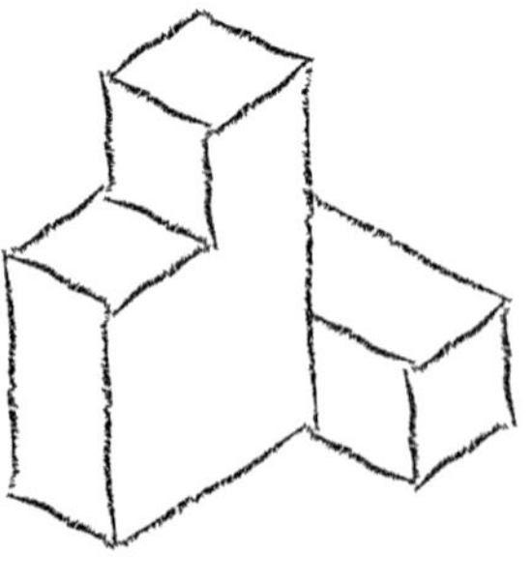

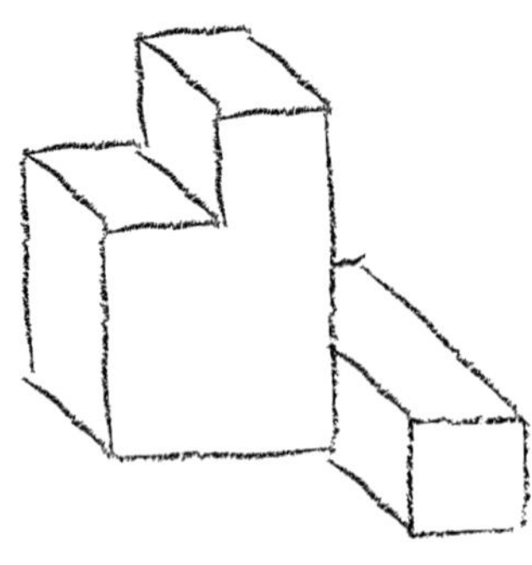

FIGURE 2.37. A comparison of isometric and oblique pictorials.

oblique. Notice that the rules established for isometric pictorial sketches also hold true for oblique pictorial sketches—you do not show the hidden back edges, you show lines only where two surfaces intersect to form an edge, and you show only the visible parts of partially obstructed surfaces.

When making oblique sketches, the length of the **receding dimension** is not too important. In fact, oblique pictorials typically look better when the true length of the receding dimension is not shown. When the true length of an object's receding dimension is sketched, the object often appears distorted and unrealistic. Figure 2.38a shows the true length of a cube's receding dimension (use a ruler to make sure), and Figure 2.38b shows the same cube with the receding dimension drawn at about one-half to three-fourths its true length. Notice that the sketch in Figure 2.38a appears distorted—it does not look very much like a cube—whereas the sketch in Figure 2.38b looks like a cube.

Other conventions pertain to the way the receding dimension is portrayed in an oblique sketch; you will learn about them in a later chapter. For now, you will concentrate on trying to make a sketch that looks proportionally correct.

When creating oblique pictorials, you can choose to have the receding dimension going back and to the left or back and to the right. The direction you choose should be the one that produces the fewest obstructed surfaces in the resulting sketch. Figure 2.39 shows two possible sketches of the same object—one with the receding dimension to the left and one with the receding dimension to the right. Notice that the first sketch (Figure 2.39a) is preferable since none of the surfaces are obscured as they are with the second sketch (Figure 2.39b).

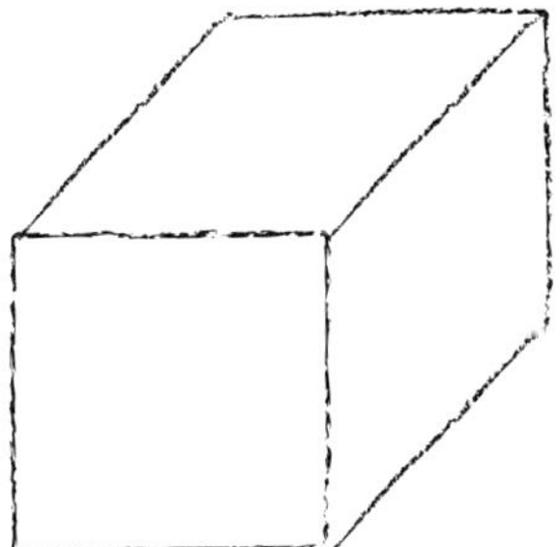

(a) Receding dimension drawn true length.

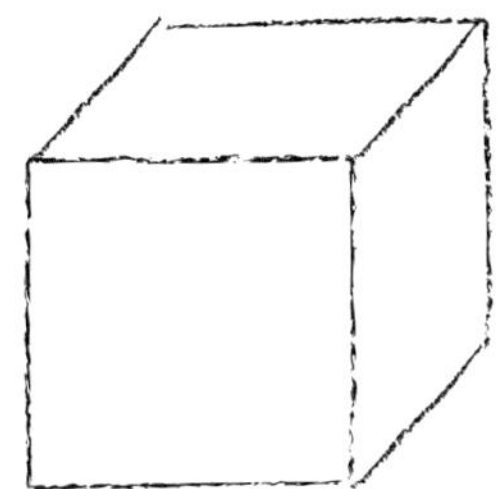

(b) Receding dimension drawn less than true length.

FIGURE 2.38. Oblique pictorials of a cube.

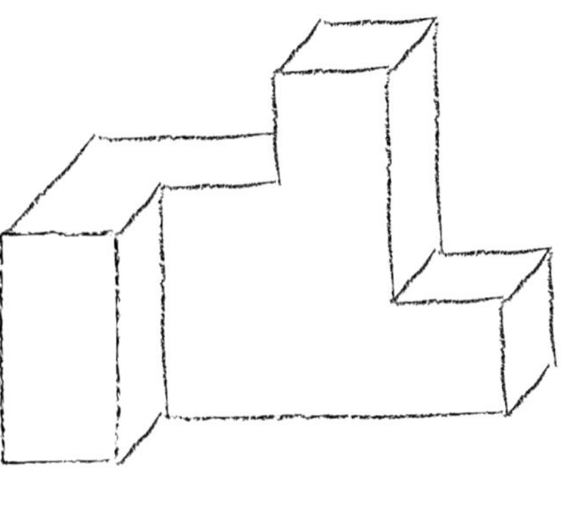

(a)

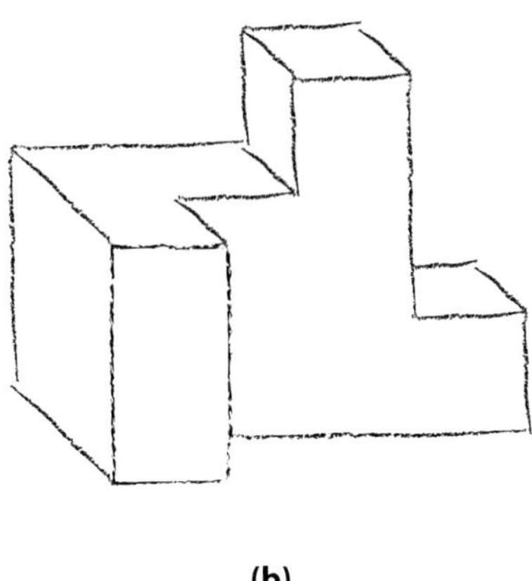

(b)

FIGURE 2.39. Two possible orientations for an oblique pictorial.

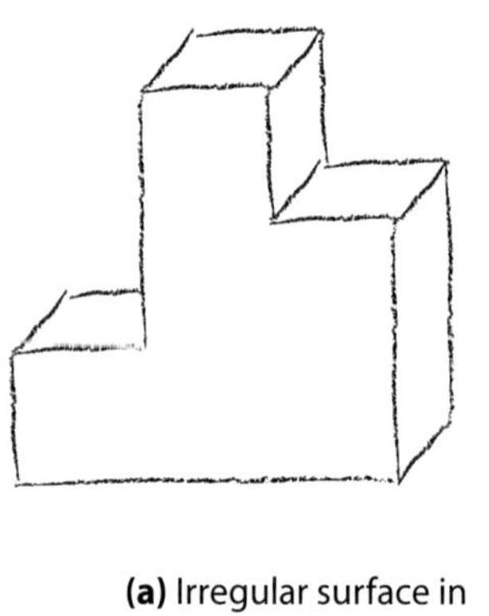

(a) Irregular surface in plane of paper.

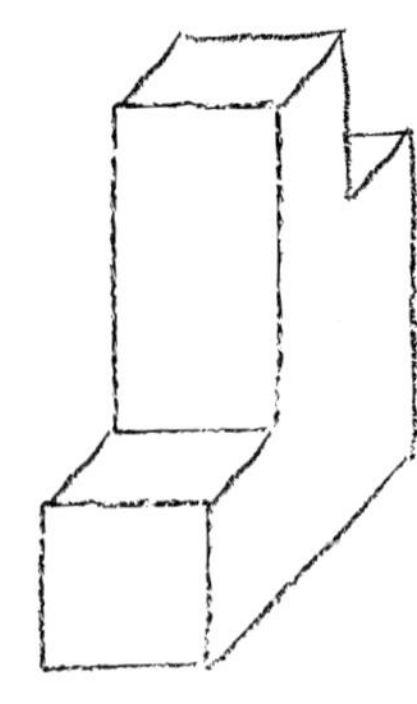

(b) Irregular surface in receding direction.

FIGURE 2.40. Two possible orientations for an oblique pictorial.

When creating an oblique pictorial, you should put the most irregular surface in the plane of the paper. This is particularly true about any surface that has a circular feature on it. Figure 2.40 shows two different oblique pictorials of the same object. In the first sketch (Figure 2.40a), the most irregular surface is placed in the plane of the paper as it should be; in the second sketch (Figure 2.40b), the irregular surface is shown in the receding dimension. Notice that the first sketch shows the features of the object more clearly than the second sketch does.

2.08.01 Circular Holes in Oblique Pictorial Sketches

When circular holes appear in an oblique pictorial sketch, as with isometric sketches, you show the partial edges of the back circle where they are visible through the hole. Once again, partial circles are visible when the object is relatively thin or when the hole has a relatively large diameter; otherwise, partial edges are not shown. Figure 2.41 shows two oblique sketches—one in which a portion of the back edge of the hole is visible and the other in which it is not.

The procedure you use to determine whether a portion of the back circle edge is visible and, if so, which portion is visible follows the procedure outlined for isometric sketches. You start by locating the center of the back edge of the hole and marking off the four radial points. You then lightly sketch the bounding box that defines the circle. Finally, as needed, you sketch the visible portions of arcs within the circular hole. Figure 2.42 shows the procedure used to sketch the visible back edges of a circular hole in an oblique pictorial.

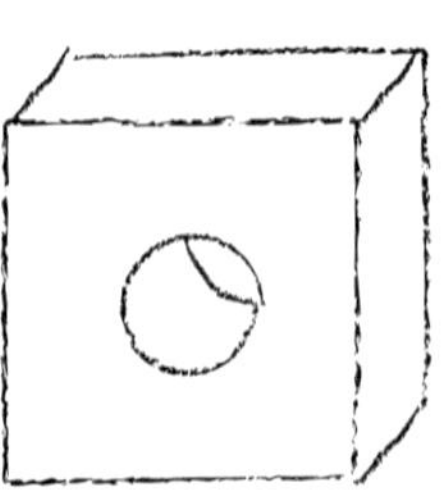

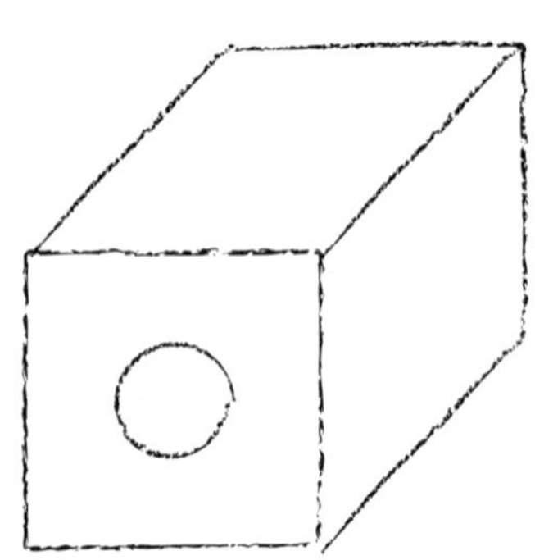

FIGURE 2.41. Oblique pictorials with circular holes in objects.

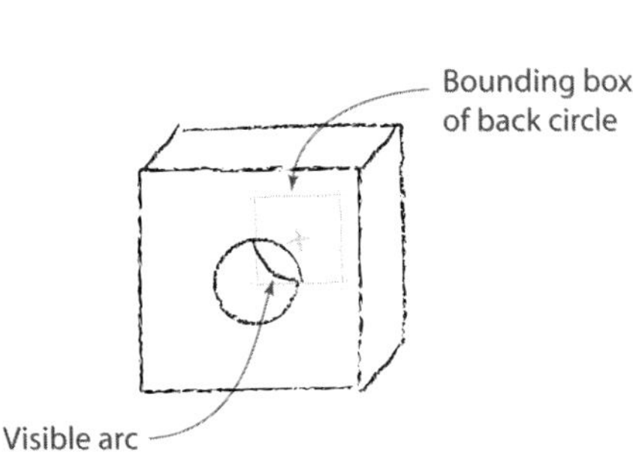

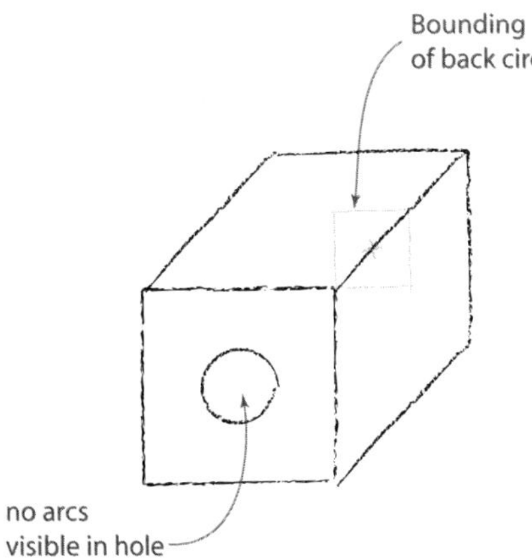

FIGURE 2.42. Determining visible back arcs in a hole.

2.09 Shading and Other Special Effects

One thing you can do to improve the quality of your pictorial sketches is to include **shading** on selected surfaces to make them stand out from other surfaces or to provide clarity for the viewer. Figure 2.43 shows an isometric sketch with all of the top surfaces shaded. Notice that the shading better defines the object for the viewer. When including shading on a pictorial sketch, try not to overdo the shading. Too much shading can be confusing or irritating to the viewer—two things you should avoid in effective graphical communication.

Another common use of shading is to show curvature of a surface. For example, the visible portion of a hole's curved surface might be shaded in a pictorial sketch. A curved surface on an exterior corner also might be shaded to highlight its curvature. Figure 2.44 shows a pictorial sketch of a simple object with curved surfaces that are shaded.

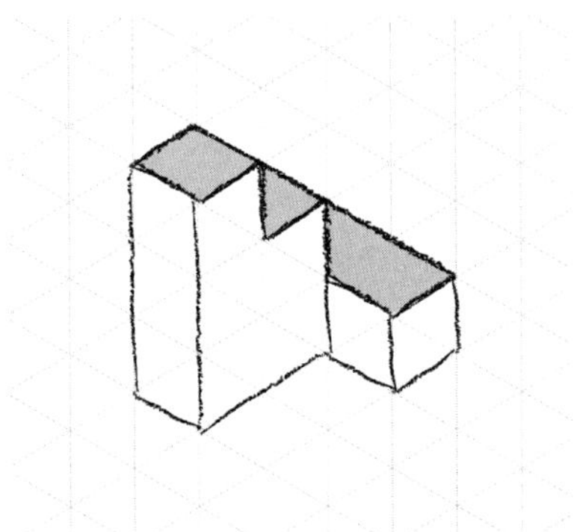

FIGURE 2.43. An object with the top surface shaded.

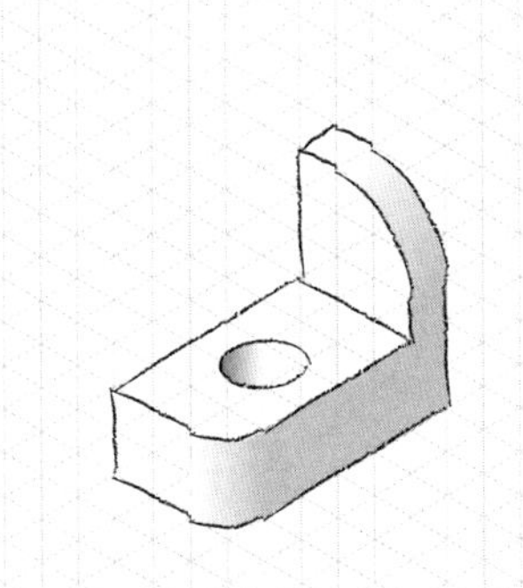

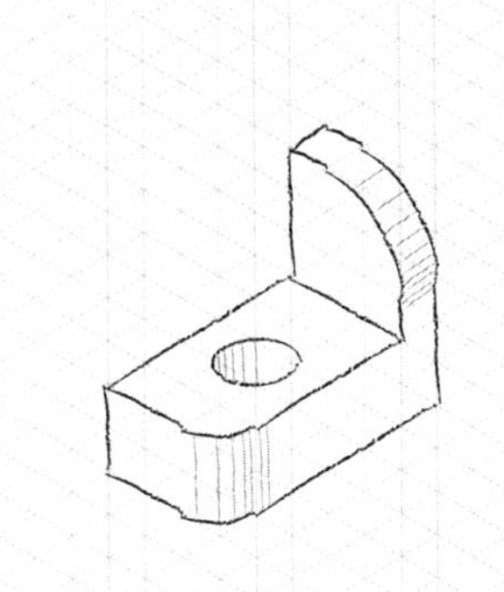

FIGURE 2.44. A simple object with two possible types of progressive shading used to emphasize the curvature of surfaces.

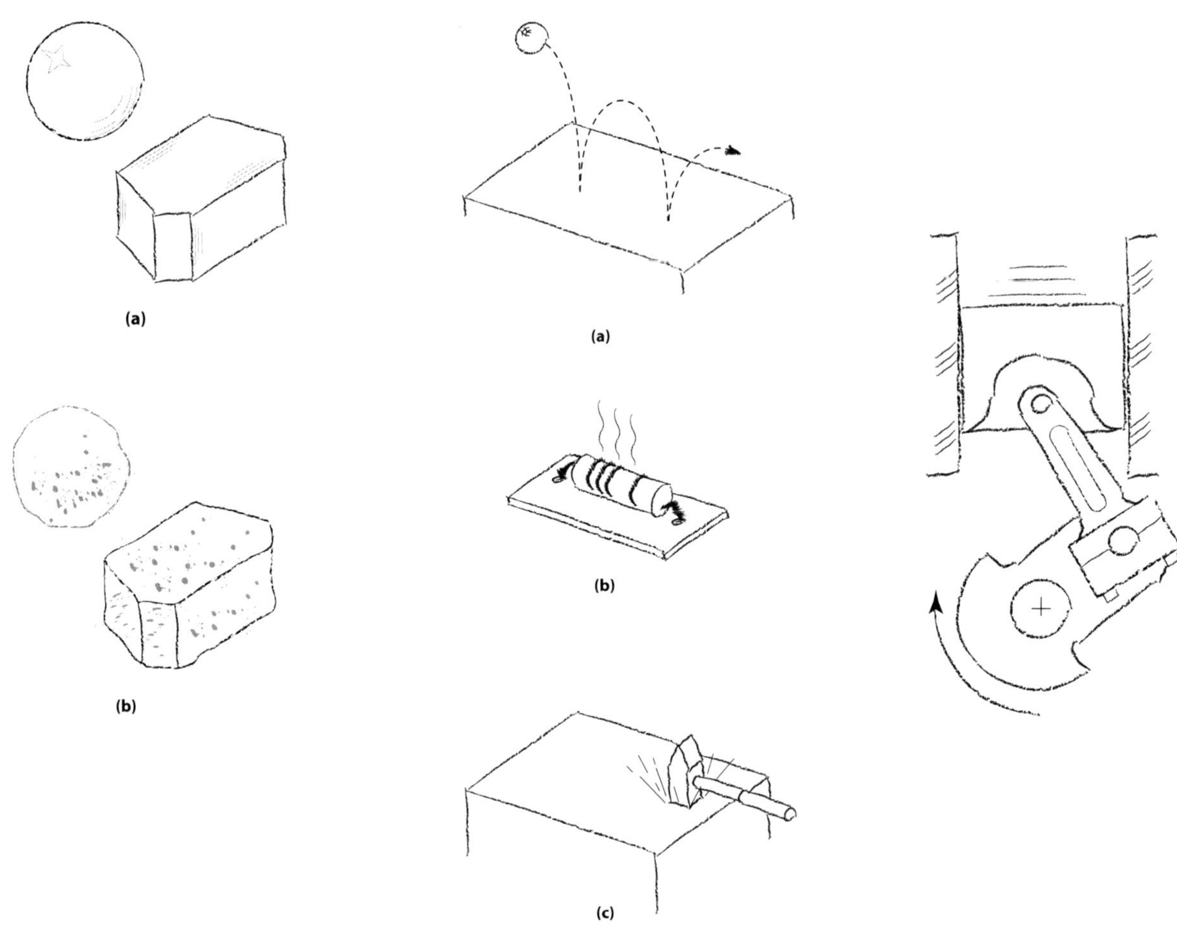

FIGURE 2.45. The addition of surface treatments to convey smooth surfaces (a) and rough surfaces (b).

FIGURE 2.46. Some sketching techniques that can be used to convey motion (a), temperature (b), and sound (c).

FIGURE 2.47. Action lines used to convey the motion of linkages.

Other sketching techniques can be used to convey features such as smooth or rough surfaces. Figure 2.45 shows different types of surface treatments that are possible for sketched objects.

You are probably familiar with techniques used in cartoons to convey ideas such as motion, temperature, and sound. Figure 2.46 shows typical cartooning lines that convey concepts not easily incorporated in a static sketch. Many of these same markings can be used in technical sketches. For example, Figure 2.47 uses action lines to convey motion for the sketch of linkages.

2.10 Sketching Complex Objects

As you refine your sketching skills, you will be able to tackle increasingly complex objects. Figures 2.48, 2.49, and 2.50 show pictorial sketches of small electronic devices. These sketches were not made to any particular scale, but were constructed so the object features appear proportionally correct with respect to one another. Notice the use of shading to enhance object appearance and to make the objects look

FIGURE 2.48. A sketch of a cell telephone.

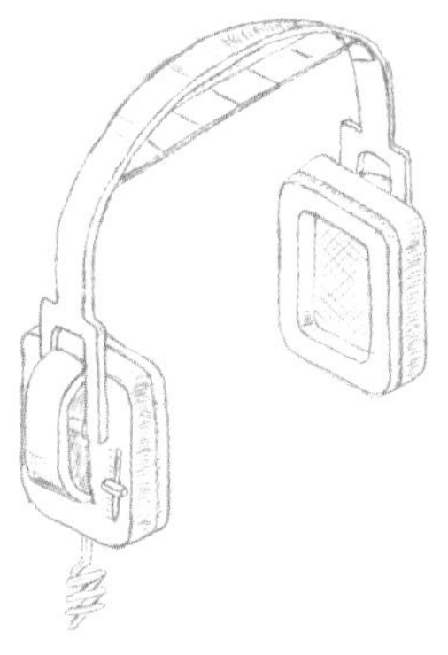

FIGURE 2.49. A sketch of a set of headphones.

FIGURE 2.50. A sketch of a camera.

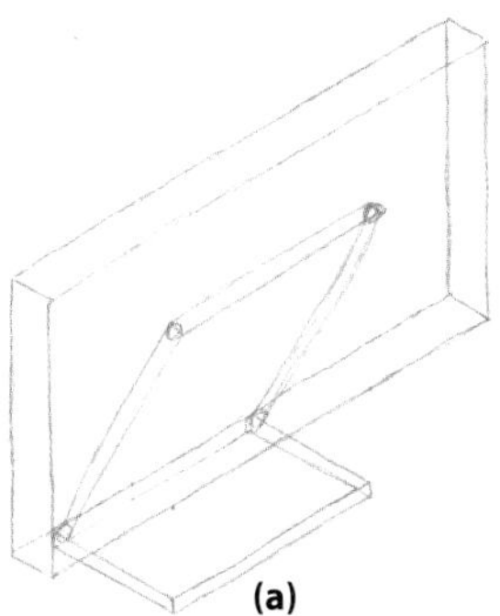

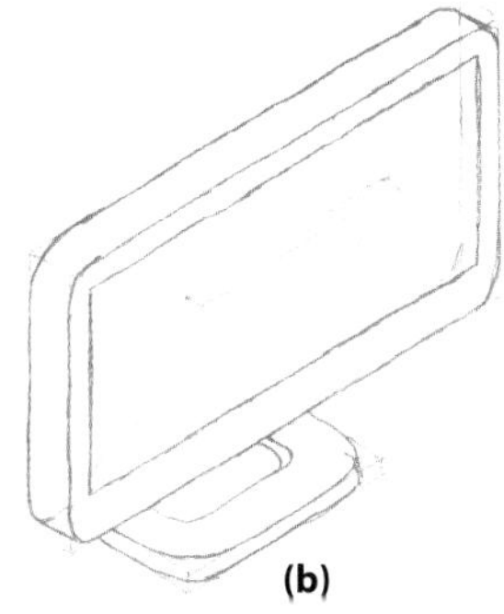

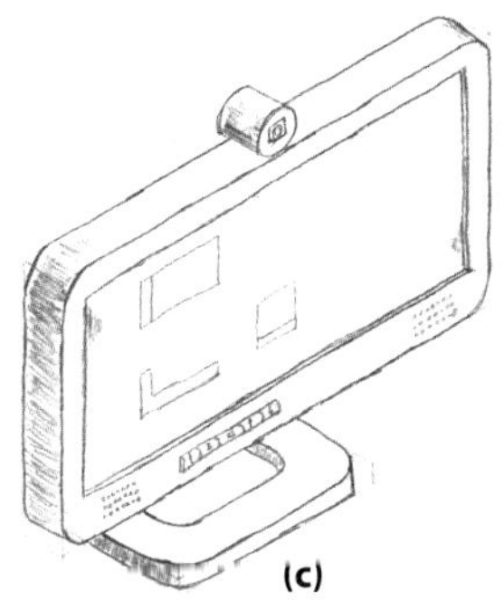

FIGURE 2.51. A sketch of a computer monitor using the method of "foundation (a), frame (b), finish (c)."

more realistic. Being able to sketch relatively complicated objects such as these will improve your ability to communicate with colleagues throughout your career. To develop this important skill, you should practice often. Do not be afraid to make mistakes—just keep trying until you get the results you want.

One way to tackle sketching a complex object is to think about it in the same way that a house is constructed—namely, "foundation, frame, finish." Using this method, you start with the "foundation" of the sketch, which usually consists of multiple guidelines and construction lines. When creating the sketch foundation, think about outlining the volume taken up by the entire object. You next "frame" the object by darkening some of the construction lines to define the basic shape of the object and its features. Once the basic frame is complete, you "finish" the sketch by adding necessary details and special features such as shading, especially on curved surfaces. Figure 2.51 shows a sketch of a flat panel computer monitor by the "foundation, frame, finish" method. Several of the exercises at the end of this chapter ask you to use this technique to develop your skills in sketching complex objects.

2.11 Strategies for Simple Pictorial Sketches

In this chapter, you learned two different ways to construct pictorial views of objects. This section outlines strategies for each type.

2.11.01 Simple Isometric Sketches

When creating an isometric pictorial from a coded plan, remember that the object "grows" up from the base according to the specified heights. You should start your sketch by drawing the visible *V* at the base of the object, as shown in Figure 2.52. You can determine the length of each side of the *V* from the coded plan. For the object defined by the coded plan in Figure 2.52, the left leg of the *V* is 2 units long and the right leg is 3 units long. The remaining bottom edges of the coded plans are hidden from view in the sketch and, therefore, are not included in the first drawing stages.

After you have created the base *V*, sketch the corner of the object at the correct height of the apex. Note that this corner will be the edge that is closest to you, the viewer. For the object shown in Figure 2.52, the height of this corner is 2 units as defined by the coded plan. The start of the isometric sketch including this corner is shown in Figure 2.53.

Starting at the "top" of this corner, move back and to the left the number of squares that are at this same height. If a change in object height is specified in the coded plan, move up or down (as shown in the coded plan) where the change occurs. When you reach the back corner, draw a vertical line back to the tip of the *V* you first sketched. This procedure is illustrated in Figure 2.54.

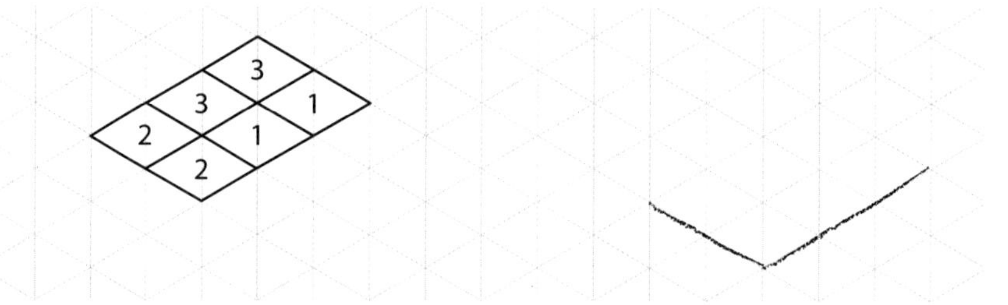

FIGURE 2.52. A coded plan with the *V* for the isometric sketch drawn.

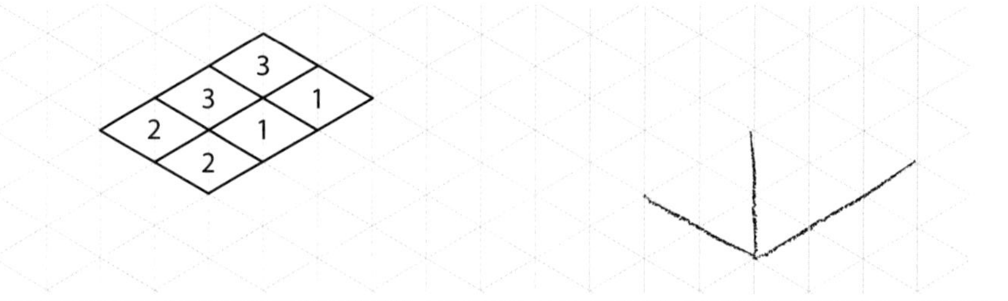

FIGURE 2.53. An isometric sketch with the nearest corner included.

FIGURE 2.54. An isometric sketch with the first side of the surface drawn.

Follow the same procedure for the surface going off to the right from the apex of the *V*, as shown in Figure 2.55.

Complete the sketch by drawing the missing top and side surfaces of the object as shown in Figure 2.56. When adding these final features, make sure you do not include lines on surfaces—only *between* surfaces. Also, include only the visible portions of surfaces that are partially obscured.

Some of the objects you sketch may not form a simple *V* at a point nearest the viewer; instead, they will have a jagged edge along the bottom. You can use a similar procedure to sketch these objects, again starting at the bottom and outlining the shape of the object from the coded plan, as shown in Figure 2.57.

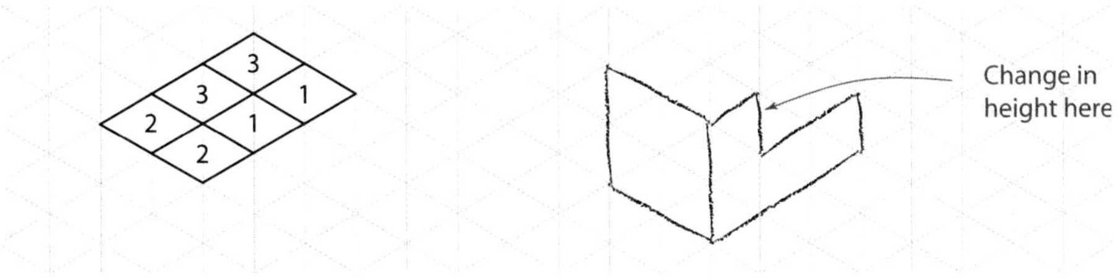

FIGURE 2.55. An isometric sketch with two side surfaces drawn.

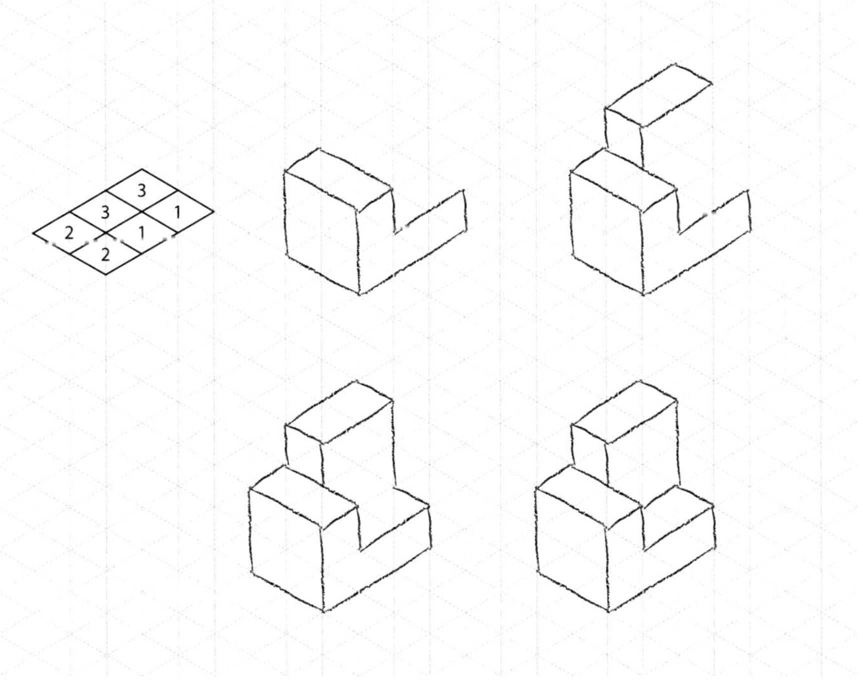

FIGURE 2.56. Completion of an isometric sketch.

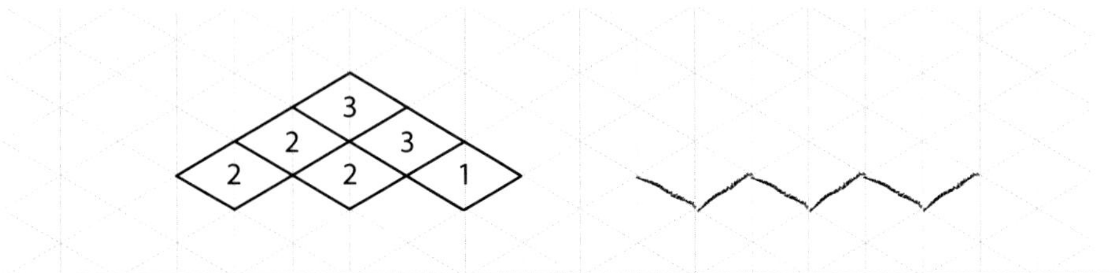

FIGURE 2.57. A jagged *V* from a coded plan.

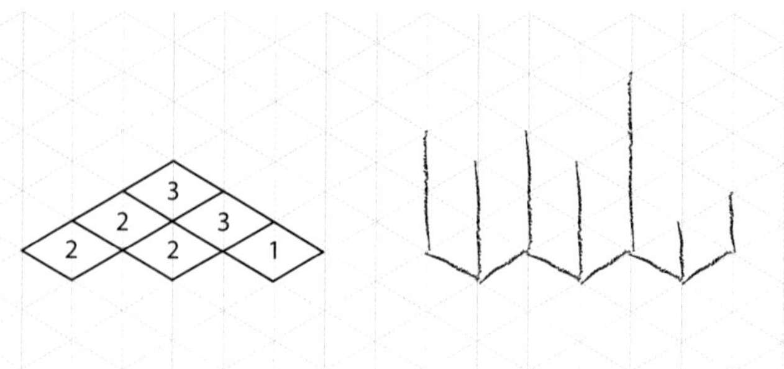

FIGURE 2.58. Heights at each corner included.

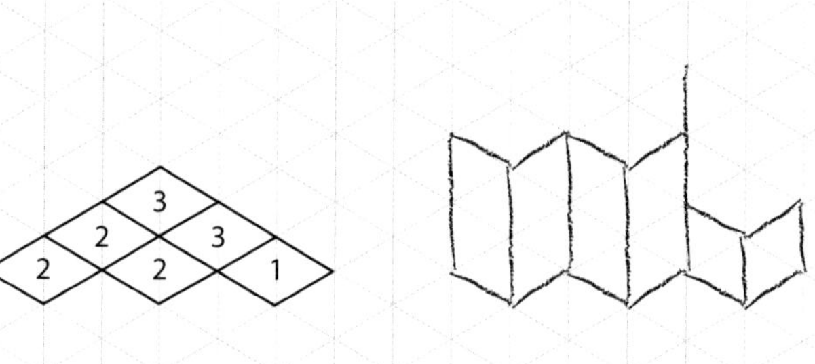

FIGURE 2.59. Side surfaces sketched.

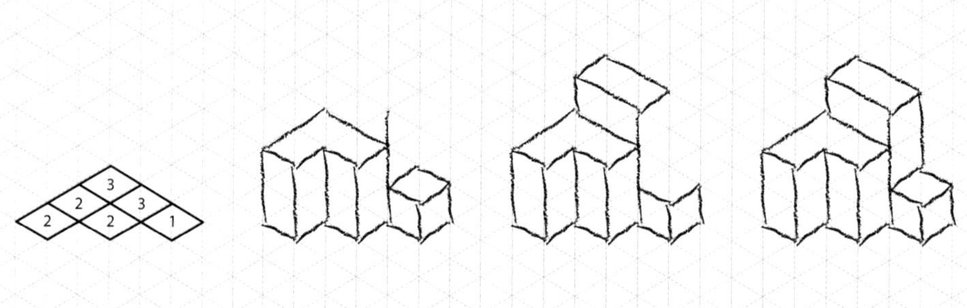

FIGURE 2.60. A completed isometric sketch.

Then you can sketch the lines that represent the height at each corner, similar to the way you sketched the height from the apex of the single *V* (see Figure 2.58). Complete the sketch by including the side and top surfaces of the object as illustrated in Figure 2.59 and Figure 2.60. For the object shown in the example, note that the final step involves erasing a portion of one of the first lines drawn (the corner at a height of 3). You need to remove part of this line so a line does not appear on the jagged side of the object.

2.11.02 Oblique Sketches

To begin your oblique sketch, you need to determine which surface on the object is closest to the viewer. Figure 2.61 shows an isometric sketch of an object with an arrow denoting the direction of the desired oblique pictorial. For this object and viewing direction, the surface labeled *A* is the one closest to the viewer in the oblique pictorial sketch.

Sketch the closest surface (in this case, surface A) in its true shape and size and decide whether you want the third dimension on the object receding back and to the left or back and to the right. Draw the visible edges receding back from each corner of

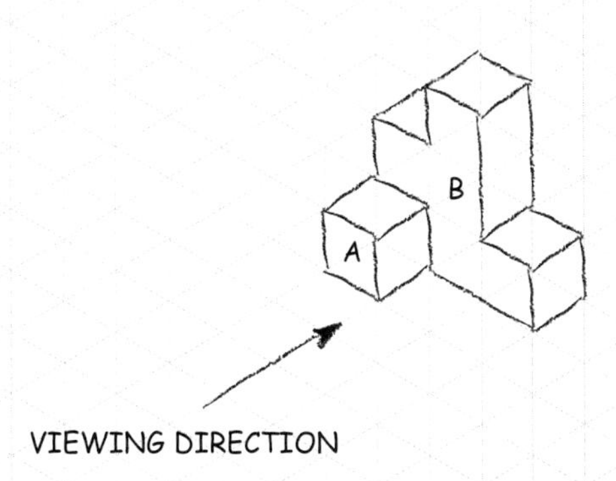

FIGURE 2.61. An isometric pictorial of an object and a viewing direction for an oblique sketch.

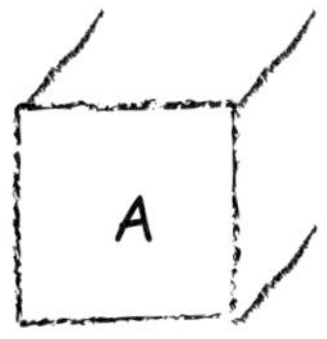

FIGURE 2.62. Surface A with receding dimensions sketched.

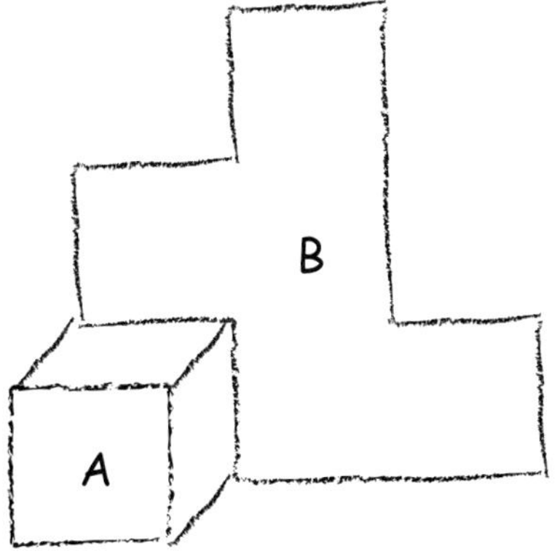

FIGURE 2.63. Surface B included in pictorial.

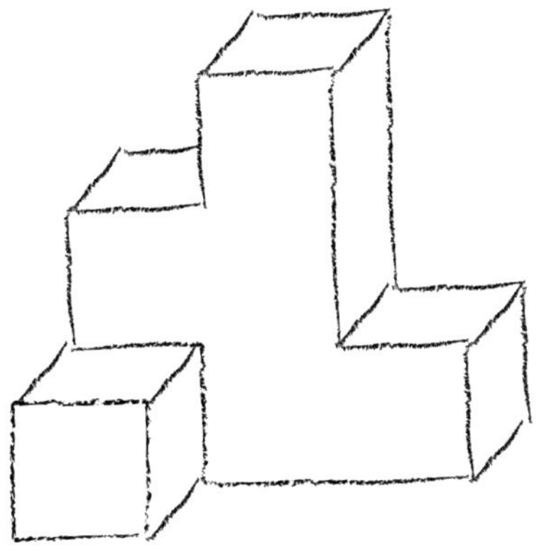

FIGURE 2.64. A completed oblique pictorial sketch.

the surface. Note that at least one corner on the surface will not have a receding line extending back from it—the receding edge will not be visible in the sketch. Figure 2.62 shows surface A with the receding edges sketched in place.

Now sketch the next surface that is parallel to the plane of the paper. For the object shown in Figure 2.63, the next closest surface is the one labeled *B*. Notice that by sketching this surface, you are connecting the endpoints of the lines drawn receding from the corners of the initial surface and, thus, are defining the side and top surfaces of the object in the pictorial. Figure 2.63 shows the result from including surface B in the oblique pictorial sketch.

Repeat these steps as often as necessary until the pictorial sketch is finished. Note that the final step is to include the back edges of the object (connecting the ends of the last set of receding lines drawn) to complete the sketch as shown in Figure 2.64.

CAUTION

When creating isometric pictorials of simple objects, remember the general rules presented earlier in this chapter—that lines are included only at the intersection between surfaces, that no hidden lines are shown, and that only the visible portion of partially obscured surfaces are sketched. One common error novices are prone to make is to include extra lines on a single surface of an object, especially when there are several changes in the object's height. Figure 2.65a shows an improper isometric pictorial sketch. Notice the extra lines included on the sketch. Figure 2.65b shows the sketch after it has been cleaned up to remove the unnecessary lines.

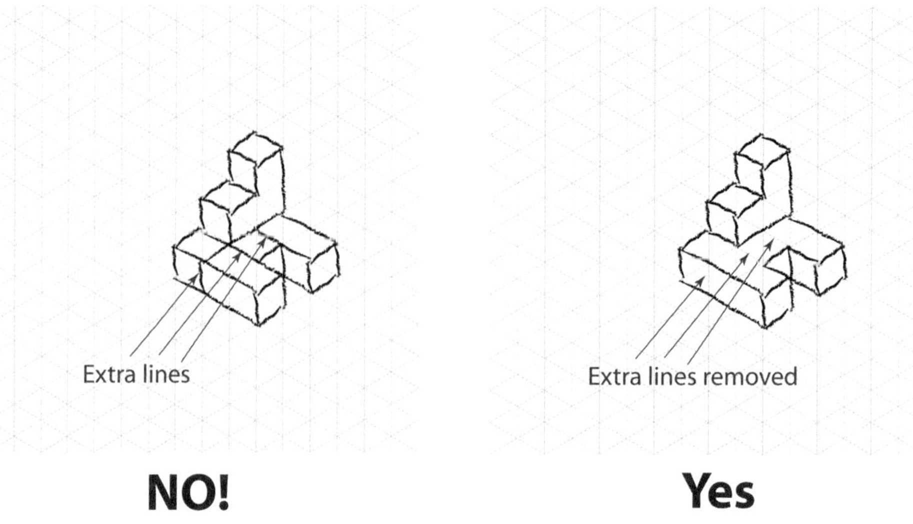

FIGURE 2.65. Isometric pictorials with and without extra lines.

Students make other common mistakes when sketching holes in isometric pictorials. One of those mistakes involves including the "back" edges of holes, even when they are not visible. Figure 2.66 shows an isometric pictorial with a hole in the object that goes all the way through. An arc representing the back edge of the hole is shown improperly in the visible part of the hole. Including the arc implies that the hole does not go all the way through the object, but stops part-way back. (Such holes are referred to as blind holes.) To avoid confusion in your isometric pictorial sketches, show only the back edge if it is visible—do not include a back edge every time you sketch an object with a hole.

Sometimes students use grid points improperly to mark off the bounding box for an isometric circular hole. Those novices fail to remember that in order to set the radial points, they need to move in the directions of the edges of the face of the object. Consider a simple box in which you want to include a circular hole emanating from the top surface. Figure 2.67a shows the four radial points incorrectly located from the center of the circle, and Figure 2.67b shows the resulting incorrect hole. Figure 2.68a shows the radial points located properly, and Figure 2.68b shows the resulting correct circular hole.

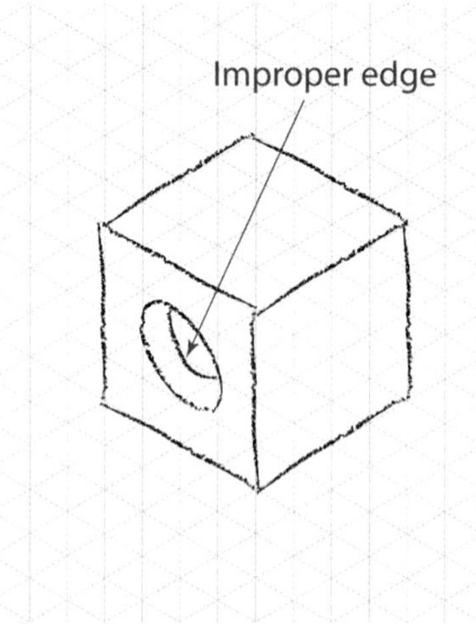

FIGURE 2.66. An isometric pictorial improperly showing the back edge of a hole.

One final error that students commonly make involves the creation of oblique pictorials. Novices sometimes forget to put the most complicated surface in the plane of the paper and show it in the receding direction instead. Figure 2.69a shows an oblique pictorial with a complex surface in the receding dimension, and Figure 2.69b shows the same object with the complex surface in the plane of the paper. Observe how the object is more understandable when you are viewing the complex surface "straight on." Also note that putting the complex surface in the plane of the paper actually makes your job easier; it is far easier to sketch the complex surface in its true size and shape than it is to sketch it as a distorted receding surface.

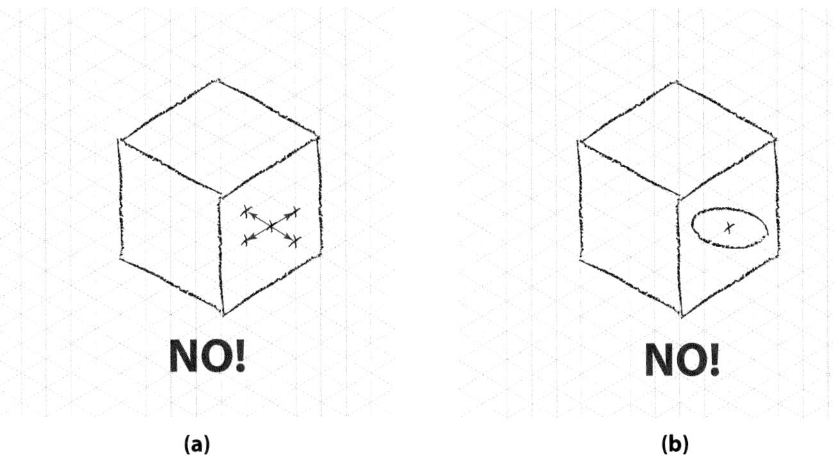

FIGURE 2.67. Improperly locating radial points on an isometric pictorial.

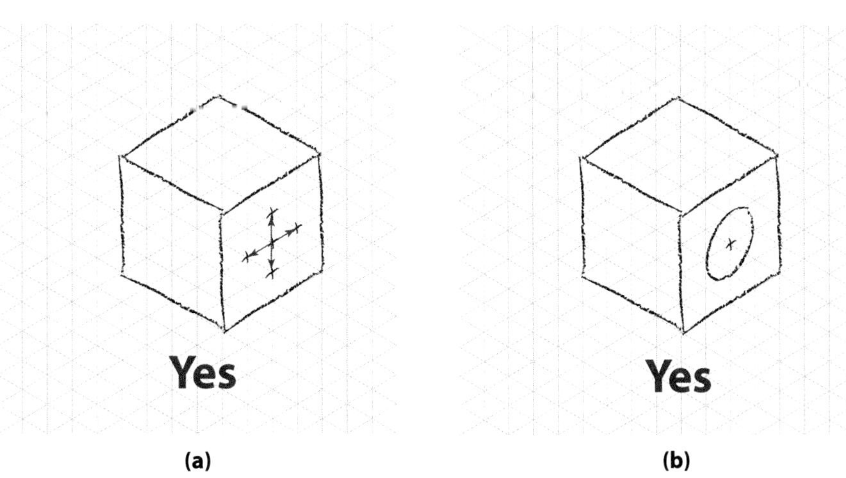

FIGURE 2.68. Properly locating radial points on an isometric pictorial.

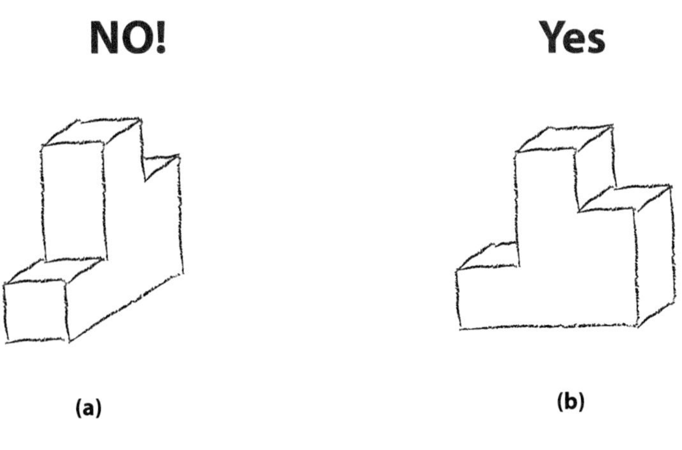

FIGURE 2.69. Oblique pictorials showing improper and proper placement of a complex surface.

2.12 Chapter Summary

In this chapter, you learned about technical sketching and about some techniques to help you master this important form of communication. Specifically, you:

- Learned about the importance of sketching for engineering professionals and the link between creativity and freehand sketching.
- Developed techniques for successfully sketching basic shapes such as lines, arcs, circles, and ellipses.
- Learned about the right-hand rule and the way it is used to define 3-D coordinate systems in space. The axes can be portrayed on paper in either isometric or oblique format.
- Discovered how to make basic isometric sketches of objects from coded plans and about some of the rules that govern the creation of these sketches. You also learned about creating ellipses in isometric to represent circular holes in objects.
- Developed techniques for creating oblique pictorials. You also learned that for this type of pictorial, you should not show the receding dimension of the object true to size in order to avoid a distorted image.

2.13 glossary of key terms

arc: A curved entity that represents a portion of a circle.

bounding box: A square box used to sketch circles or ellipses.

circle: A closed curved figure where all points on it are equidistant from its center point.

construction line: A faint line used in sketching to align items and define shapes.

ellipse: A closed curve figure where the sum of the distance between any point on the figure and its two foci is constant.

isometric axes: A set of three coordinate axes that are portrayed on the paper at 120 degrees relative to one another.

isometric dot paper: Paper used for sketching purposes that includes dots located along lines that meet at 120 degrees.

isometric grid paper: Paper used for sketching purposes that includes grid lines at 120 degrees relative to one another.

isometric pictorial: A sketch of an object that shows its three dimensions where isometric axes were used as the basis for defining the edges of the object.

left-handed system: Any 3-D coordinate system that is defined by the left-hand rule.

line: Shortest distance between two points.

oblique axes: A set of three coordinate axes that are portrayed on the paper as two perpendicular lines, with the third axis meeting them at an angle, typically 45 degrees.

oblique pictorial: A sketch of an object that shows one face in the plane of the paper and the third dimension receding off at an angle relative to the face.

receding dimension: The portion of the object that appears to go back from the plane of the paper in an oblique pictorial.

right-hand rule: Used to define a 3-D coordinate system whereby by pointing the fingers of the right hand down the x-axis and curling them in the direction of the y-axis, the thumb will point down the z-axis.

right-handed system: Any 3-D coordinate system that is defined by the right-hand-rule.

shading: Marks added to surfaces and features of a sketch to highlight 3-D effects.

3-D coordinate system: A set of three mutually perpendicular axes used to define 3-D space.

tick mark: A short dash used in sketching to locate points on the paper.

2.14 questions for review

1. What is the role of sketching in engineering design? In creativity?
2. Describe which procedure you should use to sketch straight lines. (Are you right- or left-handed?)
3. How do circles appear on an isometric pictorial? On an oblique pictorial?
4. What is a bounding box?
5. How are construction lines used in sketching?
6. Why is it important to know the right-hand rule?

2.15 problems

1. For each of the coordinate axes shown below, indicate whether they are isometric or oblique and whether they represent right-handed or left-handed systems.

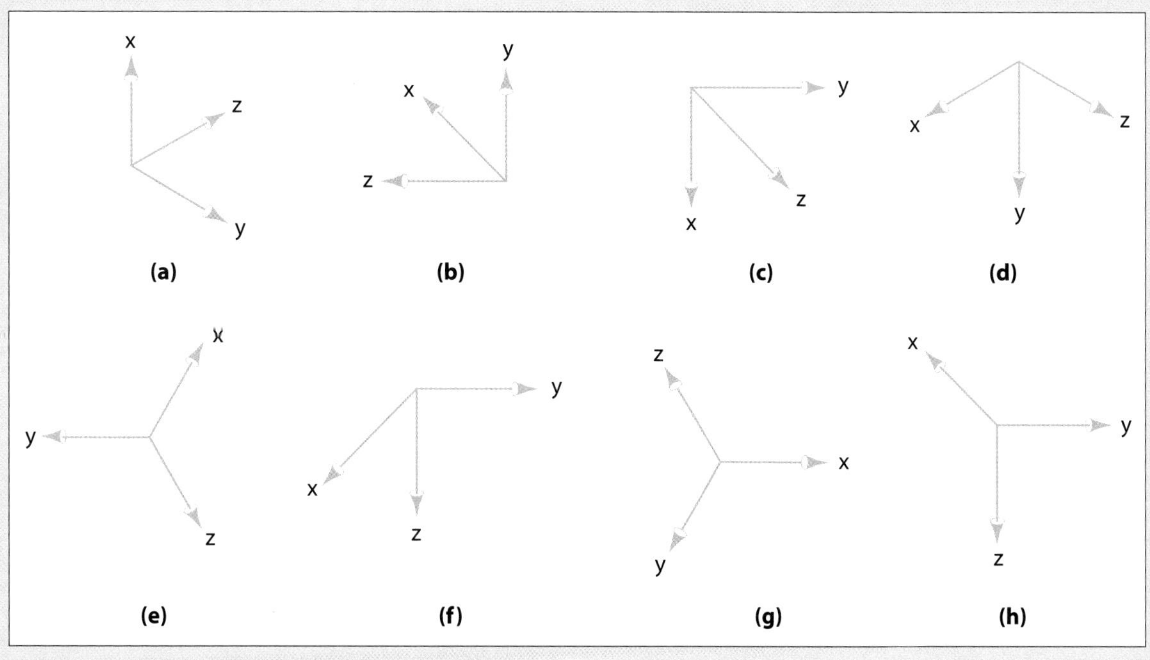

FIGURE P2.1.

2.15 problems (continued)

2. Label the third axis in each of the following figures to define a right-handed system.

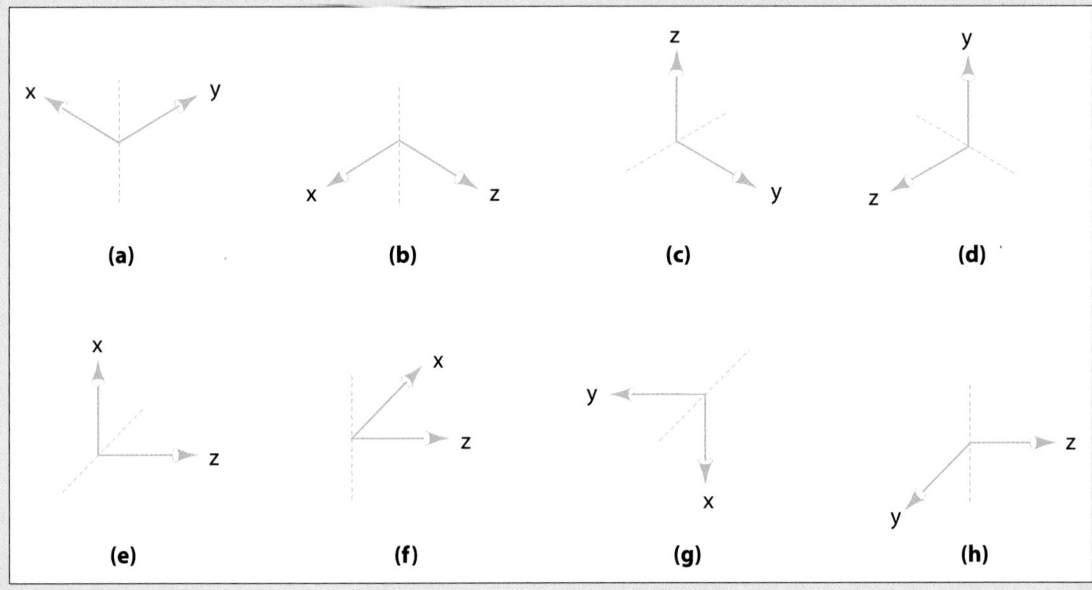

FIGURE P2.2.

2.15 problems (continued)

3. Create isometric sketches from the coded plans shown below.

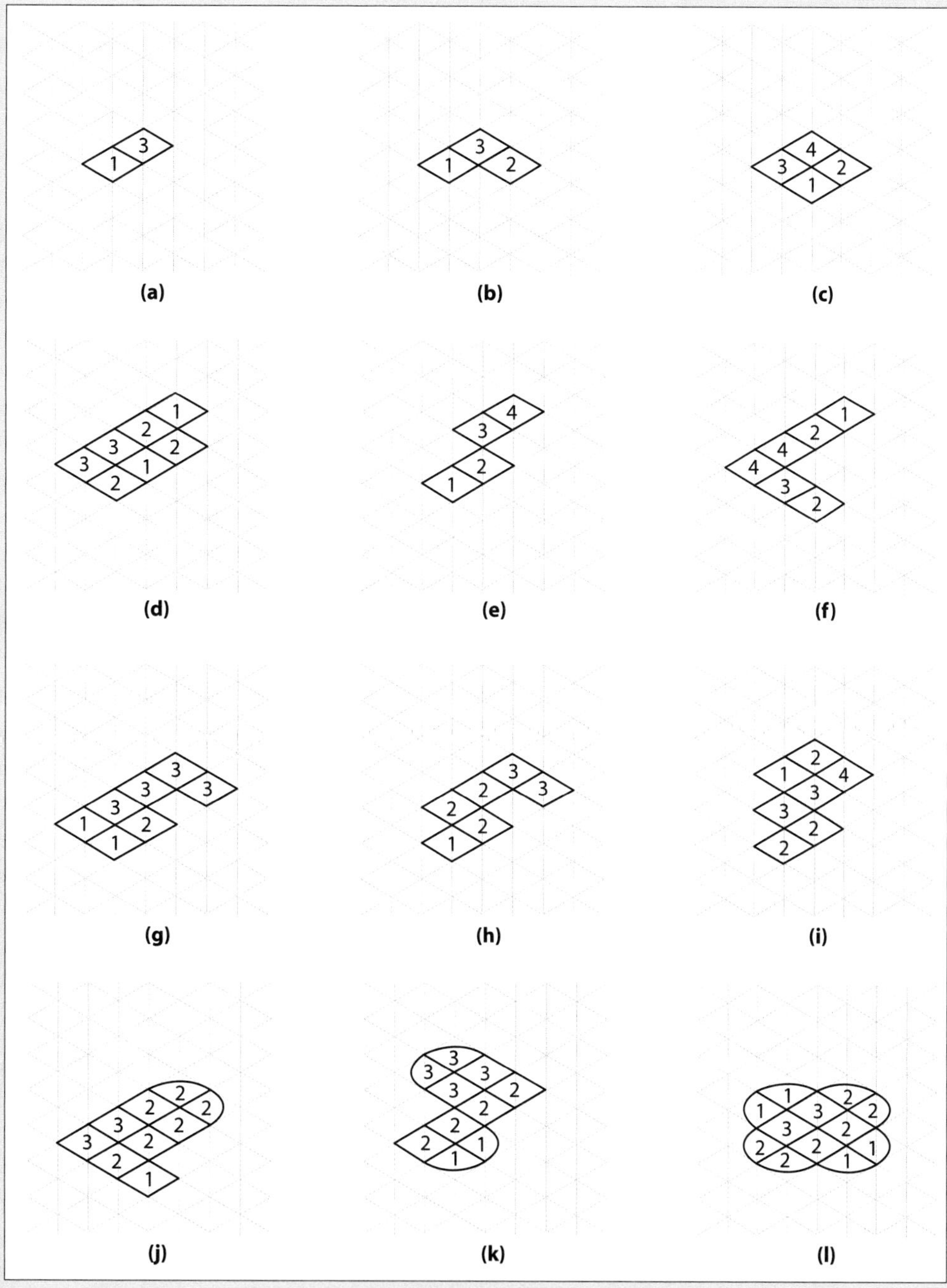

FIGURE P2.3.

4. Sketch a $6 \times 6 \times 2$ block in isometric. On the 6×6 side, sketch a hole of diameter 4, making sure you include back edges of the hole as appropriate. Also create an oblique pictorial of the block.

5. Sketch a $6 \times 6 \times 2$ block in isometric. On the 6×6 side, sketch a hole of diameter 2, making sure you include back edges of the hole as appropriate. Also create an oblique pictorial of the block.

2.15 problems (continued)

6. Sketch a 6 × 6 × 4 block in isometric. On the 6 × 6 side, sketch a hole of diameter 2, making sure you include back edges of the hole as appropriate. Also create an oblique pictorial of the block.
7. From the isometric pictorials and viewing directions defined in the following sketches, create oblique pictorial sketches that look proportionally correct.
8. Use the "foundation, frame, finish" method to create sketches of the following:

   a. stapler c. coffee mug e. calculator

   b. speedboat d. bicycle f. laptop computer

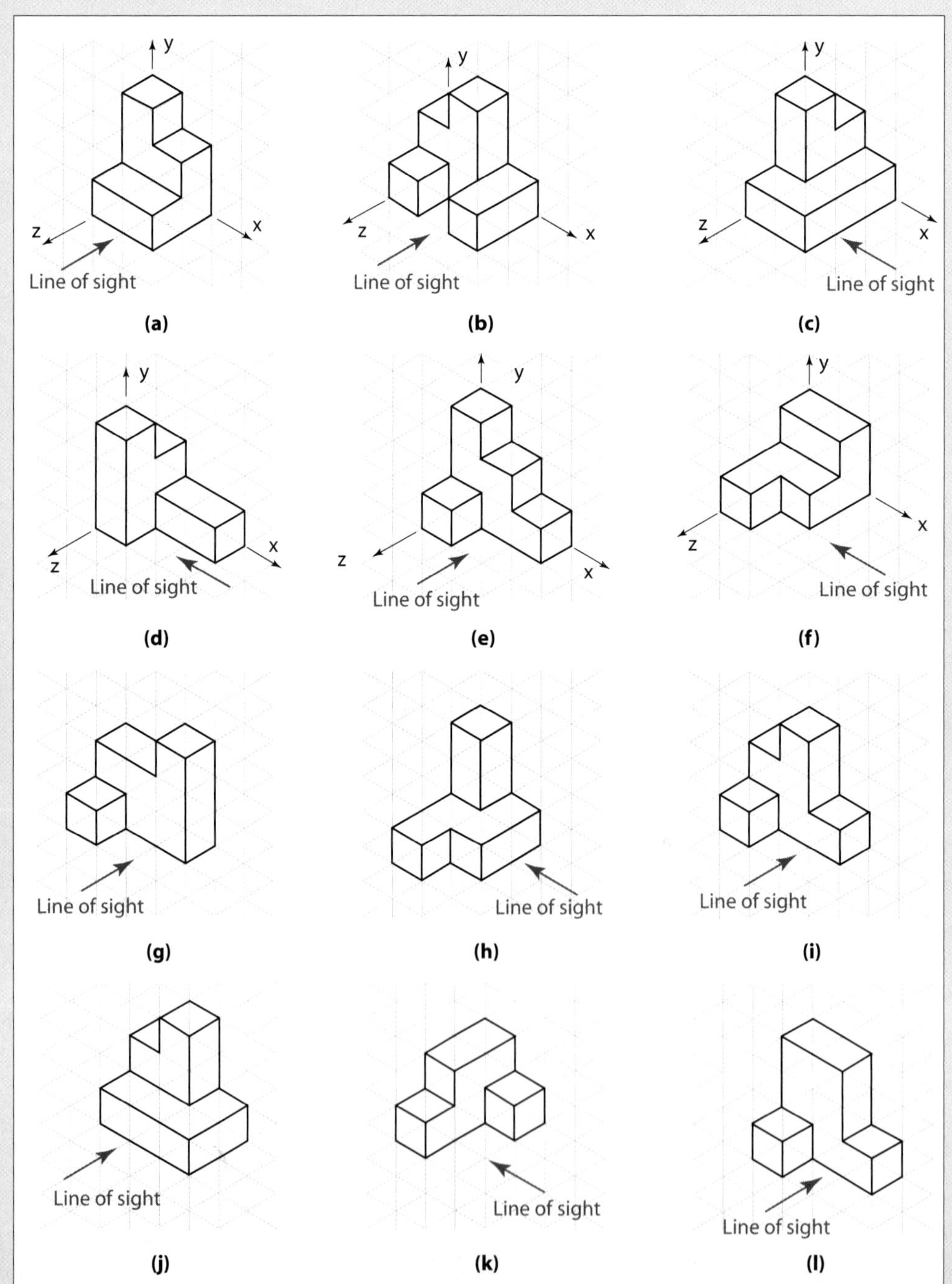

FIGURE P2.4.

3 Visualization

objectives

After completing this chapter, you should be able to

- Recognize that 3-D spatial skills are necessary for success in engineering
- Describe how a person's spatial skills develop as they age
- Examine the types of questions used to assess a person's spatial skill level
- Show how you can improve your 3-D spatial skills through techniques that include
  - Drawing different corner views of an object.
  - Rotating objects about one or more axes.
  - Sketching object reflections and making use of symmetries.
  - Considering cross sections of objects.
  - Combining two objects to form a third object through Boolean operations.

3.01 introduction

When you start your first job in the real world, an engineer or a technologist is likely to hand you a drawing and expect you to understand what is on the page. Imagine your embarrassment if you have no clue what all of the lines and symbols on the drawing mean. One of the fundamental skills you need to understand that drawing is the ability to visualize in three dimensions. The ability to visualize in three dimensions is also linked to creativity in design. People who think creatively are able to "see" things in their minds that others cannot. Their imaginations are not confined by traditional boundaries.

In this chapter, you will learn about the different types of three-dimensional (3-D) spatial skills and ways they can be developed through practice. The chapter will begin with an introduction to the background research conducted in education and to 3-D spatial skills. Then the chapter will take you through several types of visualization activities to further develop your 3-D skills through practice.

3.02 Background

Beginning in the early part of the twentieth century, IQ testing was developed to categorize a person based on his or her intelligence quotient. Anyone who took the IQ test was defined by a number that identified a level of intelligence. IQ scores over 140 identified geniuses; scores below 100 identified slow thinkers. Beginning in the 1970s, scholars began to perceive problems with this one-number categorization of a person's ability to think. One scholar in particular, Howard Gardner, theorized that there were multiple human intelligences and the one-number-fits-all theory did not accurately reflect the scope of human thought processes. Although some of his theories might be subject to scrutiny, they have gained acceptance within the scientific and educational communities. Gardner theorized that there are eight distinct human intelligences; he identified them as:

- Linguistic—the ability to use words effectively in speaking or in writing.
- Logical-Mathematical—the ability to use numbers effectively and to reason well.
- Spatial—the ability to perceive the visual-spatial world accurately and to perform transformations on those perceptions.
- Bodily-Kinesthetic—the capacity of a person to use the whole body to express ideas or feelings and the facility to use the hands to produce or transform things.
- Musical—the capacity to perceive, discriminate, transform, and express musical forms.
- Interpersonal—the ability to perceive and make distinctions in the moods, intentions, motivations, and feelings of other people.
- Intrapersonal—self-knowledge and the ability to act adaptively on the basis of that knowledge.
- Naturalist—the ability to recognize plant or animal species within the environment.

You may be acquainted with someone who has a high level of linguistic intelligence but a low level of musical intelligence. Or you might know someone who has a high level of logical-mathematical intelligence but who lacks interpersonal intelligence relationships. You may even have a friend who is generally smart but who lacks intrapersonal intelligence and attempts stunts that are beyond his or her limitations.

Most people are born with one or more of the intelligences listed. As a child, Tiger Woods was gifted with a natural ability in bodily-kinesthetic intelligence. Mozart was born with a high level of musical intelligence. However, just because a person naturally has a high level of intelligence in one area, does not mean that he or she cannot learn and improve his or her abilities in weaker areas. A person might naturally have strength in linguistics and musical intelligences, but he or she can still learn and improve in logical-mathematical endeavors. The goal of this chapter is to help those of you not born with a high level of spatial intelligence as defined by Gardner.

Learning in general and spatial skills in particular have been the subject of education research studies over the past several decades. The following are a few important questions that the research raised in the area of spatial intelligence:

- How does a person develop spatial skills?
- Why does a person need well-developed spatial skills?
- How are spatial skills measured?

The next few sections will examine researchers' answers to these questions.

3.03 Development of Spatial Skills

As a child grows, the brain develops in ways that enable the child to learn. If you think of each of the eight intelligences described by Gardner, you can understand how these skills and abilities develop as a child grows to maturity. Consider kinesthetic intelligence. Newborn infants cannot move on their own during the first few weeks of life. Within a few months, they can hold up their heads without support. By the age of four months, they can roll over; by six months, they can crawl; by one year, they can walk. Children learn to run, skip, and jump within the next year or so. Eventually, they usually develop all sorts of kinesthetic abilities that enable them to enjoy physical activities such as basketball, swimming, ballet, and bike riding. Nearly every child goes through this natural progression. However, some children develop more quickly than others; some even skip a step and go directly from rolling over to walking without ever crawling. As with most types of intelligence, some individuals—such as professional athletes—have exceptional kinesthetic skills, while others have poorly developed skills and struggle to perform the simplest tasks. However, even people who naturally have little kinesthetic ability can improve through practice and perseverance.

The remaining intelligences (mathematics, verbal, etc.) also have a natural progression; for example, to develop your mathematical intelligence, you have to learn addition before you can learn algebra. Children also acquire spatial skills through a natural progression; however, you may not be as aware of that progression of development as you are of the progressions for the other intelligences. Educational psychologists theorize that there are three distinct stages of development for spatial skills.

The first stage of development involves 2-D spatial skills. As children develop these skills, they are able to recognize 2-D shapes and eventually are able to recognize that a 2-D shape has a certain orientation in space. If you watched Sesame Street as a child, you may remember the game where four pictures of 2-D objects are shown on the TV screen—three objects are identical; the fourth is different in some way. A song urges you to pick out the object that does not belong with the other three. A child who can accomplish this task has developed some of the spatial skills at the first stage. You also may remember playing with a toy similar to a Tupperware ShapeSorter, shown in Figure 3.01. The toy is a ball that is half red and half blue with ten holes in it, each hole a different shape. A child playing this game not only has to recognize that the star-shaped piece corresponds to the star-shaped hole but also has to turn the piece to the correct orientation to fit the piece through the hole. This game challenges different 2-D skills found at the first stage of development of spatial intelligence—a child must recognize the 2-D shape of the object and then must be able to recognize its orientation in 2-D space to complete the task.

Three-dimensional spatial skills are acquired during the second stage of development. Children at this stage can imagine what a 3-D object looks like when it is rotated in space. They can imagine what an object looks like from a different point of view, or they can imagine what an object would look like when folded up from a 2-D pattern.

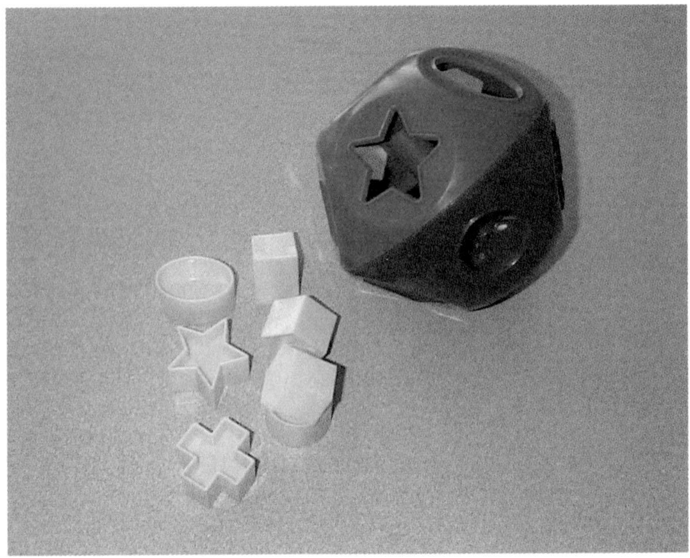

FIGURE 3.01. A Tupperware ShapeSorter toy.

People who are adept at solving the Rubik's Cube puzzle have well-developed 3-D spatial skills. Computer games such as 3D Tetris require well-developed 3-D spatial skills to perform the manipulations required to remain "alive." Soccer players who can imagine the trajectory that puts the ball in the goal from any angle on the playing field typically have well-developed 3-D spatial skills. Children have usually acquired 3-D spatial skills by the time they are in middle school. For some children, it may take a few more years, depending on their natural predisposition toward spatial intelligence and their childhood experiences.

People at an advanced stage of the development of spatial intelligence can combine their 3-D skills with concepts of measurement. Assume you are buying sand for a turtle-shaped sandbox. You go to the local gravel pit where an employee loads the sand in the back of your pickup using a big "scoop." How many scoopfuls will you need? If you can successfully visualize the volume of sand as it is transformed from the 3-D volume of one full scoop to the 3-D volume of the turtle-shaped sandbox, you have acquired this advanced 3-D visualization skill.

Many people never develop the advanced level in spatial intelligence, just like the many people who never achieve advanced skill levels in mathematics or kinesthetic intelligence. Not achieving advanced levels in some of the intelligence areas is not likely to hamper your ability to become a productive, well-adjusted member of society. However, just as a lack of basic development in verbal intelligence is likely to hurt your chances professionally, a lack of basic skills in spatial intelligence may limit your ability to be successful, especially in engineering or a technical field.

Schools help students develop most of the intelligence types, although schools do not usually provide formal training to develop spatial intelligence. You began learning mathematics in kindergarten and are likely continuing your education in math at the present time. If you get a graduate degree in a technical area, you will probably be developing your mathematical intelligence for many years thereafter. The focus on developing spatial skills from an early age, continuing through high school and beyond, is typically absent in the U.S. educational system. Developing spatial intelligence is largely ignored in schools for a variety of reasons; however, those reasons are not the subject of this text.

The lack of prior spatial training may not be a problem for you—you developed your spatial skills informally through everyday experiences or you naturally have a high level of ability in spatial intelligence. However, poorly developed 3-D spatial skills may hinder your success in fields such as engineering and technology. This is especially true as you embark on a journey through an engineering graphics course. Poorly developed spatial skills will leave you frustrated and possibly discouraged about engineering graphics. The good news if you do not have a natural ability in 3-D spatial skills is that you can develop them through practice and exercise.

3.04 Types of Spatial Skills

According to McGee (1979), spatial ability is "the ability to mentally manipulate, rotate, twist, or invert pictorially presented visual stimuli." McGee identifies five components of spatial skills:

- **Spatial perception**—the ability to identify horizontal and vertical directions.
- **Spatial visualization**—the ability to mentally transform (rotate, translate, or mirror) or to mentally alter (twist, fold, or invert) 2-D figures and/or 3-D objects.
- **Mental rotations**—the ability to mentally turn a 3-D object in space and then be able to mentally rotate a different 3-D object in the same way.
- **Spatial relations**—the ability to visualize the relationship between two objects in space, i.e., overlapping or nonoverlapping.
- **Spatial orientation**—the ability of a person to mentally determine his or her own location and orientation within a given environment.

A different researcher (Tartre, 1990) proposed a classification scheme for spatial skills based on the mental processes that are expected to be used in performing a given task. She believes that there are two distinct categories of 3-D spatial skills—spatial visualization and spatial orientation. Spatial visualization is mentally moving an object. Spatial orientation is mentally shifting the point from which you view the object while it remains fixed in space.

Regardless of the classification scheme you choose to believe, it is clear that more than one component skill makes up the broad category of human intelligence known as spatial visualization. Thus, you cannot do just one type of activity and expect to develop equally all of the components of spatial skills. You need to do a variety of tasks to develop your spatial intelligence, just as developing linguistic intelligence requires you to speak, read, write, and listen.

3.05 Assessing Spatial Skills

As with the seven other intelligence types, standardized tests have been developed to determine your level of achievements in spatial intelligence. There are many different tests—some are for 2-D shapes, and some are for 3-D objects. Some evaluate mental rotation skills, and others measure spatial relations skills. The standardized tests usually measure only one specific component of visualization skill. If you were to take a number of different visualization tests, you might find that you have a high level of ability in one component (perhaps paper folding) relative to a low ability in a different component, such as 3-D object rotations. That is normal. Many educators and psychologists believe there is no one-size-fits-all measure of spatial intelligence, just as a single IQ number does not give a clear indication of a person's overall intelligence.

One of the tests designed to measure your level of 2-D spatial skills is the Minnesota Paper Form Board (MPFB) test. Figure 3.02 shows a visualization problem similar to those found on the MPFB test. This problem tests a person's ability to determine which set of five 2-D shapes, A through E, is the composite of the 2-D fragments given in the upper left corner of the figure. The way to solve this test is to mentally rotate or move the three pieces to visualize how to put them together to coincide with the combined shape that contains the pieces. The test may seem easy, but you should have fully developed the 2-D spatial skills needed to solve this test when you were four or five years old. During the years since then, you should have developed more advanced 2-D visualization skills. For example, you should now be able to follow a map and determine whether to make a right or a left turn without turning the map.

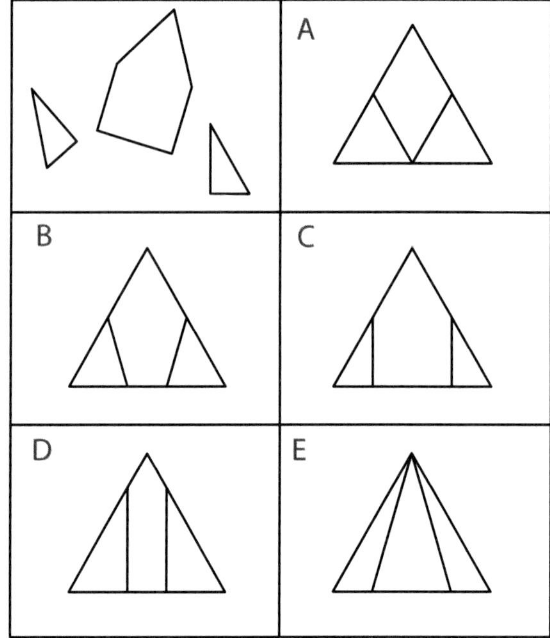

FIGURE 3.02. A problem similar to that found on the Minnesota Paper Form Board Test.

Figure 3.03 shows a visualization problem similar to what is found on the Differential Aptitude Test: Space Relations. This test is designed to measure your ability to move from the 2-D to the 3-D world. The objective is to mentally fold the 2-D pattern along the solid lines, which designate the fold lines, so the object will result in the 3-D shape. You then choose the correct 3-D object from the four possibilities shown in the figure. In your previous math classes, these 2-D patterns may have been referred to as *nets*. In engineering, the 2-D figures are called *flat patterns* or *developments*.

Mental rotations—the ability to visualize the rotation of 3-D objects—is a necessary component skill in engineering graphics and in the use of 3-D modeling software. Figure 3.4 and Figure 3.5 show problems similar to those found on two widely used 3-D spatial tests for rotations.

In the Purdue Spatial Visualization Test: Rotations, an object such as shown in Figure 3.04 is given on the top line before and after it has been rotated in 3-D space. You then have to mentally rotate a different object on the second line by the same amount and select the correct result from the choices given on the third line.

In the Mental Rotation Test, you are given an object such as shown in Figure 3.05 on the left. Of the four choices given, you pick the *two* that show correct possible rotations in space of the original object. (Note that two choices are the same object and two choices are different objects.)

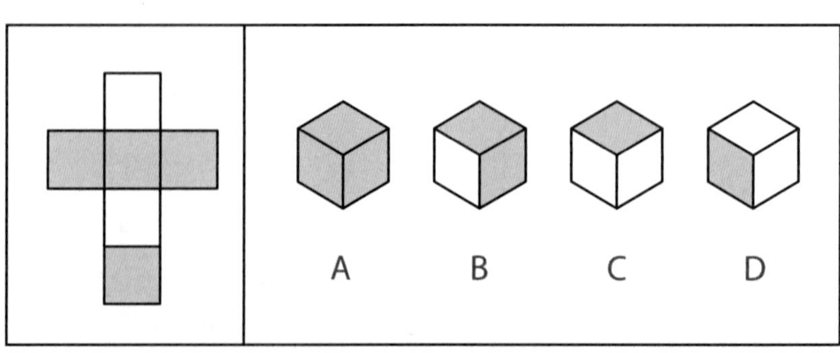

FIGURE 3.03. A problem similar to that found on the Differential Aptitude Test: Space Relations.

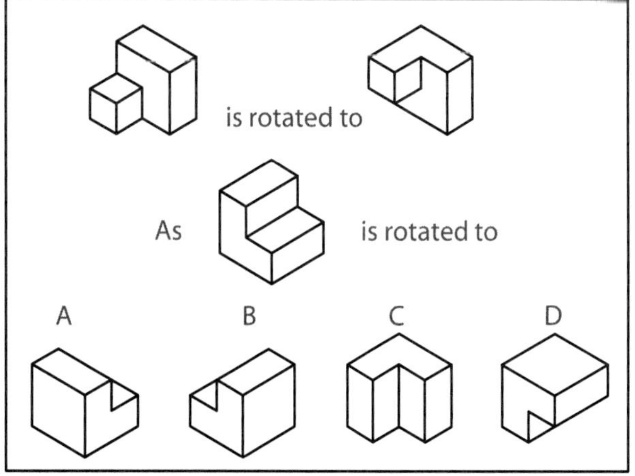

FIGURE 3.04. A problem similar to that found on the Purdue Spatial Visualization Test: Rotations.

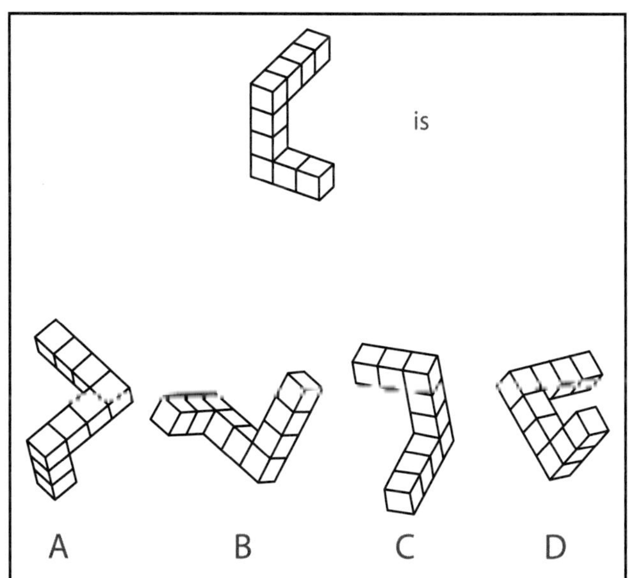

FIGURE 3.05. A problem similar to that found on the Mental Rotation Test.

Another type of spatial skill that is often tested is the ability to visualize the **cross section** that results from "slicing" a 3-D object with a **cutting plane**. One popular test of this type is the Mental Cutting Test. Figure 3.06 shows the type of problem found on this test, which challenges you to imagine the 2-D shape that is the intersection between the cutting plane and the 3-D object.

Engineers and technologists communicate with each other largely through graphical means. They use drawings, sketches, charts, graphs, and CAD models to convey ideas. Design solutions commonly have a graphical component that is backed up by pages of calculations and analysis. Your designs will not be complete without graphics. Even chemical and electrical engineers use drawings for the processes and circuits they design.

So to communicate as an engineer, you must be able to visualize and interpret the images represented in the drawings. Besides satisfying the need for effective communication, a side benefit to having well-developed 3-D spatial skills is that your brain works better when *all* parts of it are focused on solving a problem. Sketching and visualization have been shown to improve the creative process. Well-developed spatial skills contribute to your ability to work innovatively, as well as to learn to use 3-D modeling software.

FIGURE 3.06. A problem similar to that found on the Mental Cutting Test.

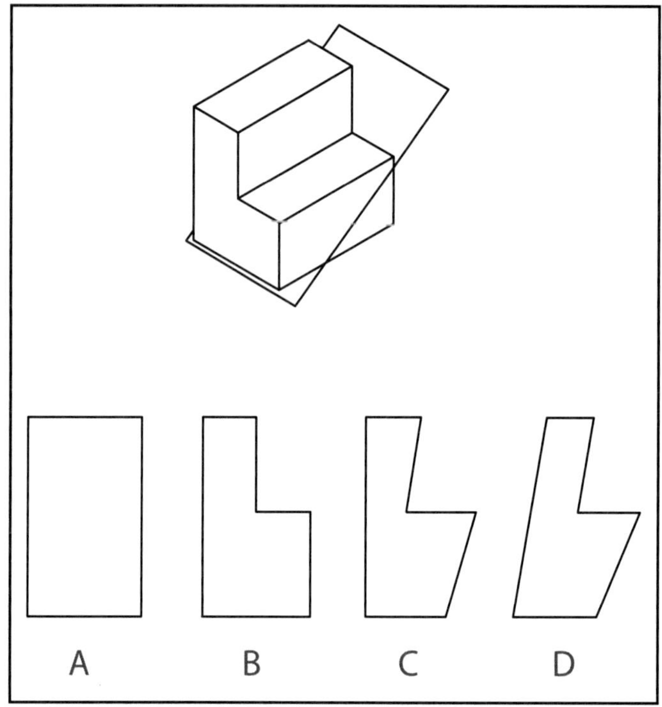

The remaining sections of this chapter will provide exercises for your brain—exercises that develop your 3-D spatial skills; exercises that help you think differently from the way you are thinking in your math and science courses; exercises that will help you improve your sketching skills.

3.06 Isometric Corner Views of Simple Objects

In Chapter 2, you learned how to create a simple isometric sketch of an object made out of blocks as specified by a coded plan. The coded plan is a 2-D portrayal of the object, using numbers to specify the height of the stack of blocks at a given location. Figure 3.07 illustrates the relationship between the coded plan, the object constructed out of blocks, and the resulting isometric sketch of the object—remember, you show edges only between surfaces on the isometric sketch.

The coded plans you viewed in Chapter 2 were constructed on isometric grid paper. The building "grew up" from the plan into the isometric grid. In the previous exercises,

FIGURE 3.07. A coded plan and its resulting isometric sketch.

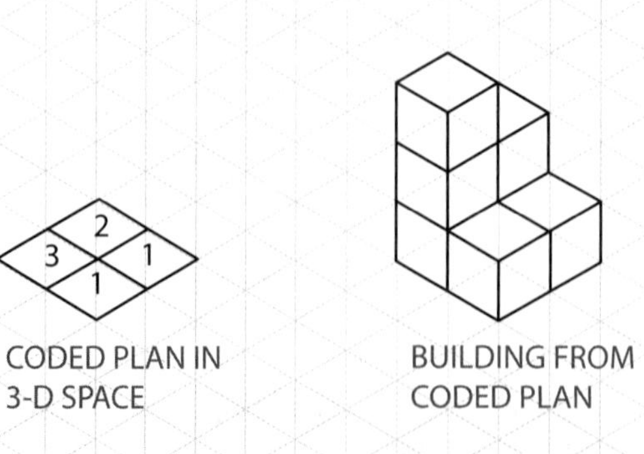

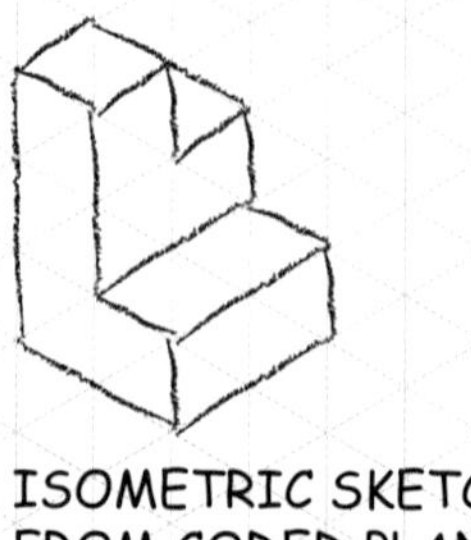

the coded plan was oriented in 3-D space on the isometric grid, which represents 3-D space. Now think about laying the coded plan flat on a 2-D sheet of paper. Figure 3.08 shows the coded plan for the object shown in Figure 3.07 laid flat in a 2-D orientation. Figure 3.9 shows the relationship between a coded plan in 2-D space, the coded plan in 3-D space, the object made of blocks, and the resulting isometric sketch.

When you orient the coded plan in 2-D space, everything you learned about these plans still applies: you "build up" from the plan. The numbers represent the height of the stack of blocks at a given location, and you show lines only where two surfaces intersect. However, now one more consideration has been introduced into the isometric sketching equation—the orientation of your "eye" with respect to the object itself. (Note that *the orientation of your eye* is often referred to as *your viewpoint.*)

Examine again the coded plans in 2-D space. Figure 3.10 shows a simple coded plan with its four corners labeled as *W*, *X*, *Y*, and *Z*. A **corner view** of the object represented by the coded plan in Figure 3.10 is the view from a given corner when the viewpoint is

FIGURE 3.08. The relationship between a coded plan in 2-D space and 3-D space.

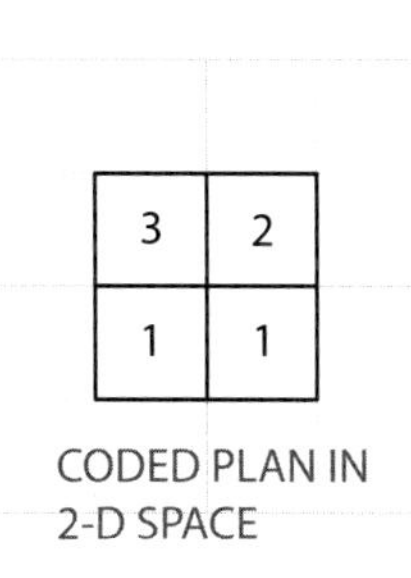

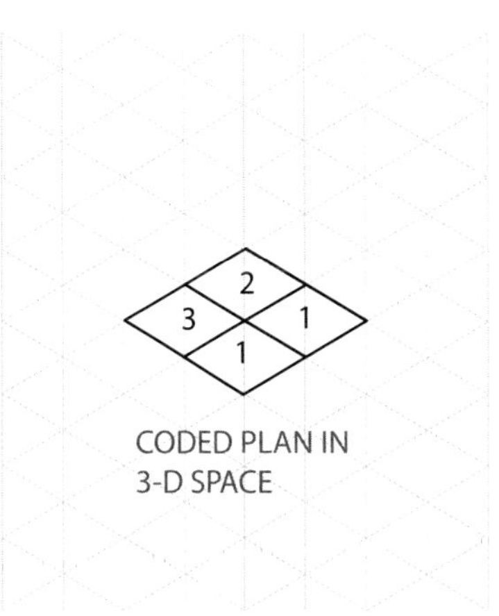

FIGURE 3.09. The relationship between coded plans, a building, and an isometric sketch.

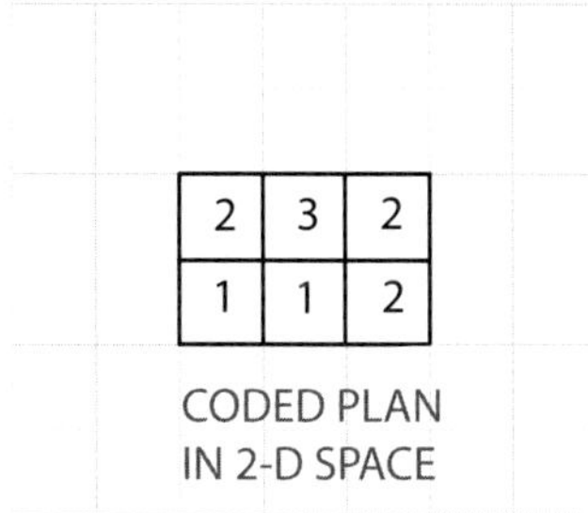

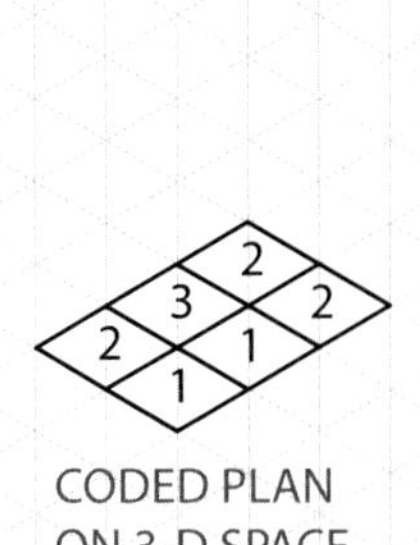

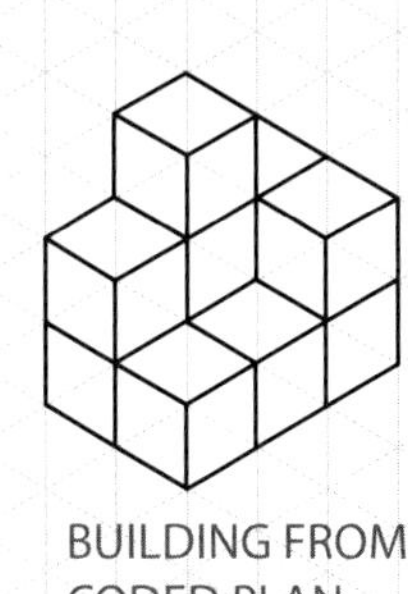

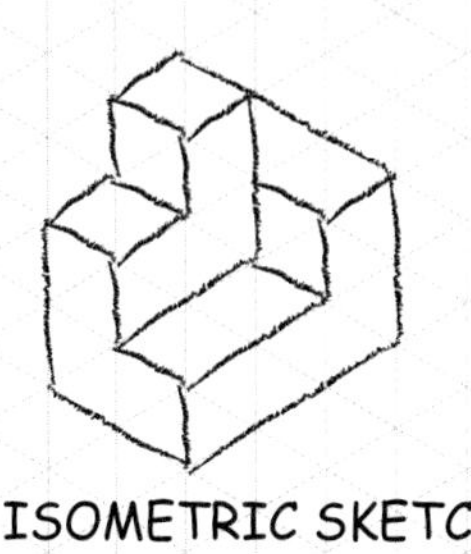

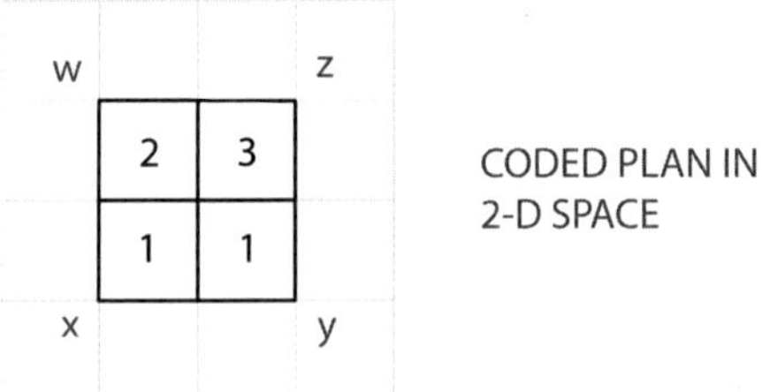

FIGURE 3.10. A simple coded plan with corners labeled.

above the object in question. This view is sometimes referred to as *the bird's-eye view*, because the viewpoint is above the object. A worm's-eye view is the viewpoint from *below* the object. Figure 3.11 shows the four corner views for the coded plan from Figure 3.10.

When the four corner views of the object are created, the object does not change—just your viewpoint of the object. The importance of viewpoint in visualization is readily apparent when you think about a complex system such as an automobile. When you are looking at a car from the front, you may have an entirely different mental image of the car than if you look at it from the side or rear. What you "see" depends largely on where your eye is located relative to the object.

With more practice, you will find it easier to make corner views from coded plans. At first, you may need to turn the paper to visualize what an object will look like from a given corner. With continued practice, however, you should be able to mentally turn the paper to sketch the object from the vantage point of any corner.

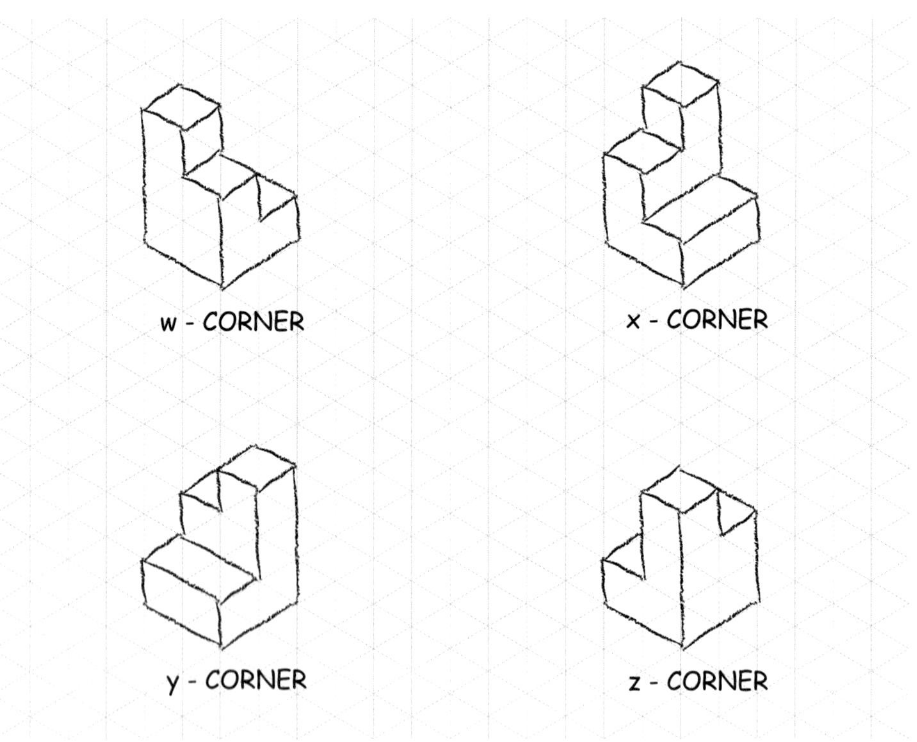

FIGURE 3.11. Sketched isometric views from the corners of the coded plan in Figure 3.10.

3.07 Object Rotations about a Single Axis

Being able to mentally visualize an object as it rotates in space is an important skill for you to acquire as an engineer or a technologist. You already have had limited exposure to the concept of rotating objects through your work with mentally rotating coded plans to obtain different corner views. In the preceding section, you started with the Y-corner view to draw the isometric. Having done that, you should be able to imagine what the object will look like from the X-corner view. If you can see in your mind what the object looks like from the X-corner view, you are mentally rotating the object in space. In this section, you will continue to work with object rotations, tackling increasingly complex objects and using increasingly complex manipulations.

You probably learned in your math classes how 2-D shapes are rotated in 2-D space about a pivot point, as illustrated in Figure 3.12. In this figure, the shape has been rotated 90 degrees counterclockwise (CCW) about the pivot point, which is the origin of the 2-D xy coordinate system. After rotation, the newly oriented shape is referred to as the "image" of the original shape. Notice that when the 2-D shape is rotated about the pivot point, each line on the shape is rotated by the same amount–in this case, 90 degrees CCW about the pivot point. Also notice that the point on the shape that was originally located at the pivot point, the origin, remains at that same location after rotation.

In Chapter 2, you learned about 3-D coordinate systems and how three axes (the x-, y-, and z-axes) can be used to describe 3-D space. When you rotate an object in 3-D space, the same principles apply as for 2-D rotations. In fact, you can reexamine the rotation of the shape in Figure 3.12 from a 3-D perspective. Figure 3.13 shows the 2-D shape drawn in 3-D space before and after it was rotated 90 degrees CCW about the pivot point, which is the origin of the xyz coordinate system.

Observe and understand how each line on the shape is rotated the same amount– 90 degrees CCW about the origin–and that the point on the shape originally in contact with the origin remains at the origin after rotation. One other thing you may notice is that the pivot point is the point view of the z-axis. The point view of a line is what you see as you look down the length of the axis. To illustrate this principle, take a pen or pencil and rotate it so you are looking directly at its point; notice that the length of the pen "vanishes" and only the "point" remains visible, as shown in Figure 3.14. As such, the original rotation of the 2-D shape, as shown in Figure 3.12, could be considered a 90 degree CCW rotation about the z-axis in 3-D space.

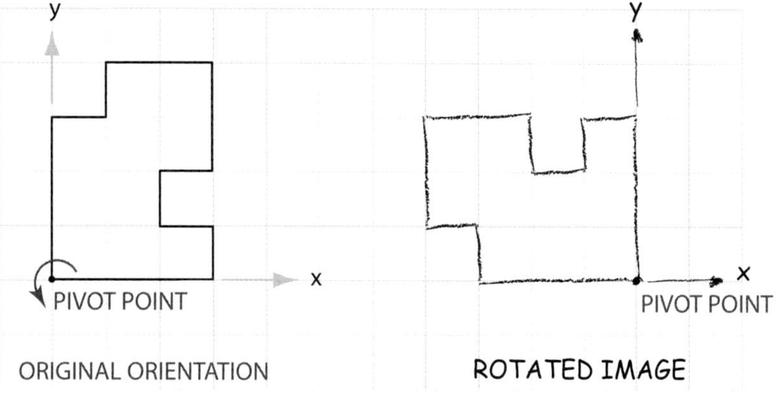

FIGURE 3.12. A shape rotated about a pivot point in 2-D space.

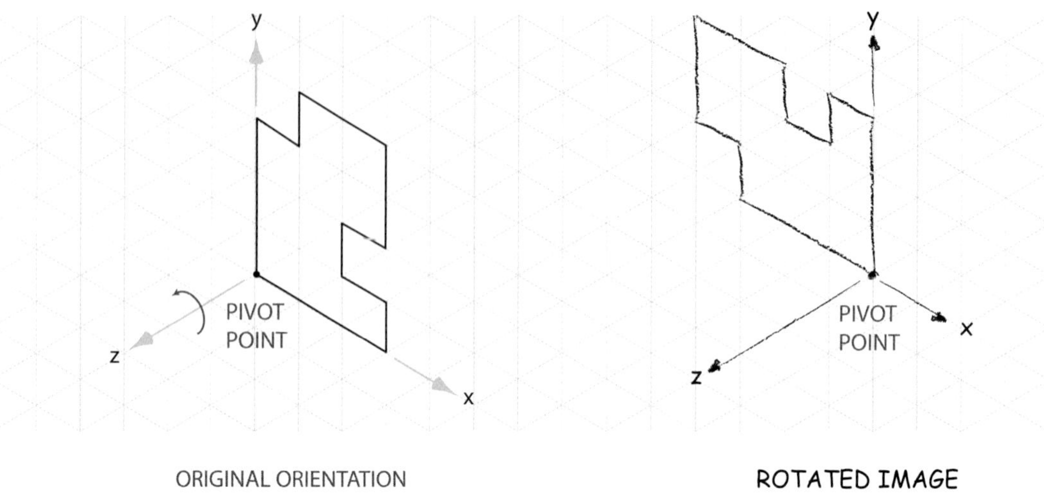

FIGURE 3.13. A 2-D shape rotated in 3-D space.

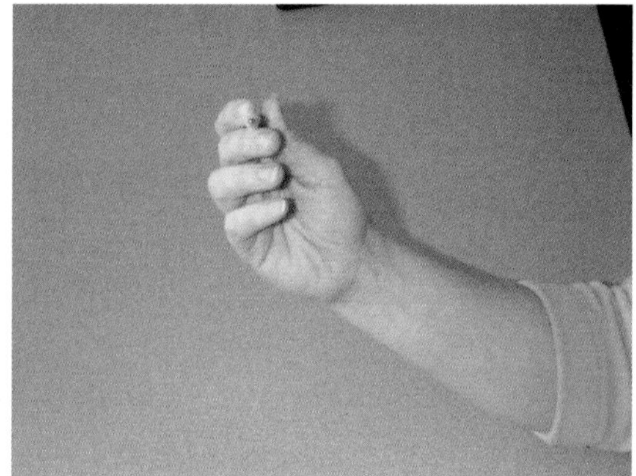

FIGURE 3.14. Looking down the end of a pencil.

Think back to what you learned in Chapter 2 about the right-hand rule. If you point the thumb of your right hand in the positive direction of the z-axis and curl your fingers, you will see that the 90 degree CCW rotation mimics the direction that your fingers curl, as illustrated in Figure 3.15. This CCW rotation of the 2-D shape represents a *positive* 90 degree rotation about the z-axis. The CCW rotation is positive because the thumb of your right hand was pointing in the positive direction of the z-axis as the shape was rotated. If you point the thumb of your right hand in the negative direction of the z-axis and the shape is rotated in the direction the fingers of your right hand curl, your fingers indicate a clockwise (CW) rotation of the shape about the z-axis, as shown in Figure 3.16. A CW rotation about an axis is defined as a negative rotation. Remember that the thumb of your right hand is pointing in the negative z-direction. Also remember that the pivot point of the shape remains at a fixed location in space as it is rotated in the negative z-direction.

You should now be ready to tackle rotations of 3-D objects in 3-D space. Imagine the 2-D shape from the past several figures is a surface view of a 3-D object. Assume you can extend the surface you have been seeing in the xy plane into the z-dimension. The result of extending that surface in the third dimension is a solid object. The terminology of 3-D CAD software says that the shape was extruded. You will learn more about extrusion later in this text. If this shape is "extruded" 3 units into the z-direction, the object will appear as shown in Figure 3.17. In this figure, notice that instead of a single point located on the axis of rotation (the z-axis in the figure), an entire edge of the object is located on that axis. The edge is hidden from sight in this view, but you can imagine it

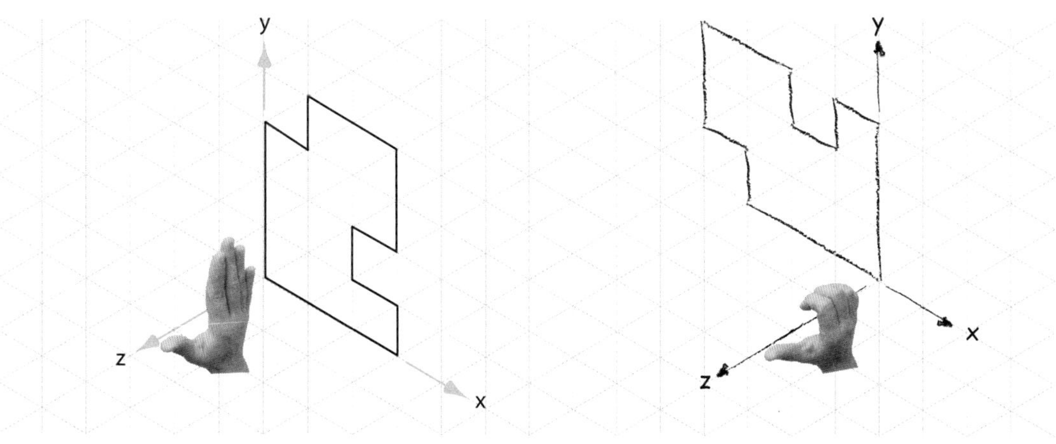

FIGURE 3.15. Positive rotation of a 2-D shape about the z-axis.

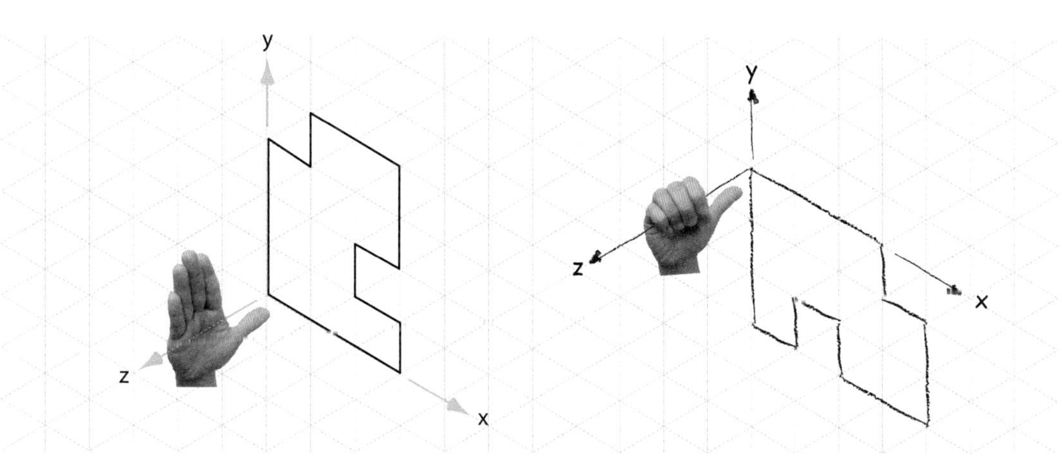

FIGURE 3.16. Negative rotation of a 2-D shape about the z-axis.

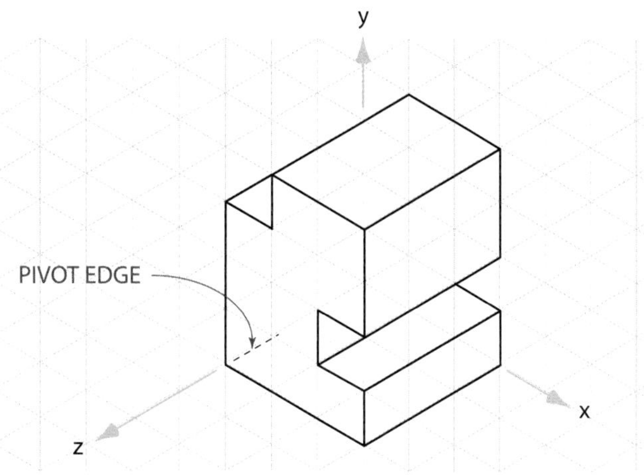

FIGURE 3.17. A 2-D shape from Figure 3.12 extruded three units in the z-direction.

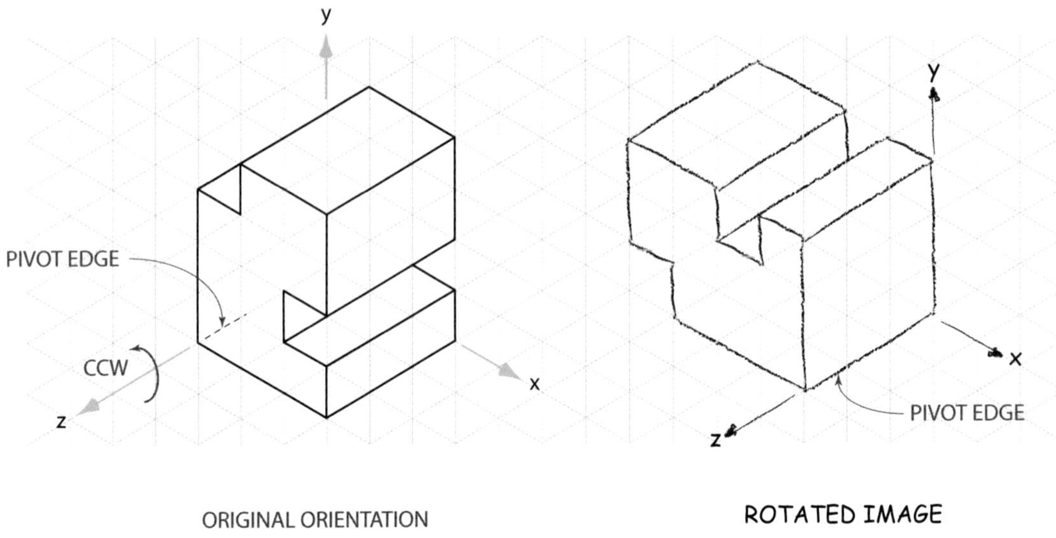

FIGURE 3.18. A 3-D object rotated 90 degrees counterclockwise about the z-axis.

nonetheless. Now think about rotating the entire object about the z-axis in a positive direction (or CCW) 90 degrees from its original position. When this happens, the image shown in Figure 3.18 appears. Instead of a single pivot point, the 3-D rotation has a pivot edge. Throughout the rotation, the edge remained in contact with the axis of rotation. All parts of the object also rotated by the same amount (90 degrees CCW about z) just as all parts of the surface were rotated when you were considering 2-D shapes.

Just as 2-D shapes can be rotated positively (CCW) or negatively (CW) about the z-axis, 3-D objects can be rotated in either direction. Figure 3.19 shows the same object after it has been rotated negative 90 degrees (CW) about the z-axis. This figure also makes clear that the pivot edge of the object remains in contact with the axis of rotation as the object is rotated.

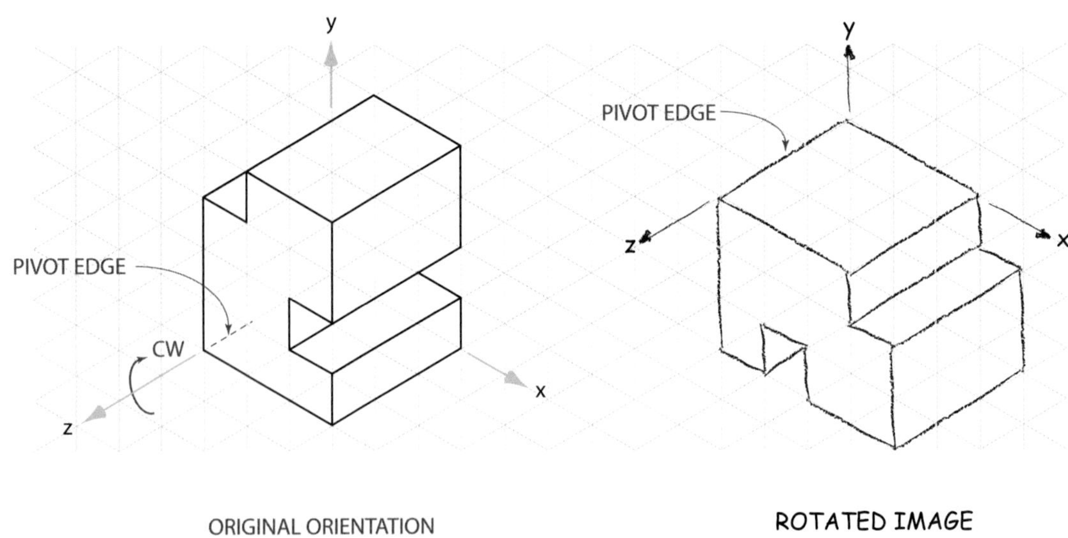

FIGURE 3.19. A 3-D object rotated 90 degrees clockwise about the z-axis.

Any object can be rotated about the x- or y-axis by following the same simple rules established for rotation about the z-axis:

1. The edge of the object originally in contact with the axis of rotation remains in contact after the rotation. This edge is called the pivot edge.
2. Each point, edge, and surface on the object is rotated by exactly the same amount.
3. The rotation is positive when it is CCW about an axis and negative when it is CW about an axis. The direction is determined by looking directly down the positive end of the axis of rotation.
4. An alternative method for determining the direction of the rotation is the right-hand rule. Point the thumb of your right hand into the axis of rotation—into either the positive or negative end of the axis of rotation—and curl your fingers in the direction the object is rotated. The direction you obtain from the right-hand rule is the same as the direction defined in number 3 above, positive is CCW and negative is CW.

Figure 3.20 and Figure 3.21 illustrate the positive and negative 90 degree rotations obtained about the x-axis and the y-axis, respectively.

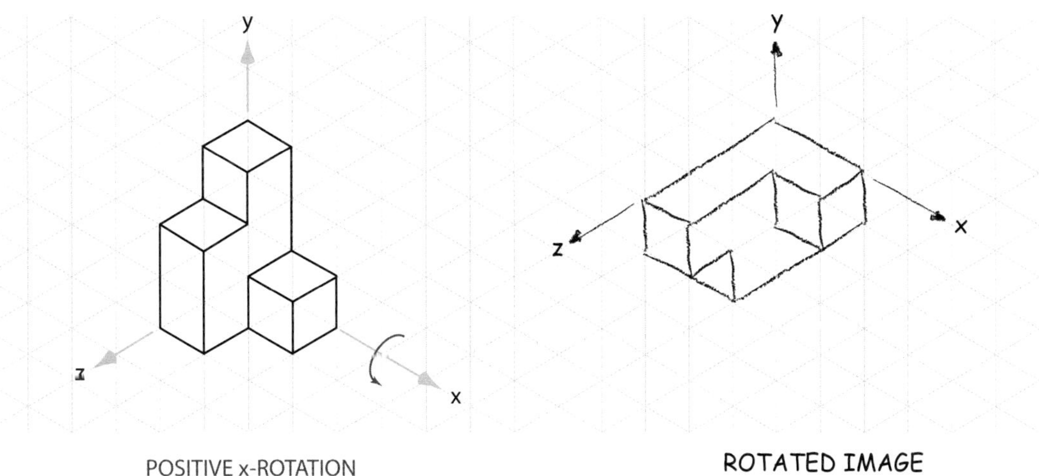

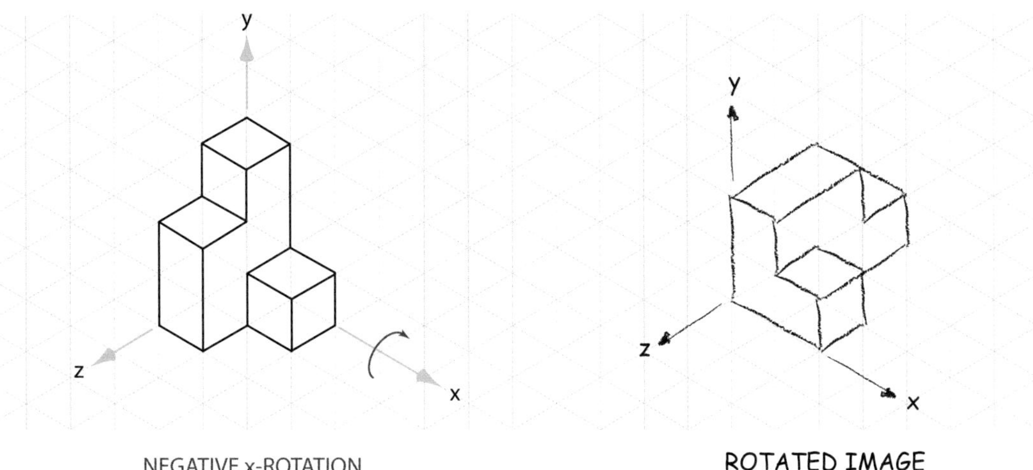

FIGURE 3.20. Positive and negative rotations about the x-axis.

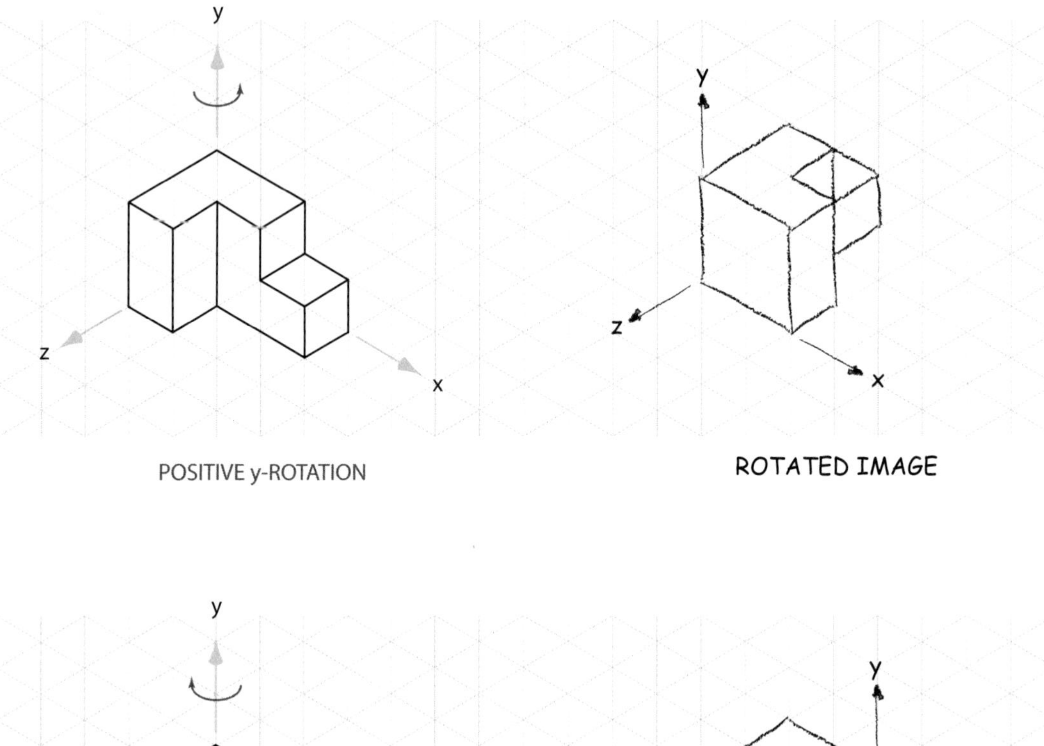

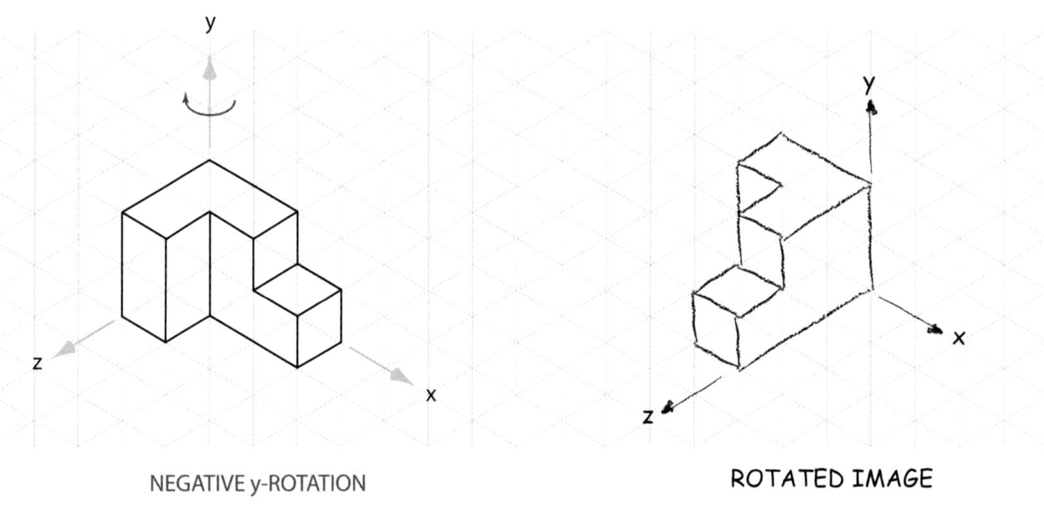

FIGURE 3.21. Positive and negative rotations about the y-axis.

3.07.01 Notation

Specifying in writing a positive, or CCW, rotation about any axis is cumbersome and time-consuming. For this reason, the following notations will be used to describe object rotations in this text:

- To denote positive rotations of an object about the indicated axis.
- To denote negative rotations of an object about the indicated axis.
- Also, for simplicity in sketching, this text will always rotate an object in increments of 90 degrees about the indicated axis. Figure 3.22 illustrates the result when you rotate the object according to the notation given.

3.07.02 Rotation of Objects by More Than 90 Degrees about a Single Axis

In all examples and figures in the preceding sections, objects were rotated exactly 90 degrees about a single axis. In reality, you can rotate objects by any number of degrees. If you rotate an object in two increments of 90 degrees about the same axis, the total rotation will be 180 degrees. Similarly, if you rotate an object in three increments,

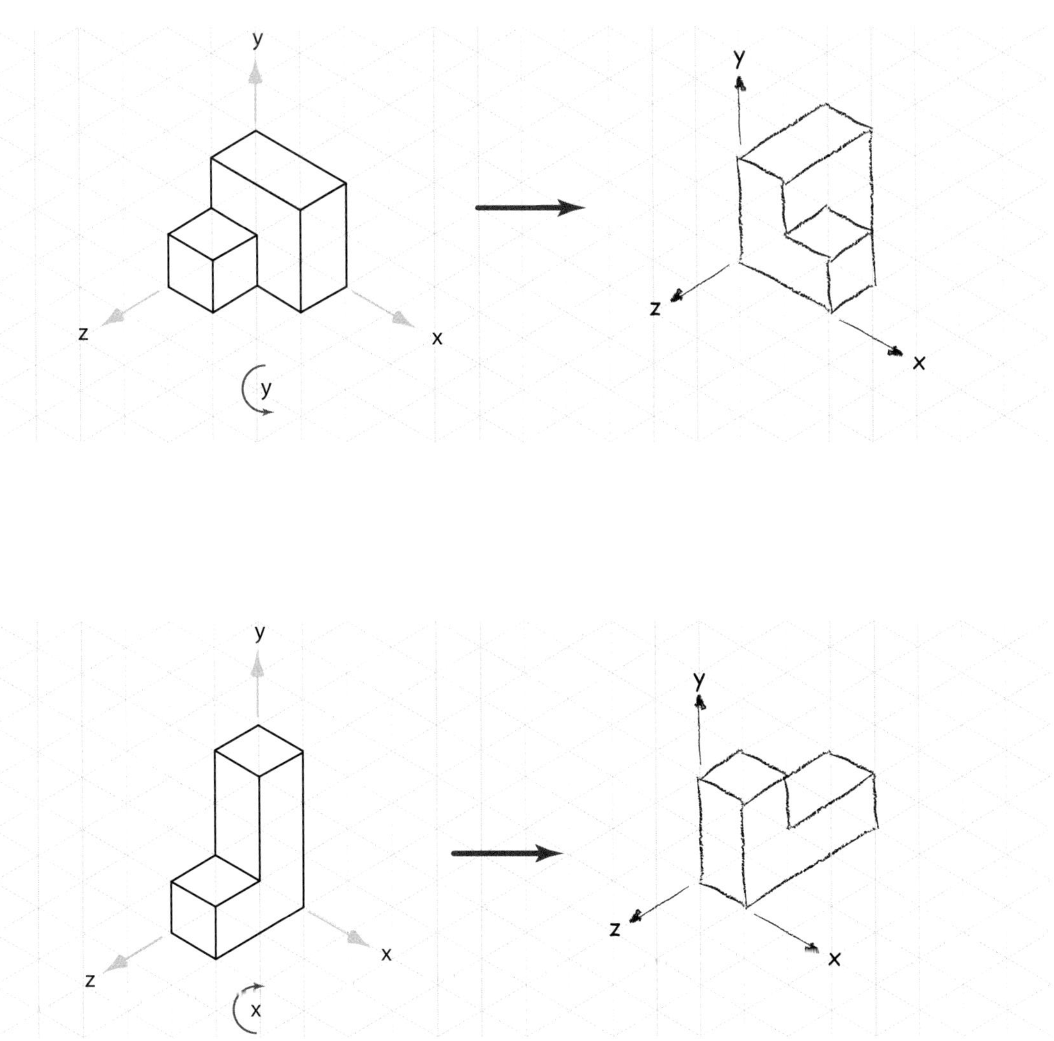

FIGURE 3.22. Object rotations specified by notation.

the total rotation will be 270 degrees. Figure 3.23 shows an object that has been rotated 180 degrees about a single axis, along with the symbol denoting the amount and direction of rotation. Notice that the two 90 degree positive x-axis rotations indicate the total 180 degree rotation achieved.

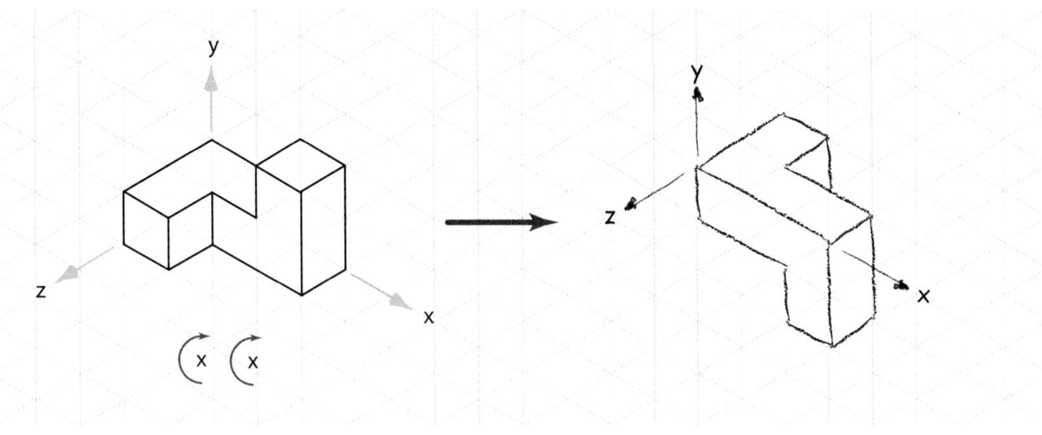

FIGURE 3.23. An object rotated 180 degrees about an axis.

Once you are free to rotate objects in multiple increments of 90 degrees, you can achieve several equivalent rotations. The term *equivalent rotations* means that two different sets of rotations produce the same result.

3.07.03 Equivalencies for Rotations about a Single Axis

When an object is rotated in multiple increments about an axis, the following equivalencies can be observed:

- A positive 180 degree rotation is equivalent to a negative 180 degree rotation.
- A negative 90 degree rotation is equivalent to a positive 270 degree rotation.
- A positive 90 degree rotation is equivalent to a negative 270 degree rotation.

These equivalencies are illustrated in Figures 3.24, 3.25, and 3.26, respectively.

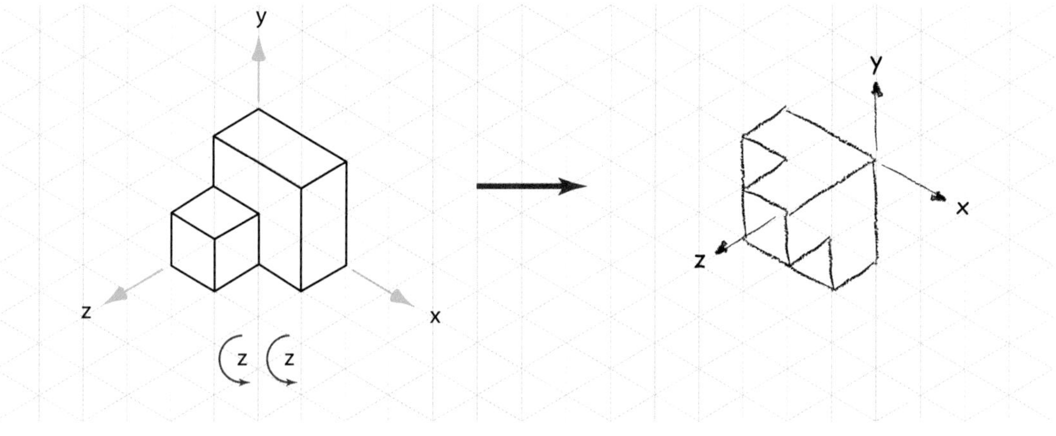

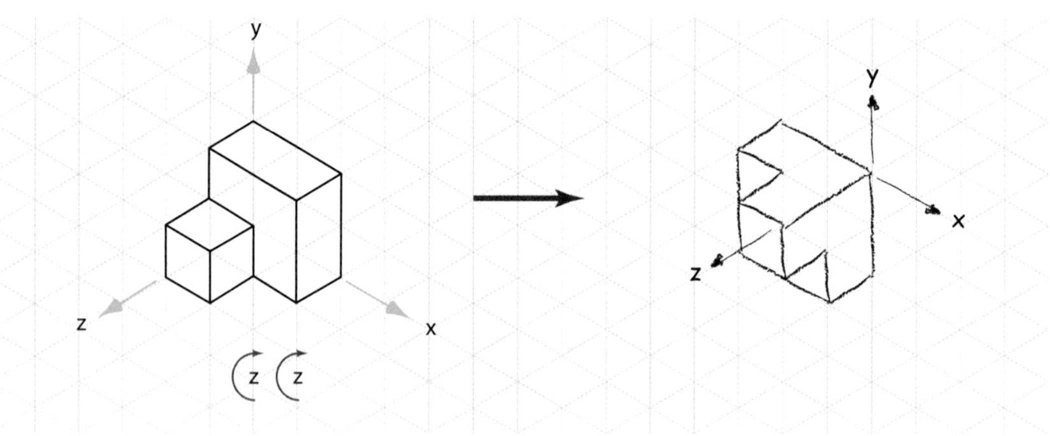

FIGURE 3.24. A positive 180 degree rotation is equivalent to a negative 180 degree rotation.

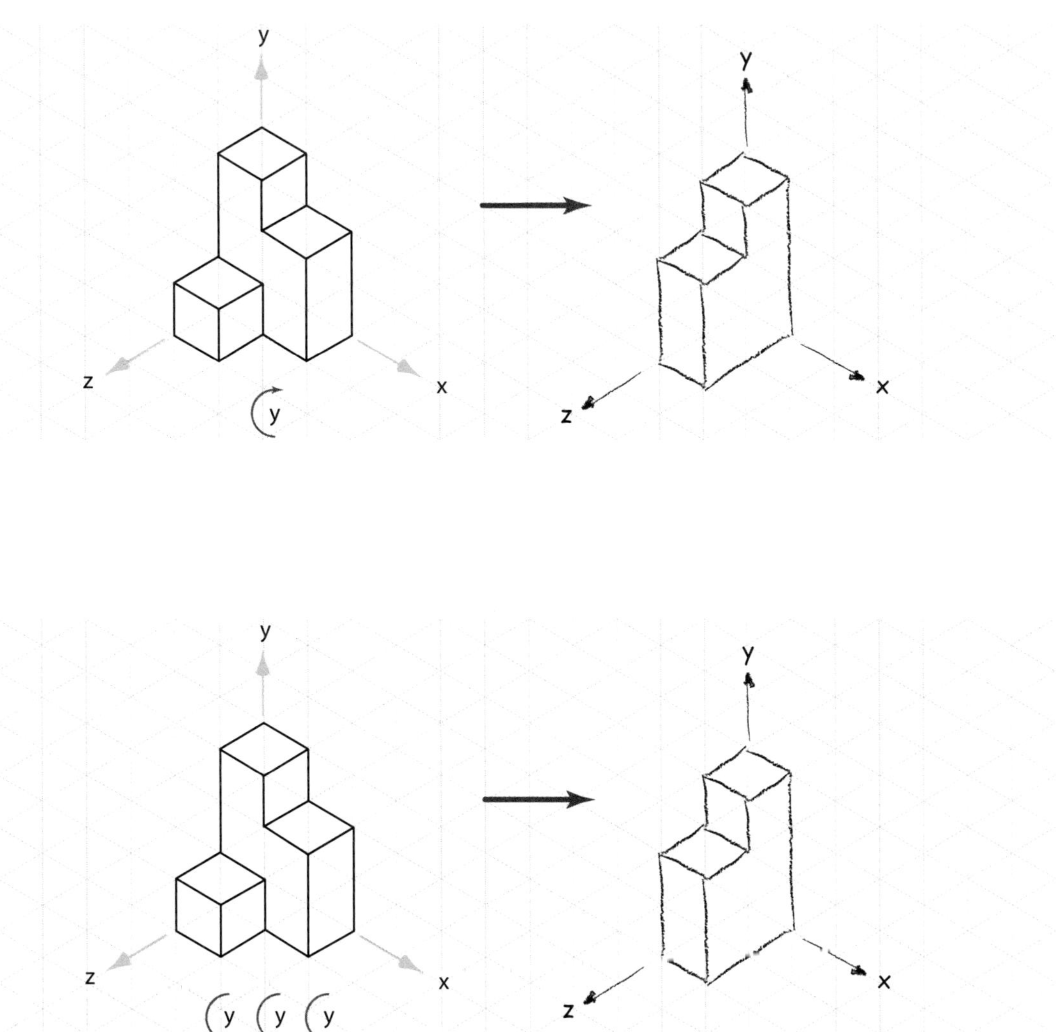

FIGURE 3.25. A negative 90 degree rotation is equivalent to a positive 270 degree rotation.

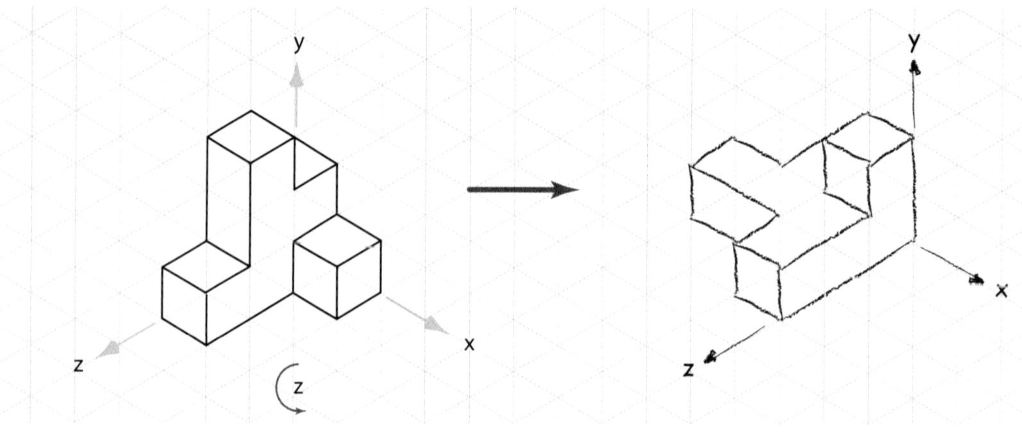

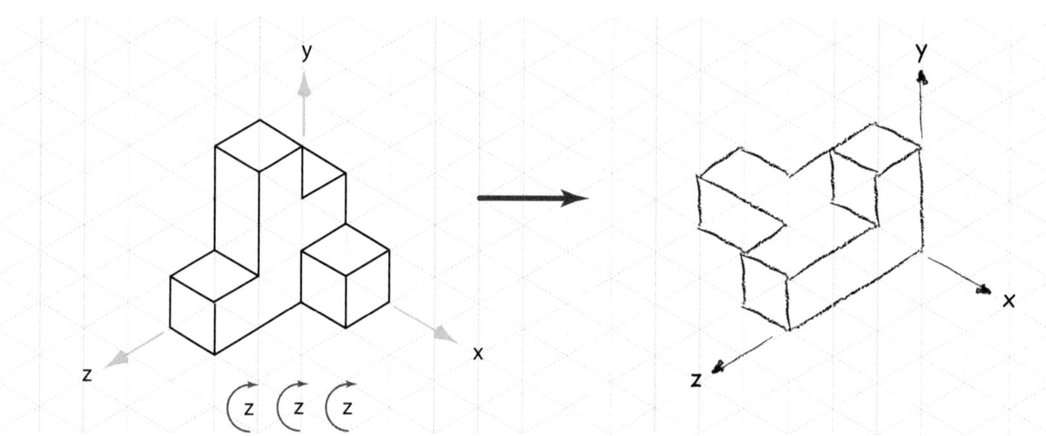

FIGURE 3.26. A positive 90 degree rotation is equivalent to a negative 270 degree rotation.

3.08 Rotation about Two or More Axes

In the same way you rotated an object about a single axis, you can also rotate the object about more than one axis in a series of steps. Figure 3.27 shows an object that has been rotated in the positive direction about the x-axis and then rotated in the negative direction about the y-axis. The rotation notation used in the figure indicates the specified two-step rotation. Figure 3.28 shows the same set of rotations, only this time they are shown in two single steps to achieve the final result. Notice that when an object is rotated about two different axes, a single edge no longer remains in contact with the axis of rotation (since there are now two of them). For rotations about two axes, only a single point remains in its original location, as shown in Figure 3.27 and Figure 3.28.

When rotating an object about two or more axes, you must be careful to perform the rotations in the exact order specified. If the rotations are listed such that you rotate the object CCW in the positive direction about the x-axis and then rotate it CW in the negative direction about the z-axis, you must perform the rotations in that order. Object rotations are not commutative. (Remember that the commutative property in math states that $2 + 3 = 3 + 2$.) For object rotations, rotating about the x-axis and then rotating about the y-axis is *not* the same as rotating about the y-axis and then rotating about the x-axis.

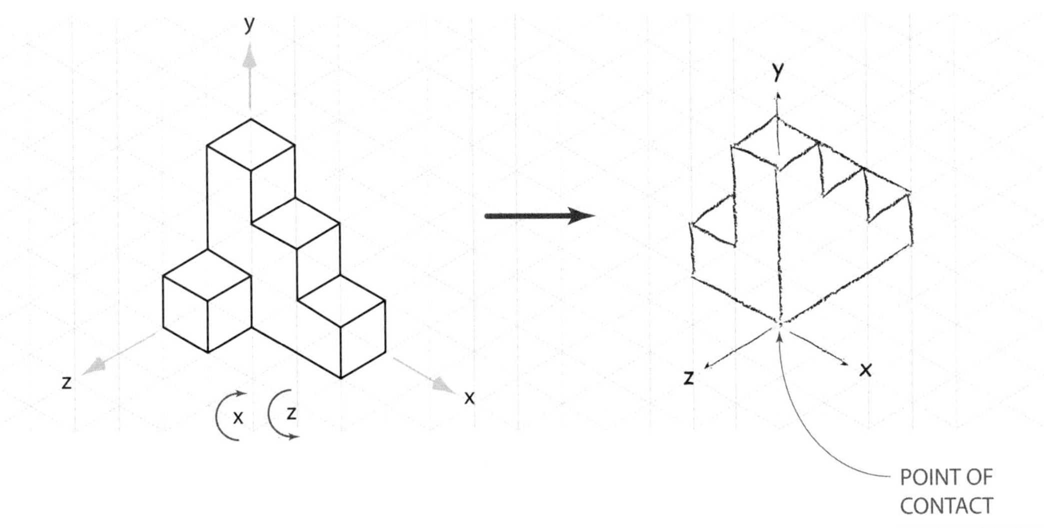

FIGURE 3.27. An object rotated about two axes.

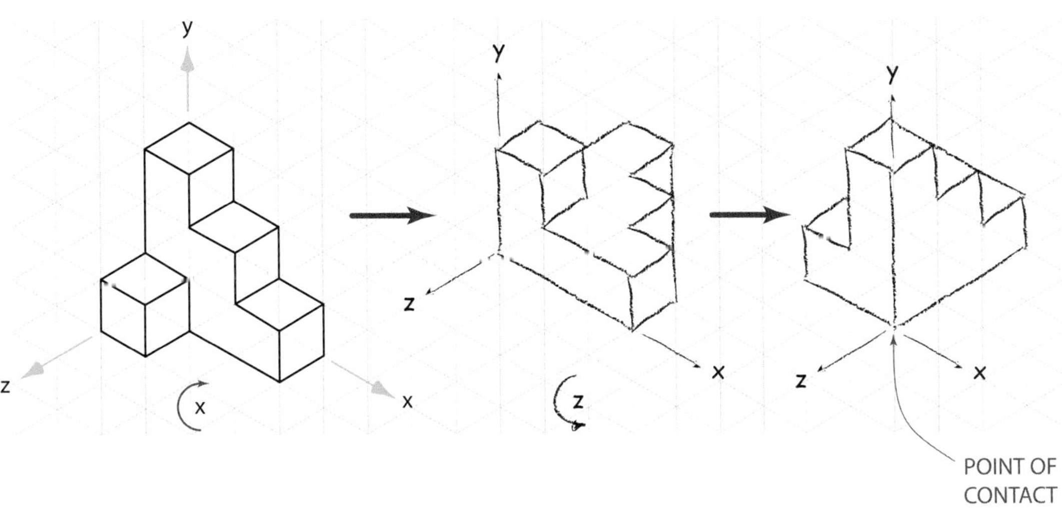

FIGURE 3.28. An object rotated in two steps.

In the top portion of Figure 3.29, the object has been rotated about positive y and then rotated about negative z to obtain its image. In the bottom portion of the figure, the object has been rotated about negative z and then rotated about positive y to obtain a new image of the rotated object. The second image is obtained by reversing the order of the rotations. The resulting images are not the same when the order of rotation is changed. Why? Because with the first set of rotations, the edge of the object on the y-axis serves as the pivot line for the first rotation, which is about positive y. For the second set of rotations, the edge of the object on the z-axis serves as the pivot edge for the first of the two rotations. When you rotate first about negative z, you are using an entirely different object edge than the initial pivot line; hence, the difference in rotated images.

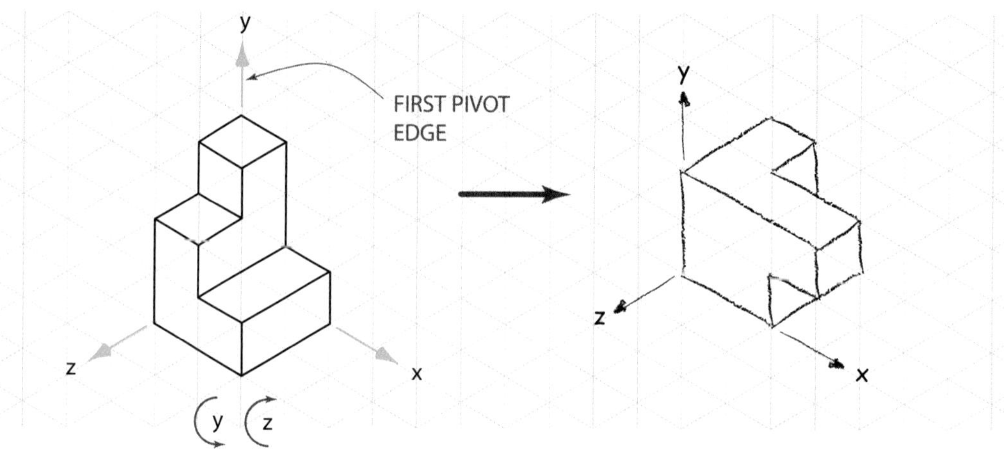

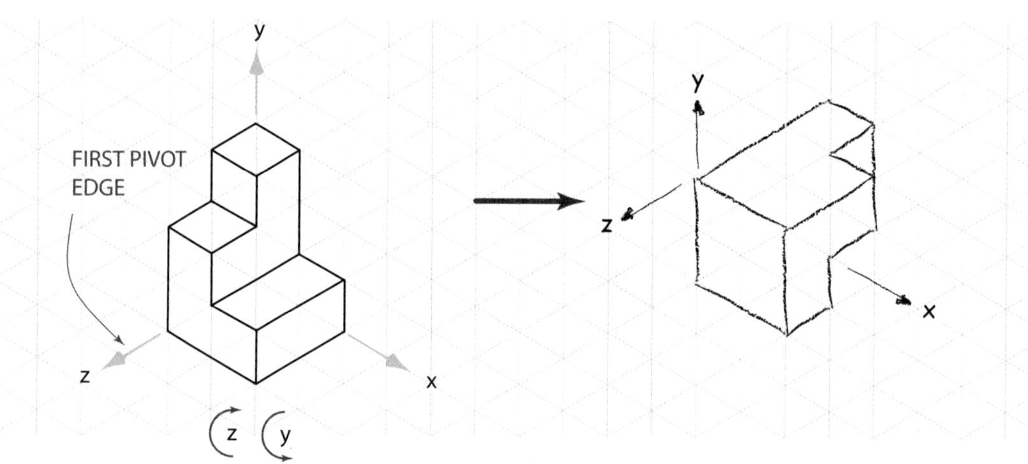

FIGURE 3.29. Object rotations about two axes—order not commutative.

3.08.01 Equivalencies for Object Rotations about Two or More Axes

Just as there are equivalencies for rotations of an object about a single axis, there are equivalencies for object rotations about two axes. Figure 3.30 shows one pair of rotational equivalencies. Can you find another set? How about positive x and then negative z? No! Or positive y and then positive z? Yes! There are several possibilities for each pair of rotations. But it is impossible to come up with simple rules for equivalency, as in the previous discussion of equivalent rotations about a single axis. Equivalent rotations for objects about two or more axes are likely to be determined through trial and error and a great deal of practice.

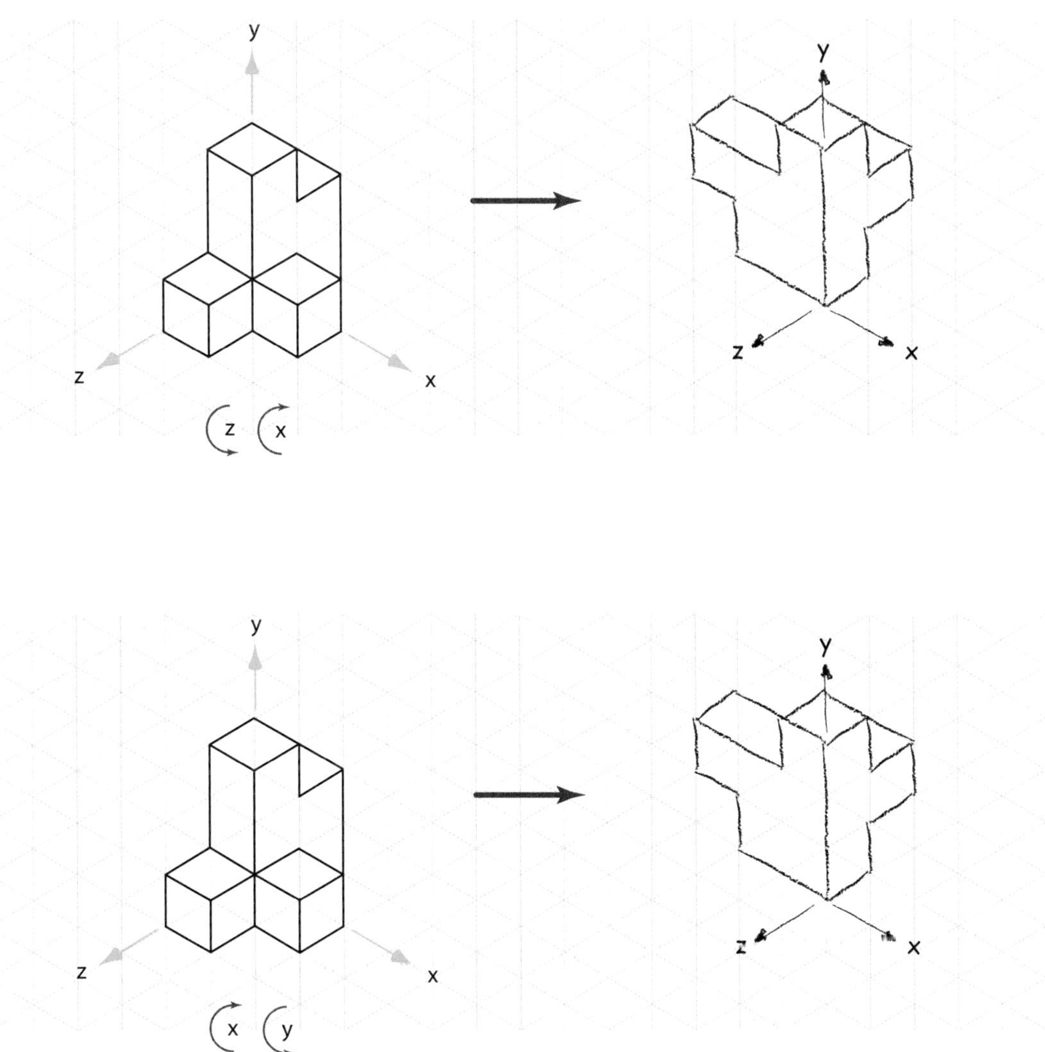

FIGURE 3.30. Equivalent rotations about two axes.

3.09 Reflections and Symmetry

Now that you know the basics of how to visualize an object rotated about an axis, you are ready to move on to visualizing reflections and symmetry. Visualizing planes of symmetry, for example, could save you a great deal of computation time when you are using analysis tools such as FEA. You will learn more about FEA in later chapters of this text.

You are probably familiar with the concept of **reflections** because you are used to looking at your image reflected back to you from a mirror. With a mirror, you see a reflected 2-D image of your face. If you have a mole on your right cheek, you will see the mole on the right cheek of the reflection. Even though your face is three-dimensional, your face in the mirror is a 2-D reflection—as if your face were projected onto a 2-D plane with your line of sight perpendicular to the plane. You may be able to see somewhat in the third dimension from this mirror plane; however, your depth perception will be a bit off because the image is only two-dimensional. Three-dimensional reflection of objects is different from 2-D reflections with mirrors. For one thing, you reflect a 3-D object *across* the plane so that a 3-D image ends up on the other side of the reflection plane.

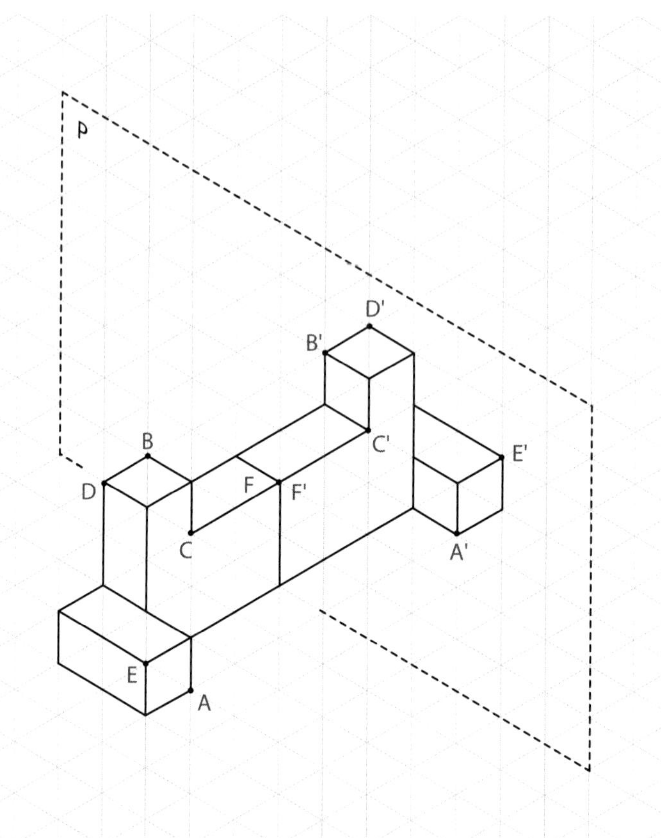

FIGURE 3.31. An object and its 3-D reflection.

Formally stated, in the case of 3-D object reflections, such as shown in Figure 3.31, each point A of the object is associated with an image point A´ in the reflection such that the plane of reflection is a perpendicular bisector of the line segment AA´. What this means is the distance between a point on an object and the reflection plane is equal to the distance between the corresponding point on the image and the reflection plane. The distances are measured along a line perpendicular to the plane of reflection. Figure 3.31 shows a simple object and its reflection across a reflection plane.

In this figure, several points on the original object are labeled, as well as their corresponding points on the reflected image. In this case, the plane of reflection coincides with one planar end of the original object; therefore, the corresponding planar end of the reflected image also coincides with the reflection plane. If you measure the distance between point A on the object and the reflection plane, you will find that it is 3 units. Then if you measure the distance between A´ and the reflection plane, you will find the distance to be 3 units again. It is also possible to reflect an object across a plane when the object is located some distance from the reflection plane, as illustrated in Figure 3.32.

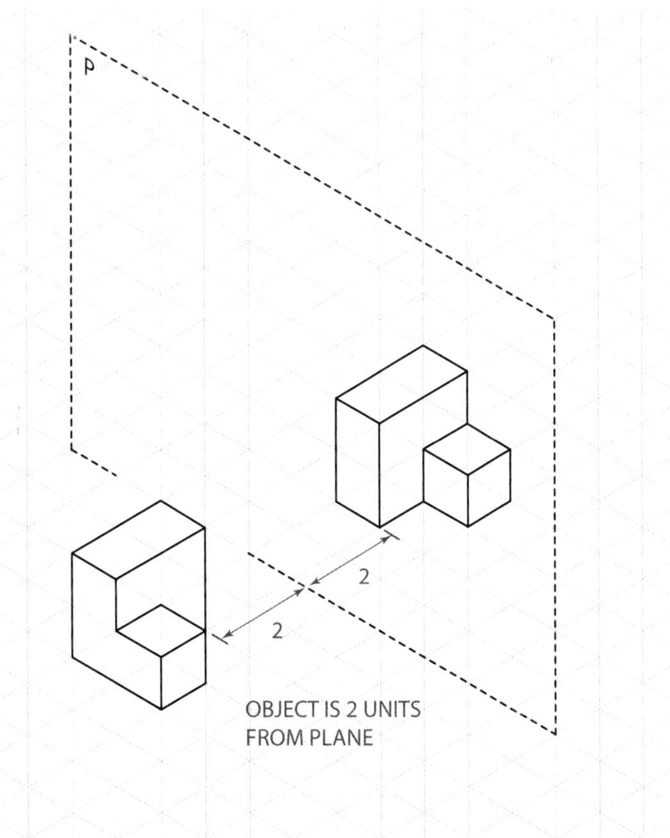

FIGURE 3.32. An object located at a distance from the plane and its reflection.

3.09.01 Symmetry

Your job as an engineer may be easier if you can recognize planes of symmetry within an object. A plane of **symmetry** is an imaginary plane that cuts through an object such that the two parts, one on either side of the plane, are reflections of each other. Not all objects have inherent symmetry. The human body is roughly symmetrical and has one plane of symmetry—a vertical plane through the tip of the nose and the belly button. The left side is a reflection of the right side. Some objects contain no planes of symmetry, some contain only one plane of symmetry, and still others contain an infinite number of planes of symmetry. Figure 3.33 shows several objects and their planes of symmetry: one object contains no planes of symmetry, one object has just one plane of symmetry, one object has two planes of symmetry, and the last object contains an infinite number of planes of symmetry.

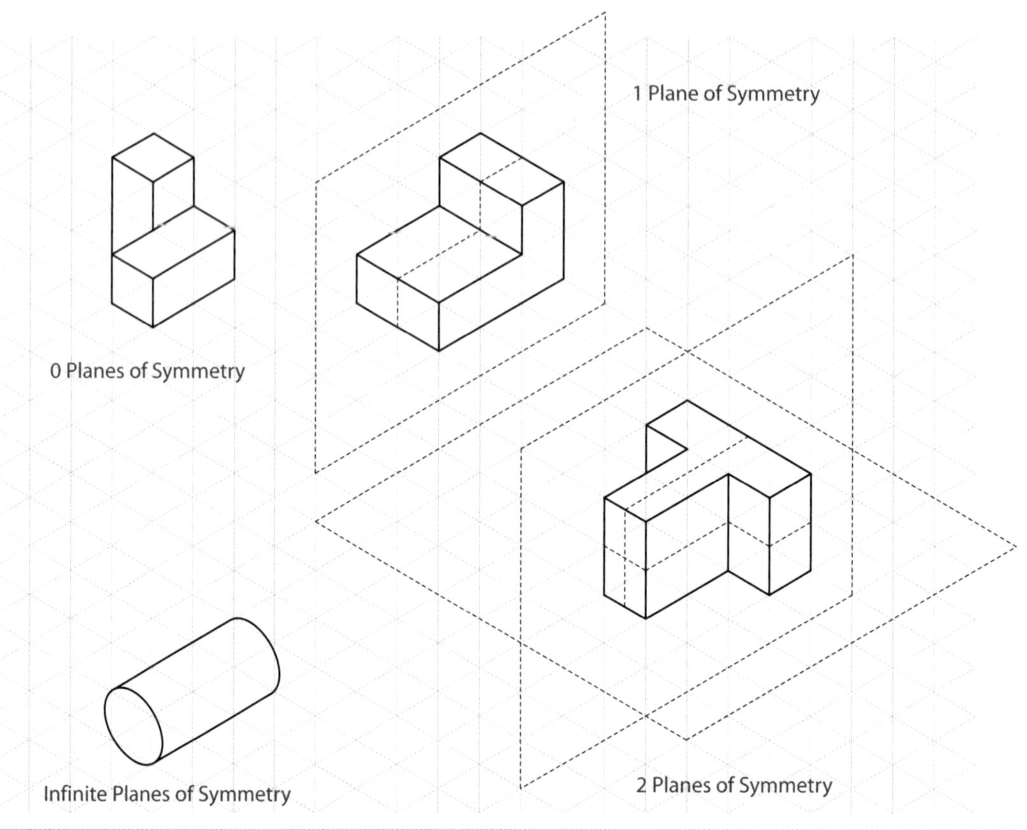

FIGURE 3.33. Objects and their planes of symmetry.

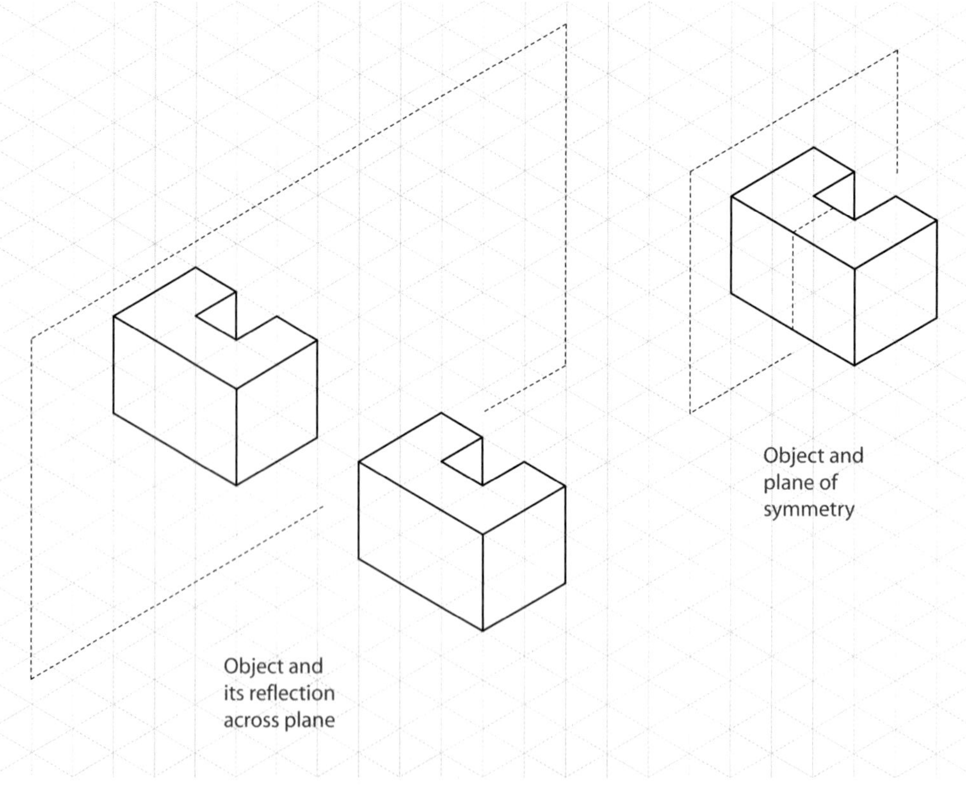

FIGURE 3.34. A comparison of object reflection and symmetry.

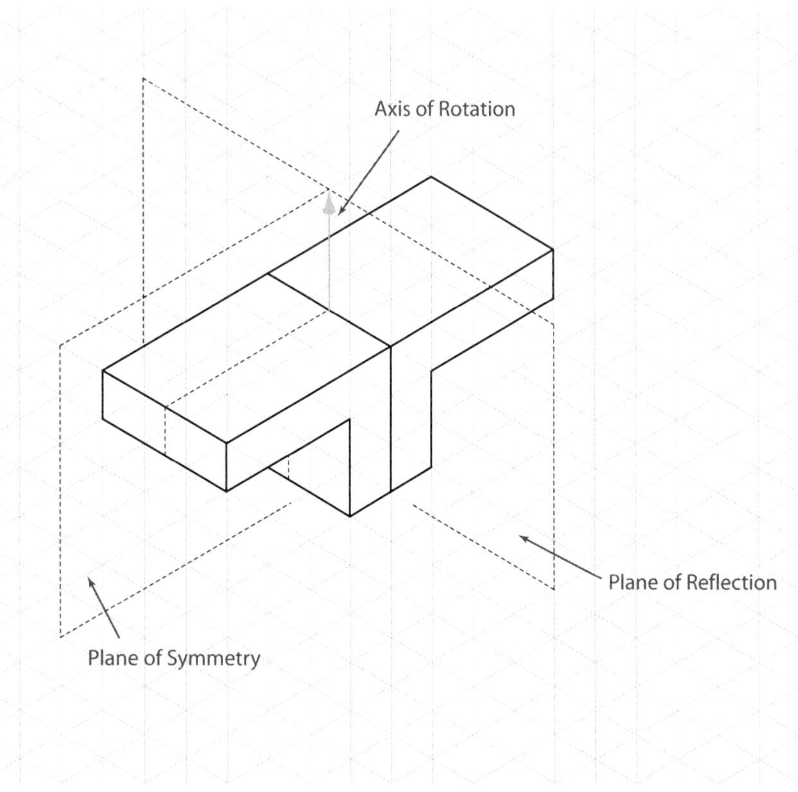

FIGURE 3.35. Object reflection through rotation.

There is one major difference between object reflection and object symmetry. With reflections, you end up with two separate objects (the original and its reflected image); with symmetry, you have a single object that you imagine is being sliced by a plane to form two symmetrical halves. Figure 3.34 illustrates the difference between the two.

For an object that is symmetrical about a plane, you can sometimes obtain its reflection by rotating the object 180 degrees. To do this, the axis of rotation must be the intersection between the plane of reflection and the plane of symmetry (two planes intersect to form a line). This concept is illustrated in Figure 3.35. Note that a reflection of an object that is not symmetrical cannot be achieved through a simple 180 degree rotation of the object. Hold up your hands in front of you to obtain an object (left hand) and its reflected image (right hand). Note that because your hands have no planes of symmetry, it is impossible to rotate one of them in space to obtain the other one.

3.10 Cross Sections of Solids

Visualizing cross-sections enables an engineer to figure out how a building or a mechanical device is put together. Visualizing cross-sections enables an electrical engineer to think about how circuit boards stack together within the housing that contains them. Chemical engineers and materials engineers think about the cross sections of molecules and the way those molecules combine with other molecules. Geological engineers and mining engineers visualize cross sections of the earth to determine where veins of rock and ore may be located. Most of the skills described in these examples are at an advanced level; in this section, you will learn about cross-sections of solids from a fundamental level. Then you can apply the principles to the visualization of more complex parts and systems in later courses and, of course, in your professional work.

Simply stated, a cross-section is defined as "the intersection between a solid object and a cutting plane." Because a plane is infinitely thin, the resulting intersection of the two planes is a planar section. The limits of the cross-sectional plane are the edges and

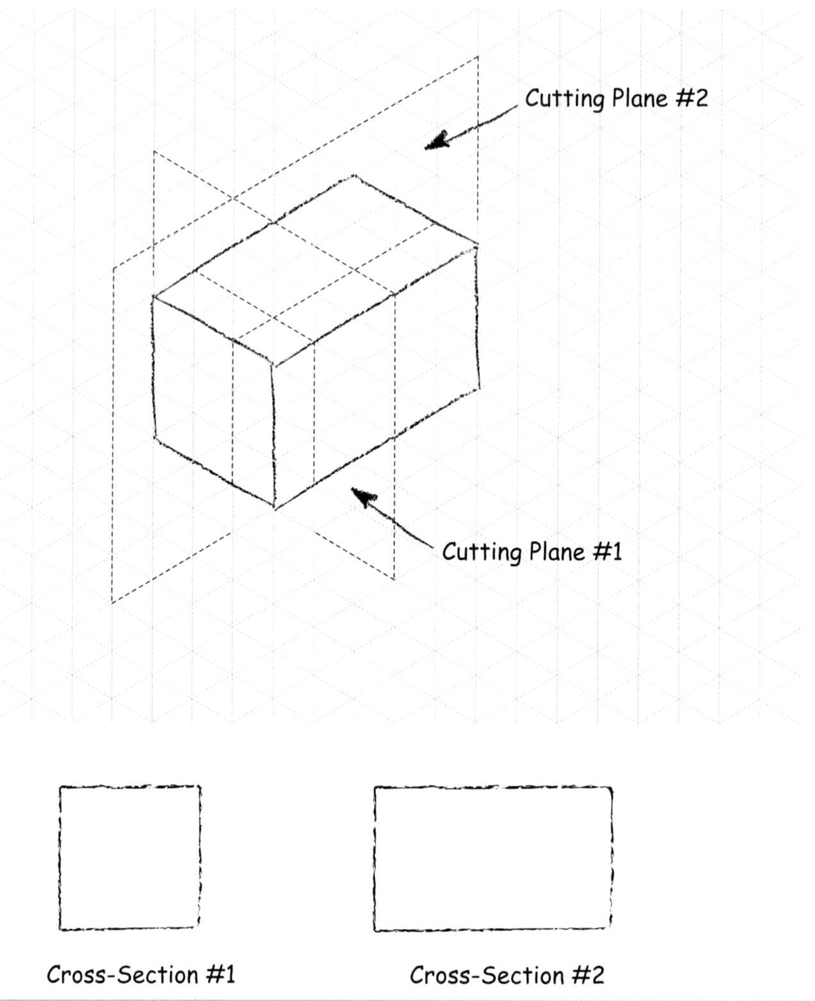

FIGURE 3.36. Cross sections from a square prism.

the surfaces where the plane cuts through the object. Consider a loaf of bread. Imagine a single slice of infinitely thin bread. One slice of bread would represent the cross section obtained by slicing a vertical plane through the loaf. Because most loaves of bread are not "constant" in shape along their lengths, the cross section changes as you go along the loaf. You know from experience that the cross sections, or slices, on the ends of the loaf are typically smaller than the slices in the middle.

The cross-section obtained by intersecting a cutting plane with an object depends on two things: (1) the orientation of the cutting plane with respect to the object and (2) the shape of the original object.

Consider the square prism shown in Figure 3.36. It is cut first by a vertical cutting plane perpendicular to its long axis to obtain the square cross section shown. If the cutting plane is rotated 90 degrees about a vertical axis, the result is the rectangular cross section shown in the figure. The two cross-sections are obtained from the same object. The difference in the resulting cross-sections is determined by changing the orientation of the cutting plane with respect to the object.

Now consider the cylinder shown in Figure 3.37. If a cutting plane is oriented perpendicular to the axis of the cylinder, a circular cross -section results; if the plane is located along the axis of the cylinder, a rectangular cross-section is obtained. Observe that this rectangular cross section through the cylinder is identical to the cross section obtained by slicing the rectangular prism along its long axis in Figure 3.36.

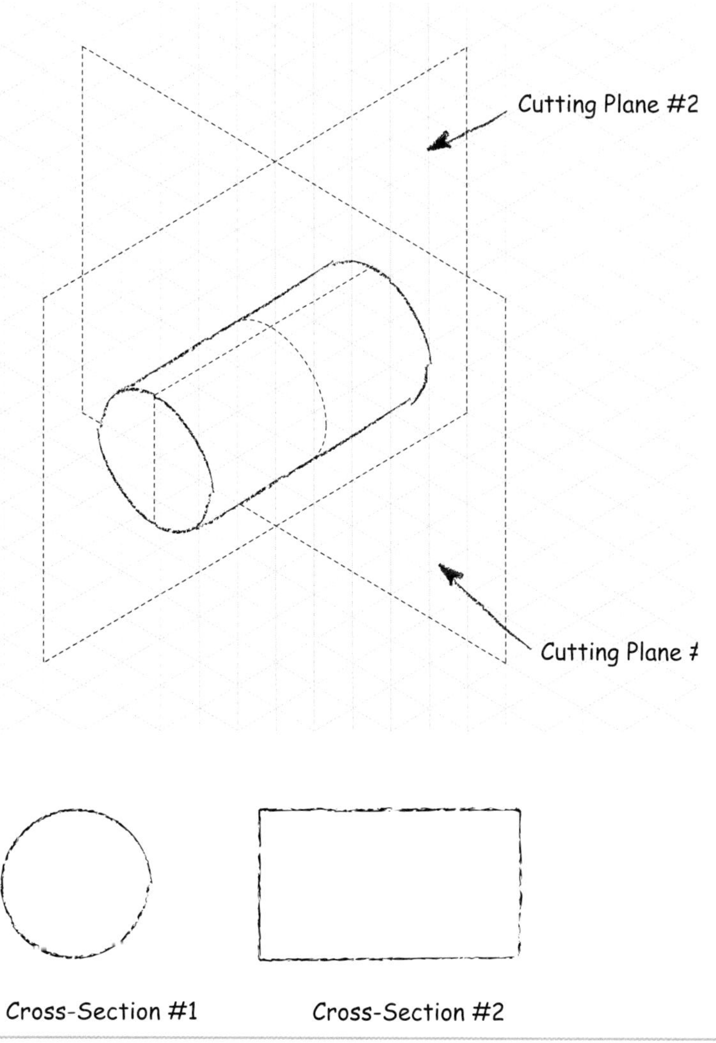

FIGURE 3.37. Cross sections of a cylinder.

Because a resulting cross-section through an object depends on the orientation of the cutting plane with respect to the object, most objects have several cross-sections associated with them. Figure 3.38 shows a cylinder with four possible cross sections. Can you imagine the orientation of the cutting plane with respect to the cylinder for each cross-section?

You already know that the first two cross-sections, rectangle and circle, were obtained by orienting the cutting plane perpendicular to and along the long axis of the cylinder, respectively.

What about the third cross-section? It was obtained by orienting the cutting-plane at an angle with respect to the axis of the cylinder.

The fourth cross-section was also obtained by angling the cutting plane with respect to the cylinder axis, but the angle was such that a portion of the cutting plane went through the flat circular end surface of the cylinder.

Figure 3.39 shows several cross-sections obtained by slicing a cube with cutting planes at different orientations.

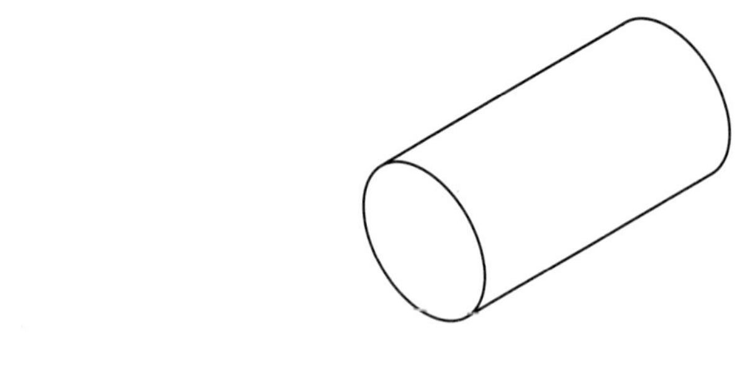

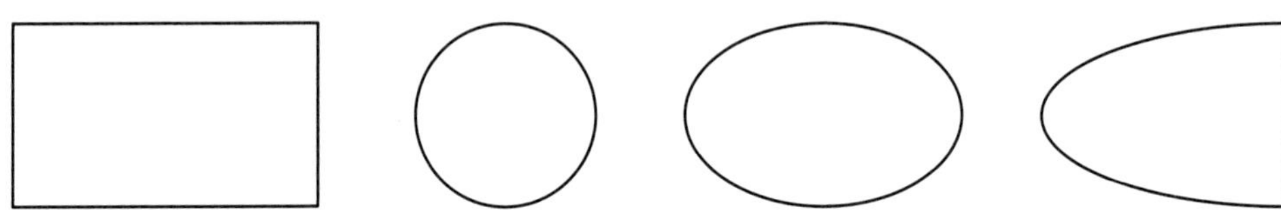

FIGURE 3.38. Various cross sections of a cylinder.

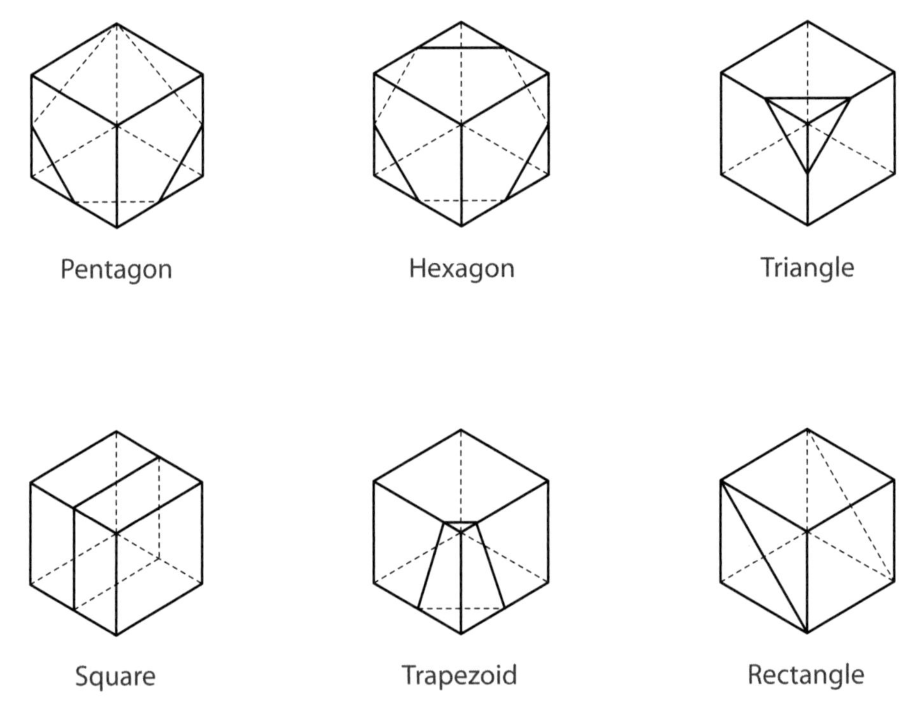

FIGURE 3.39. Various cross sections of a cube.

3.11 Combining Solids

Another skill that will be helpful to you as an engineer is the ability to visualize how two solids combine to form a third solid. The ability to visualize **combining solids** will be helpful as you learn how to use solid modeling software. In early versions of 3-D CAD software, commands used to combine solids were sometimes known as **Boolean operations**. This terminology was borrowed from mathematics set theory operations, called Booleans, where basic operations include unions, intersections, and complements between sets of numbers. Boolean logic is now the foundation of many modern innovations. In fact, if you have performed a search on the Web using an AND or an OR operator, you have used Boolean logic to help you narrow or expand your search. In terms of 3-D CAD, the Boolean set operations typically correspond to software commands of Join, Intersect, and Cut. To help you become familiar with the terminology since you probably will be building 3-D computer models, this section will use the same terminology.

Two overlapping objects can be combined to form a third object with characteristics of each original object apparent in the final result. To perform any Cut, Join, and Intersect operations to combine objects, the objects must be overlapping initially. What is meant by overlapping is that they share a common volume in 3-D space—called the **volume of interference**. Figure 3.40a shows two objects that overlap; Figure 3.40b shows the volume of interference between the two objects. Notice that the volume of interference takes shape and size characteristics from each of the two initial objects.

When two objects are **joined**, the volume of interference is absorbed into the combined object. The result is a single object that does not have "double" volume in the region of interference. The Boolean Join operation is illustrated in Figure 3.41.

When two objects are combined by **intersecting**, the combined object that results from the intersection is the volume of interference between them, as shown in Figure 3.42.

In the **cutting** of two objects, the combined object that results from the cutting depends on which object serves as the cutting tool and which object is cut by the other object. The result of a cutting operation is that the volume of interference is removed from the object that is cut, as illustrated in Figure 3.43.

FIGURE 3.40. Overlapping objects and volume of interference.

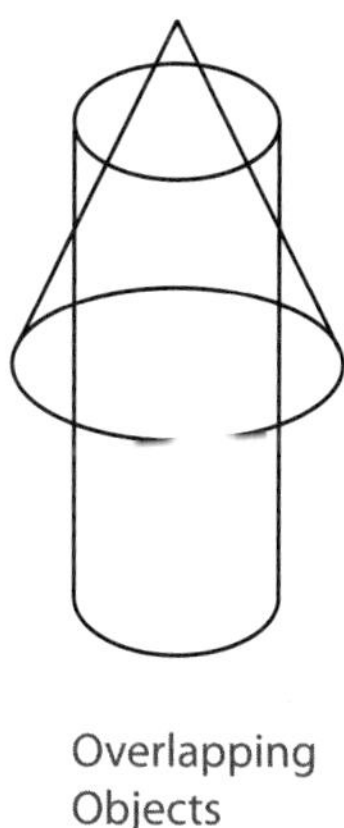

FIGURE 3.41. Result of two objects joined.

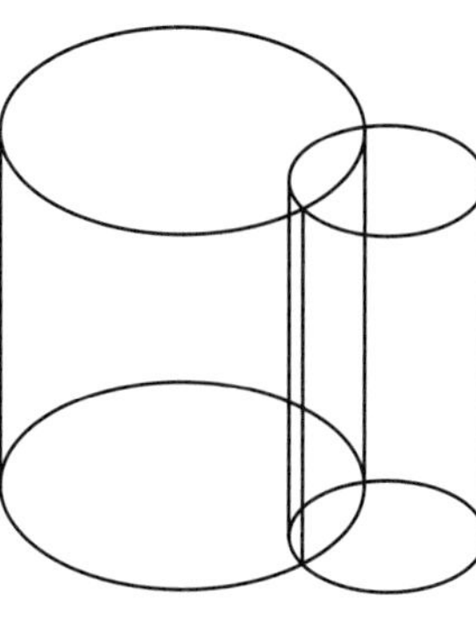

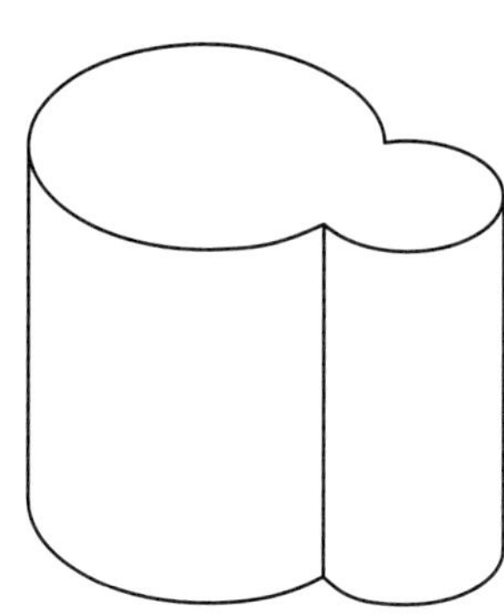

FIGURE 3.42. Result of two objects intersected.

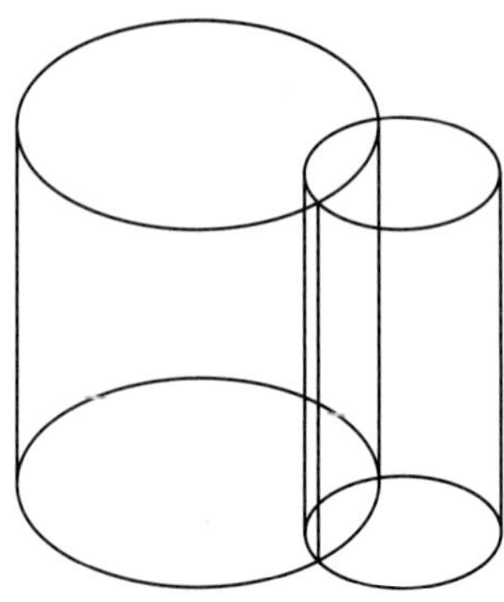

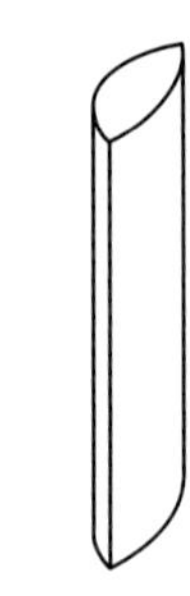

FIGURE 3.43. Result of two objects cutting.

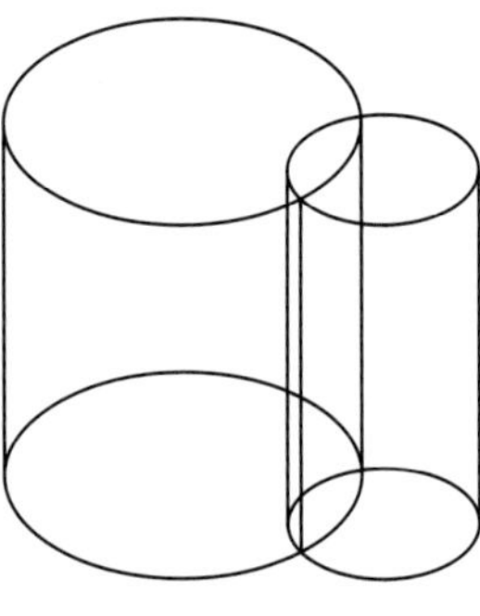

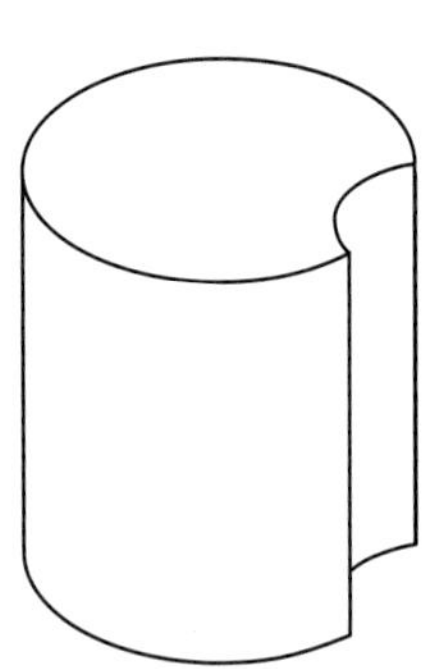

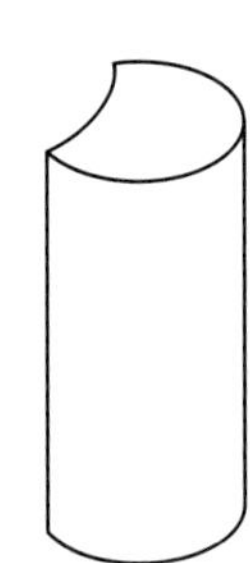

3.12 Strategies for Developing 3-D Visualization Skills

In this chapter, you learned about several different types of exercises you might want to tackle in order to develop your 3-D visualization skills. In this section, you will learn how to get started with those exercises.

3.12.01 The Sketching of Corner Views

3	3	3
3	2	1
3	2	2

FIGURE 3.44. A coded plan.

When you have a large task, such as constructing the isometric view of an object, the easiest way to get started is to complete an isometric view of a small piece first and then move on successively to other pieces of the object. You should follow this same process for almost every task in this chapter and throughout this textbook. In terms of objects, the basic building blocks are points, edges, and surfaces. To get started, you need to break down the object into its elements.

When sketching an isometric view of an object from its coded plan, you should look first at the corner from which you are sketching to determine which side of the object is on the left and which is on the right. The corner will be defined by a vertical line on the grid paper with the left and right sides emanating from there. Figure 3.44 shows a coded plan with the corners identified. If you want to sketch the Y-corner view, the arrows indicated show the left and right directions emanating from this corner. If you cannot immediately see this, turn the page so that *Y* is directly in front of you.

Next, you should sketch the height of the object from the corner you selected. As you move to the left from this corner, how does the height of the object change? For the coded plan in Figure 3.44, the height at the corner is 2; and as you go to the left, the object goes back two squares at the same height and then switches to a height of 3 for

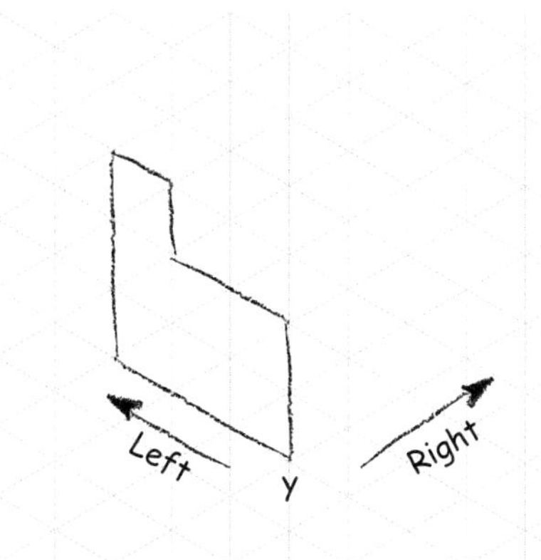

FIGURE 3.45. The left side surface of a building.

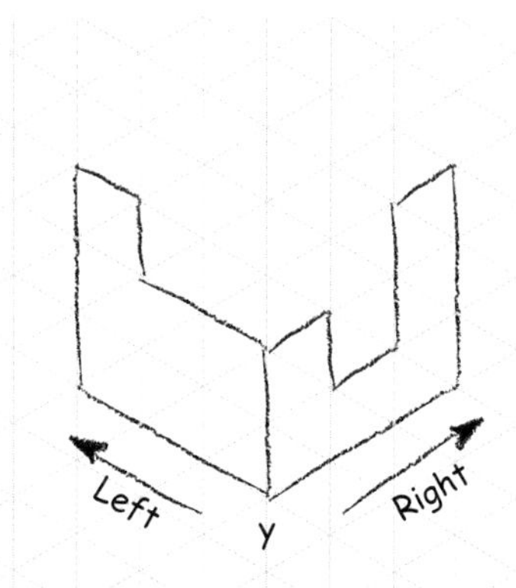

FIGURE 3.46. Left and right side surfaces of an object.

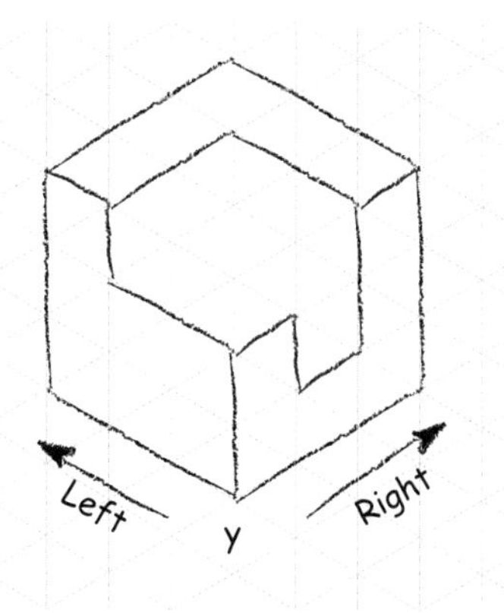

FIGURE 3.47. The top surface of the object.

one additional square. With this height information, you should be able to sketch the left-side surface of the object as shown in Figure 3.45.

You have now completed drawing one surface on the object. Go back to the original corner and think about moving in the other direction to define the surface on the object's right side. The height of the object goes from 2 to 1 to 3, one square deep at each height just indicated. Using this information, you can sketch the right-side surface going to the right from your chosen corner as illustrated in Figure 3.46.

Look again at the coded plan. You will see that the maximum height of the object is 3 units. At the same height, five blocks form an L going from the left side to the right side. You can sketch the L-shaped top surface at the given height as shown in Figure 3.47.

Look at the coded place once more, this time to determine where the object's height is 2. You will see three blocks at this height; together these three blocks trace another smaller L, as shown in Figure 3.48.

Finally, examine the sketch to determine where you need to add lines to complete the isometric. Add those lines. For the object you have been working with, you need to add lines to define the top surface that has a height of 1 and add whatever lines are needed at vertical corners to complete the isometric sketch. Figure 3.49 shows the completed sketch of the object.

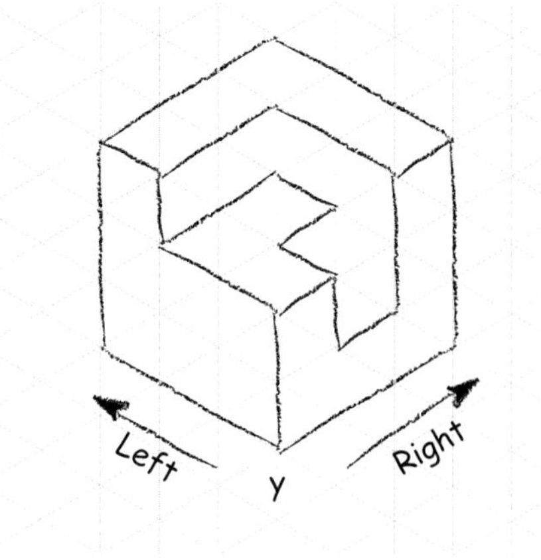

FIGURE 3.48. The second top surface of the object.

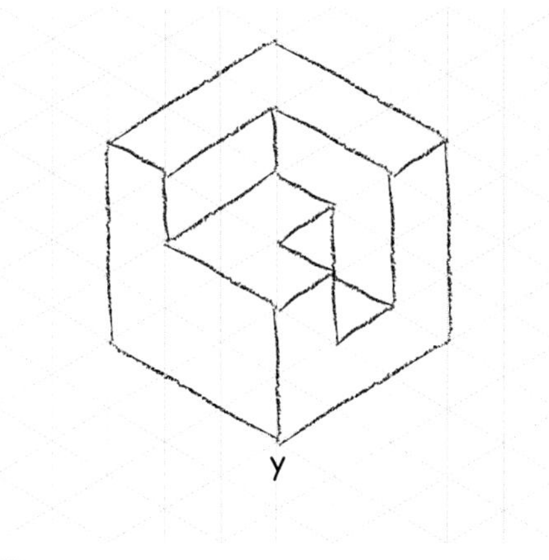

FIGURE 3.49. The completed isometric sketch.

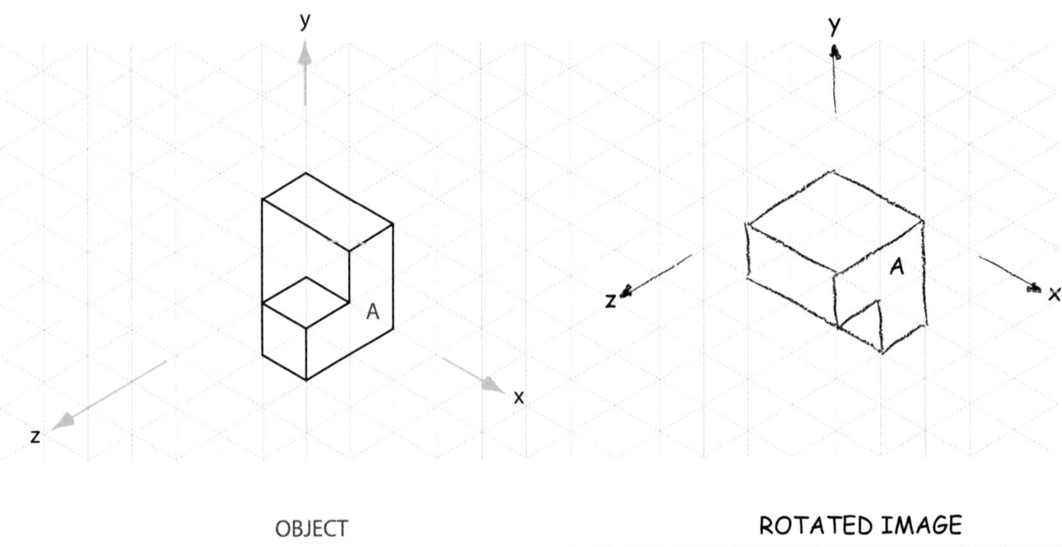

FIGURE 3.50. An object in original position and its rotated image.

3.12.02 Object Rotations about One Axis

Once again, the simplest way to visualize rotation of an object is to focus on just one of the object's surfaces. You should look at the object and mentally select one surface to serve as your focal point for the rotation. This surface is called the key surface for purposes of this exercise. When you mentally rotate the key surface, you usually can let the remainder of the object follow. Figure 3.50 shows an object and its rotated image. What axis was the object rotated about, and was it a positive or a negative rotation?

This exercise will use the L-shaped surface, labeled *A*, as the key surface. By focusing on surface A on the original object, you can see that the surface was rotated positive 90 degrees about the x-axis. By definition, when an object is rotated in space, all points, edges, and surfaces on the object are rotated the same amount. This means that the entire object from Figure 3.50 is rotated 90 degrees about the x-axis, because surface A was rotated by that amount. Note that the surface you choose as the key surface could be any of those on the object, and this surface is likely to change from problem to problem.

Just as focusing on one surface can help you *identify* an object rotation, starting with one surface also can assist you in the task of *performing* an object rotation. Look at the object in Figure 3.51. Your task is to rotate the object negative 90 degrees about the z-axis and sketch the image that results.

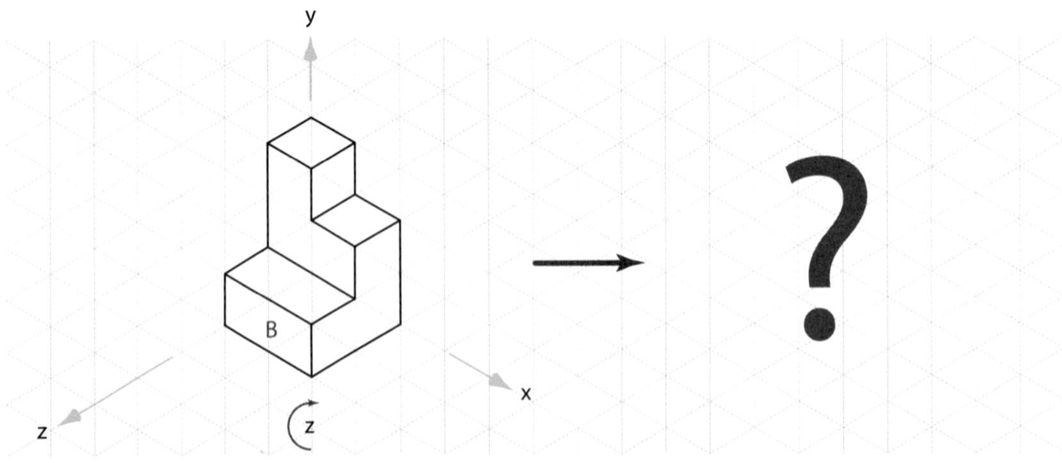

FIGURE 3.51. An object for rotation.

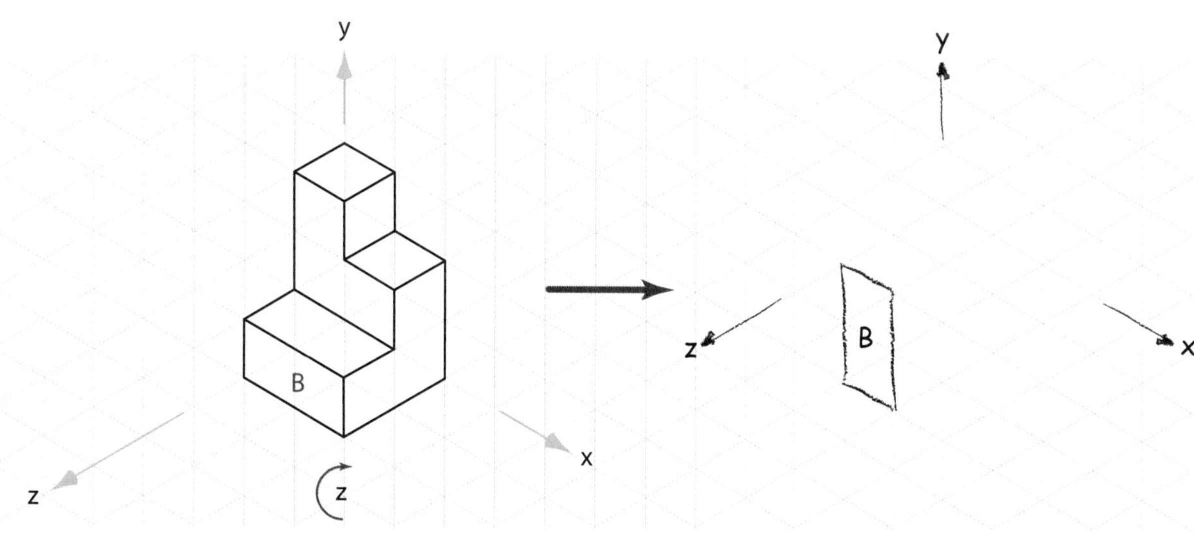

FIGURE 3.52. Rotation of surface B only.

To perform the task, you can select any surface; however, because the surfaces on the right side of the object will be hidden from view after the rotation is performed, they are probably not the best choice with which to begin this exercise. The surface of the object labeled *B* is probably the easiest one to begin with. Imagine surface B is rotated by the specified amount and sketch the surface only where it will appear after rotation. Figure 3.52 illustrates the rotation of the surface 90 degrees negatively (CW) about the z-axis.

Knowing the location of surface B after rotation, you can think about rotating surfaces C and D by the same amount as surface B and sketching the rotated surfaces C and D as shown in Figure 3.53. Notice that with 3-D rotations of an object, surfaces visible in the original view of the object are not always visible in the rotated image and surfaces that were hidden in the original view often become visible in the image of the rotated object.

Now you need to imagine what the "back" left-side surface of the object looks like after rotation so you can include it in your sketch. The back surface is hidden from view in the original orientation; but after rotation, it is the "top" surface of the object and is clearly visible. Going back to the object in its original position, you can see that this back surface starts at a height of 1 and jumps to a height of 3, meaning that it will

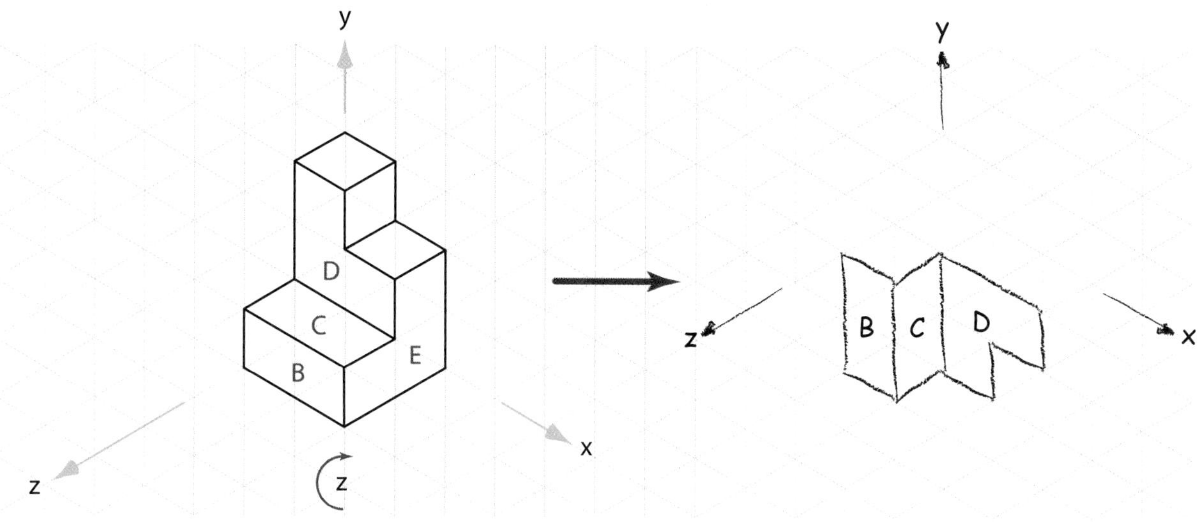

FIGURE 3.53. Rotation of surfaces C and D.

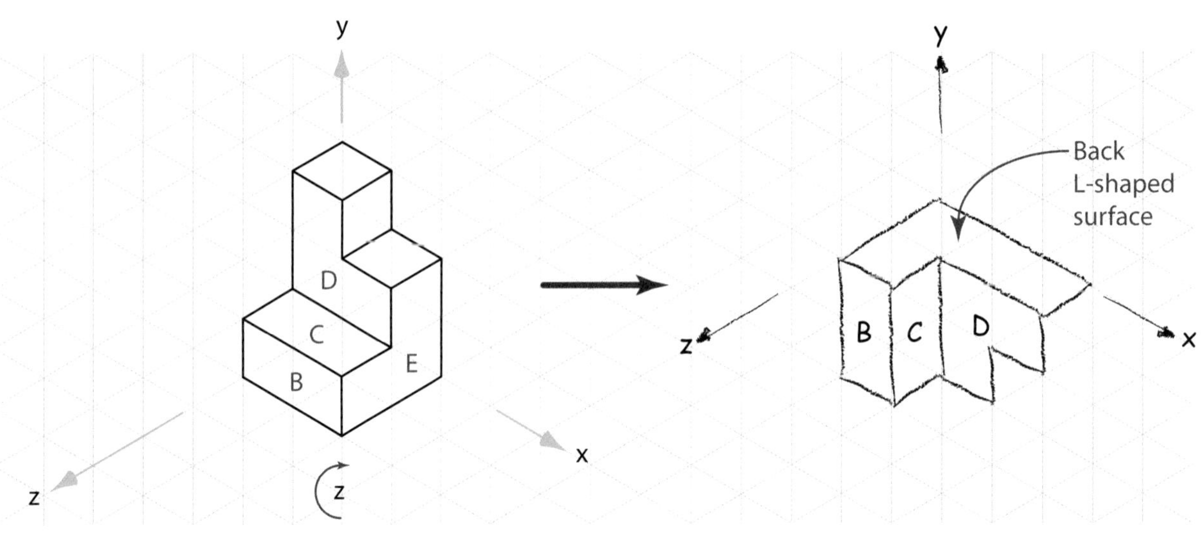

FIGURE 3.54. The back left-side surface after rotation.

appear L-shaped when rotated into position in the new view. This L-shaped surface can be sketched along with the other surfaces of the object as shown in Figure 3.54.

The last step is to clean up the drawing, adding lines as needed to complete the isometric sketch and to define the remaining surfaces on the image of the rotated object as illustrated in Figure 3.55.

3.12.03 Object Rotations about Two or More Axes

A step-by-step procedure for visualizing rotations about two or more axes is not as simple as the procedure for visualizing rotations of an object about one axis. The procedure is not simple because when rotating about two or more axes, many of the surfaces on the object that start out visible before rotation will be hidden after the first or successive rotations. One way to deal with the difficulty of visualizing rotation about two axes is to sketch the object as it appears after the first rotation and before the second rotation. For example, if an object is to be rotated positively about the x-axis and then negatively about the y-axis, sketch the intermediate step—the rotation about the x-axis—and complete the task from there. This way, the complex rotation is divided into a series of two single axis rotations. Figure 3.56 illustrates the method.

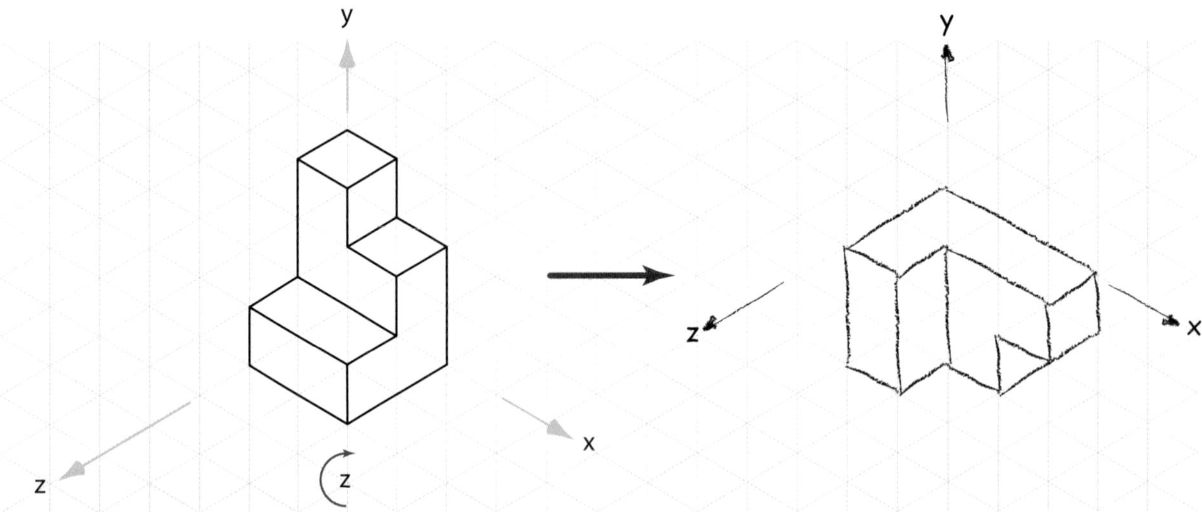

FIGURE 3.55. The completed sketch of the object after rotation.

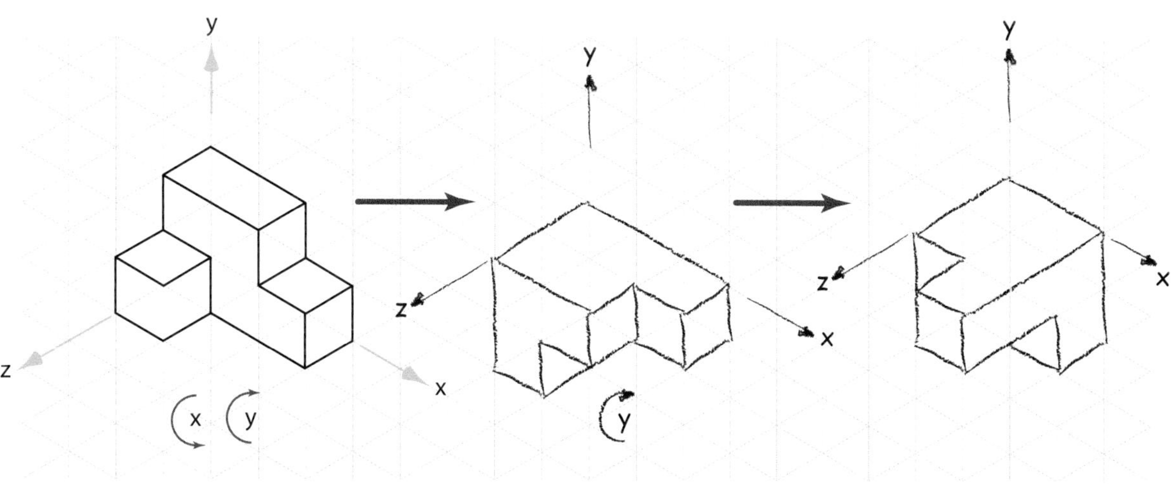

FIGURE 3.56. Intermediate rotation sketched.

You might be able to visualize the rotation about two or more axes by concentrating on a single key surface. But if you do try to do this, take care to assure that the key surface you choose to rotate remains visible after the rotation has been performed. Figure 3.57 shows an object and its rotated image. Note that surfaces A and B are no longer visible after the specified two rotations, so you would have to concentrate on surface C when mentally rotating the object.

One technique that may help you visualize rotations about two or more axes is to focus on a single surface that was not visible originally (instead of a visible one) and imagine its orientation after rotation. For example, the object shown in Figure 3.58 has a U-shaped surface on its back left side. Although you cannot see the surface in the original object orientation, the overall shape of the objects tells you that it is there. The surface of the object in its original position is hidden from view, but you should be able to visualize the surface nevertheless. Whenever you are asked to perform the rotations specified, you can think about rotating just the one surface and sketching it in its new position, similar to the way you started with the rotations about one axis. Figure 3.59 shows just the back surface rotated about both axes specified and into its final position.

When sketching the surface in the position after being rotated as specified, you can fill in the lines composing the entire object that corresponds to the same specified series of rotations as the single key surface. The complete sketch of the object after both specified rotations have been performed is shown in Figure 3.60.

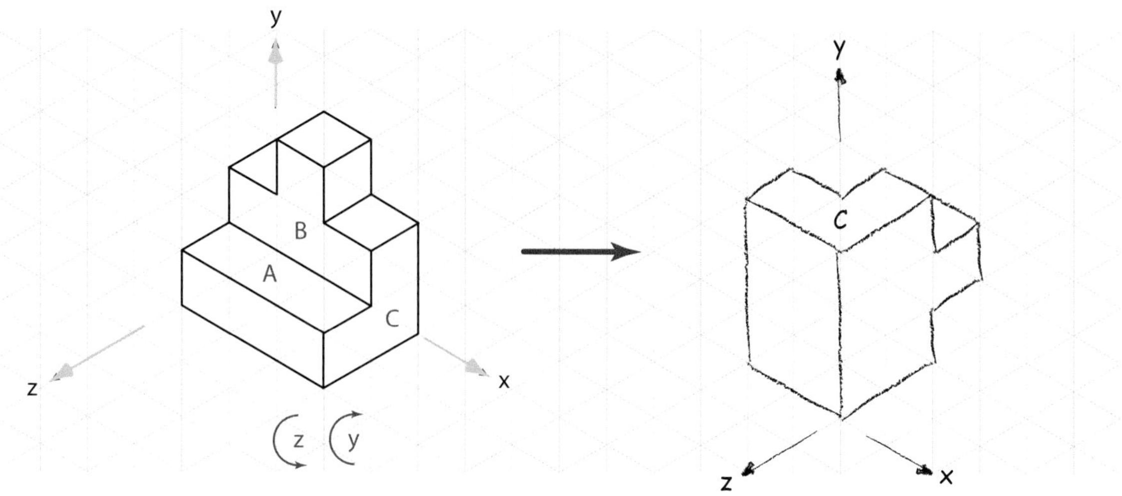

FIGURE 3.57. Visible surfaces before and after rotation.

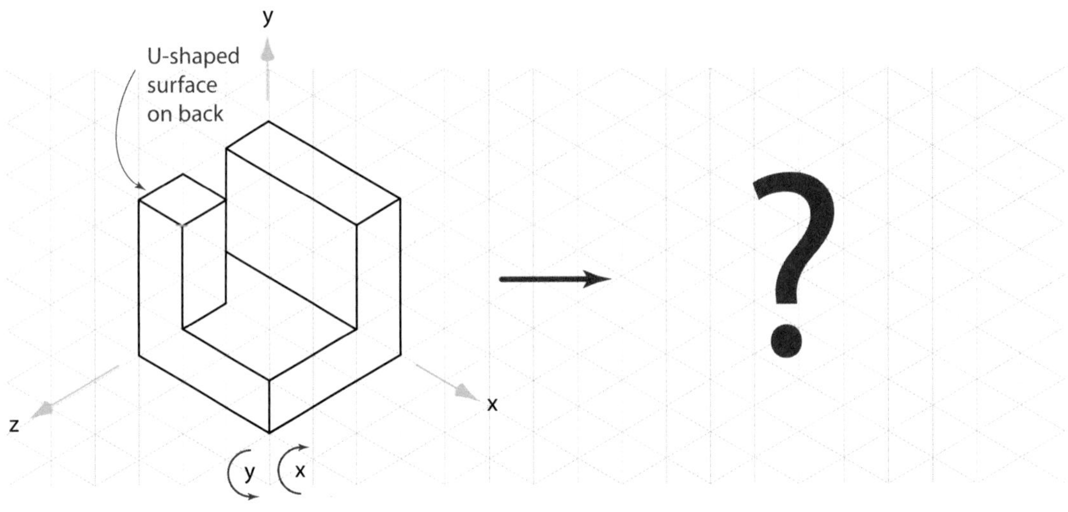

FIGURE 3.58. Original object position.

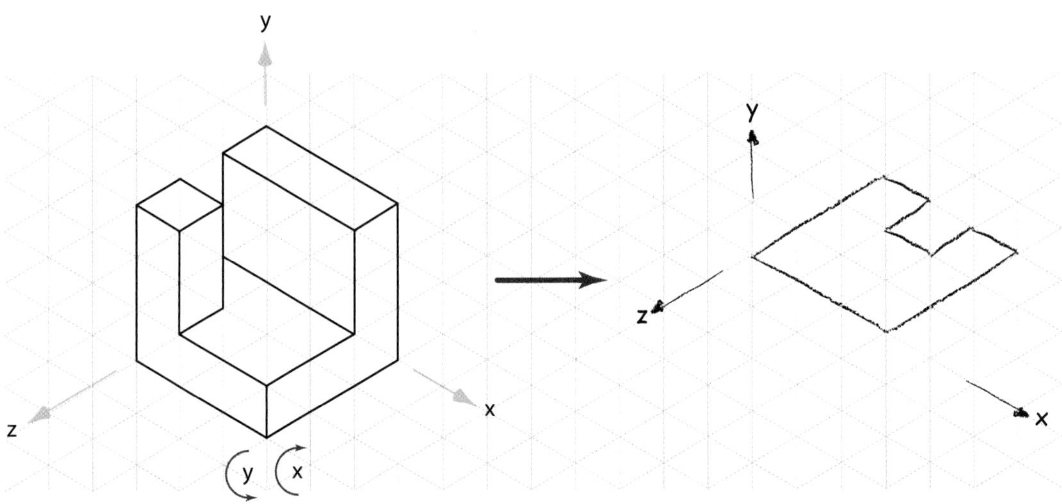

FIGURE 3.59. Rotation of U-shaped surface.

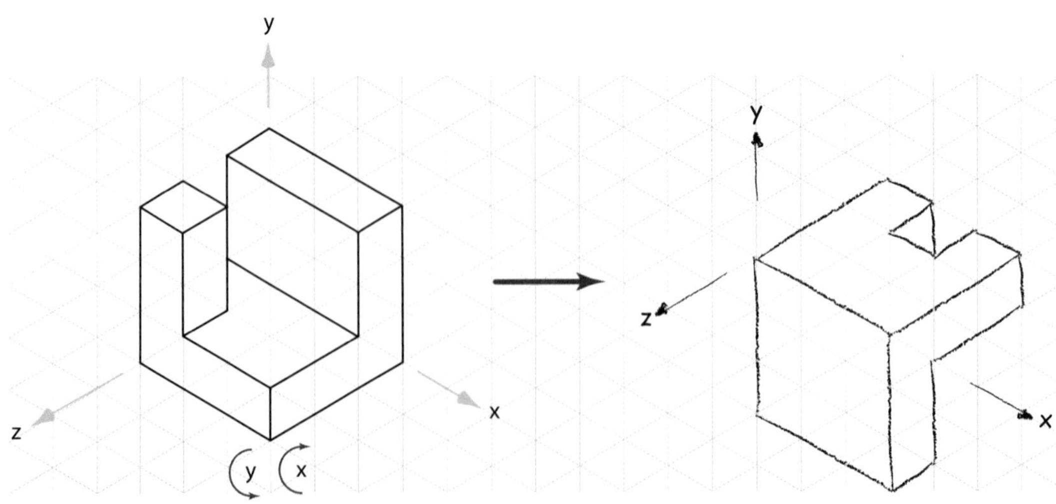

FIGURE 3.60. Completed object rotation.

3.12.04 Reflections

When creating a sketch of the image of an object as it appears on the other side of a reflection plane, you must remember that points on the object that are invisible in the original object may become visible in its reflection. The easiest way to create an object reflection is to start with one surface that is fully visible in the original view and will be fully visible in the reflection; then transfer that surface, point by point, from your view of the object to your view of the reflected image. (Usually, at least one surface on the object meets this criterion.) Since the object is the same distance from the reflection plane on either side of the plane, each point on the object and its corresponding reflected point on the image must be the same distance from the reflection plane on either side of the plane.

Figure 3.61 shows an object located 2 units from the plane of reflection. For this object, each of A, B, C, D, and E are located 2 units from the plane; another point hidden from view directly below A is the same distance from the plane of reflection. If each of these six points is reflected across the plane, they will be a total of 4 units—2 units to the plane and 2 units across the plane on the other side of it—from their location on the original object in a direction "perpendicular" to the plane. By "connecting the dots," you can sketch the surface they define, as in Figure 3.62, which shows the reflection of these points and the surface they form (the reflected surface) on the other side of the reflection plane.

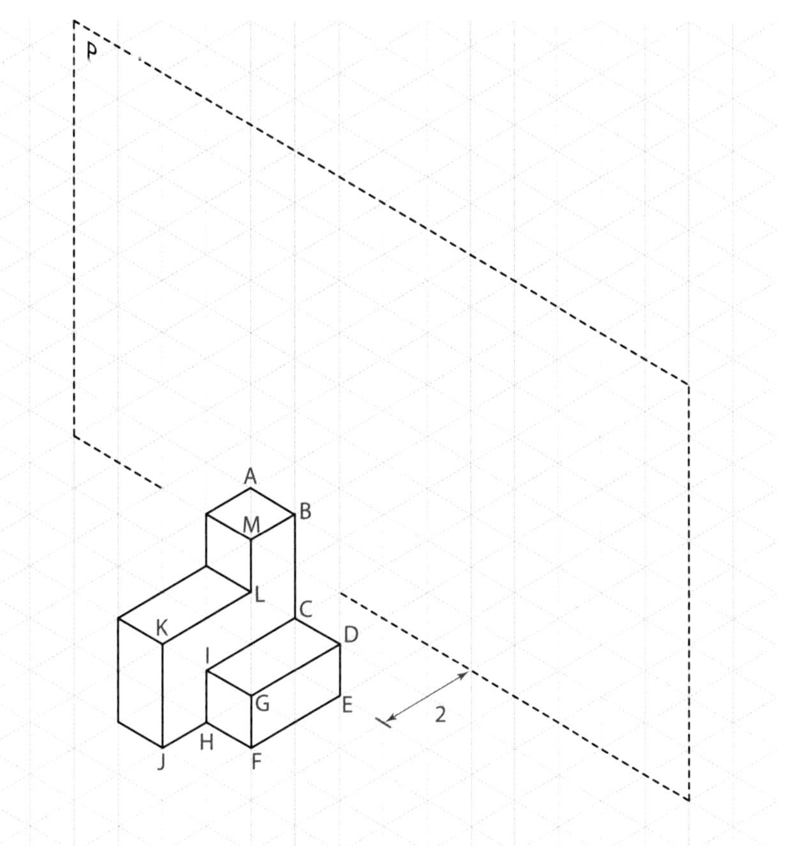

FIGURE 3.61. An object located 2 units from reflection plane.

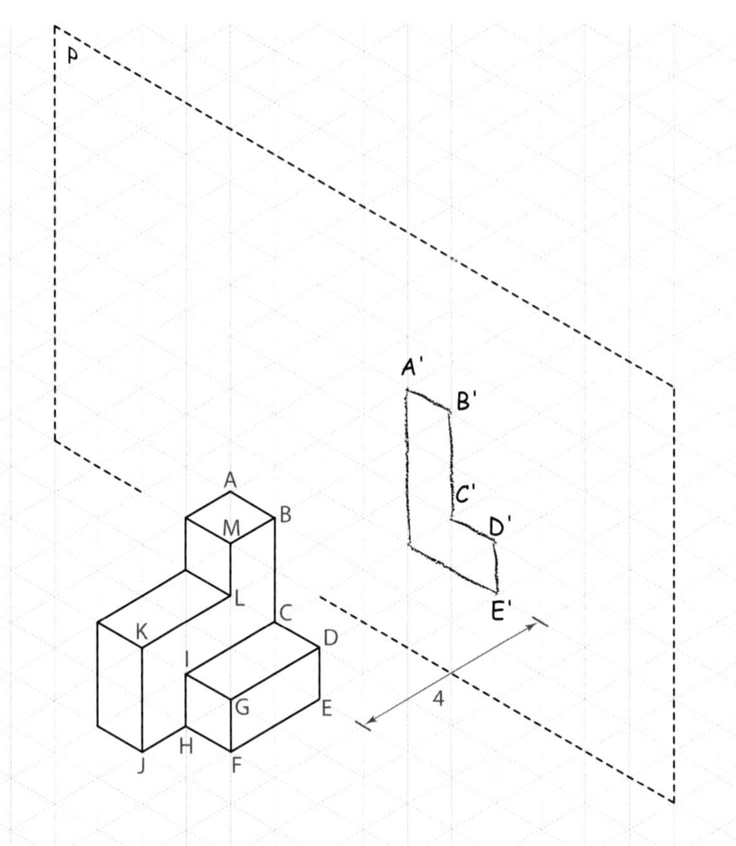

FIGURE 3.62. Reflection of points A, B, C, D, and E and the resulting surface.

For the object in this example, you can focus next on two surfaces that also will be visible in the reflection. One surface is defined by points D, E, F, and G; and the other surface is defined by points C, D, G, and I. Points C, D, and E have already been reflected, so the only points you have to consider to define the two surfaces are G, F, and I. Each point G, F, and I is located 4 units from the plane of reflection; thus, the reflection of each of these three points will be 8 units from its location on the original object in a direction perpendicular to the plane of reflection. The reflection of these points and the surfaces that result from connecting the reflected points are shown in Figure 3.63.

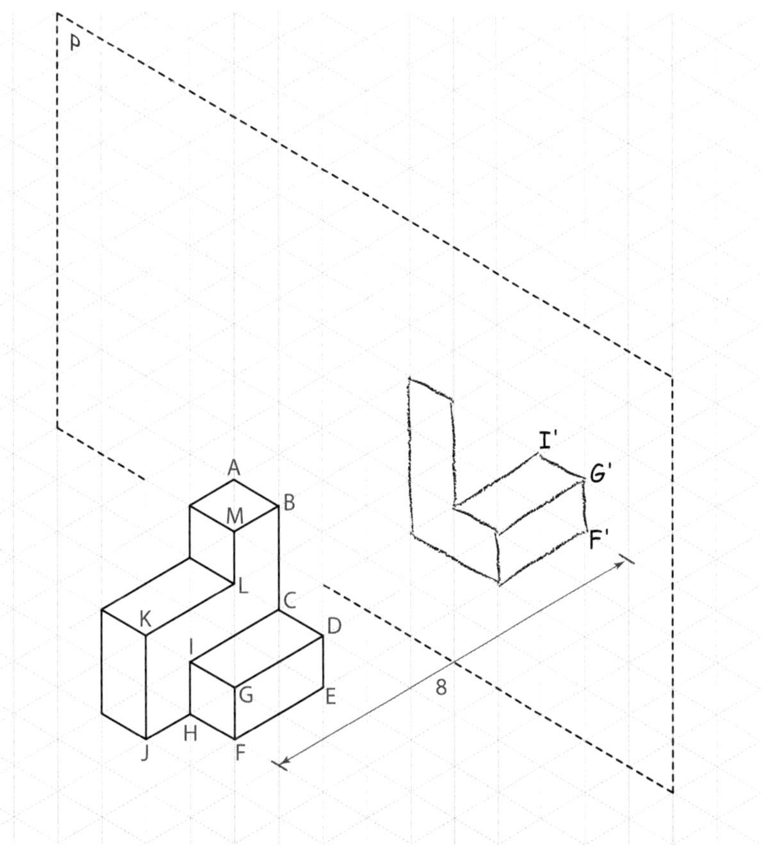

FIGURE 3.63. Reflection of two additional object surfaces.

Now consider the Z-shaped surface defined by points B, C, I, H, J, K, L, and M. You can reflect each of these points one at a time, realizing that H will be obscured from view in the reflected Z-shaped surface. Figure 3.64 illustrates the result of defining the reflected Z-shaped surface.

You now have enough of the object reflected that you can easily complete the sketch by adding the missing surfaces. The completed sketch is shown in Figure 3.65. Point labels have been excluded from the figure for clarity's sake. You could have reflected the object's other two surfaces point by point, but you can probably manage without that step.

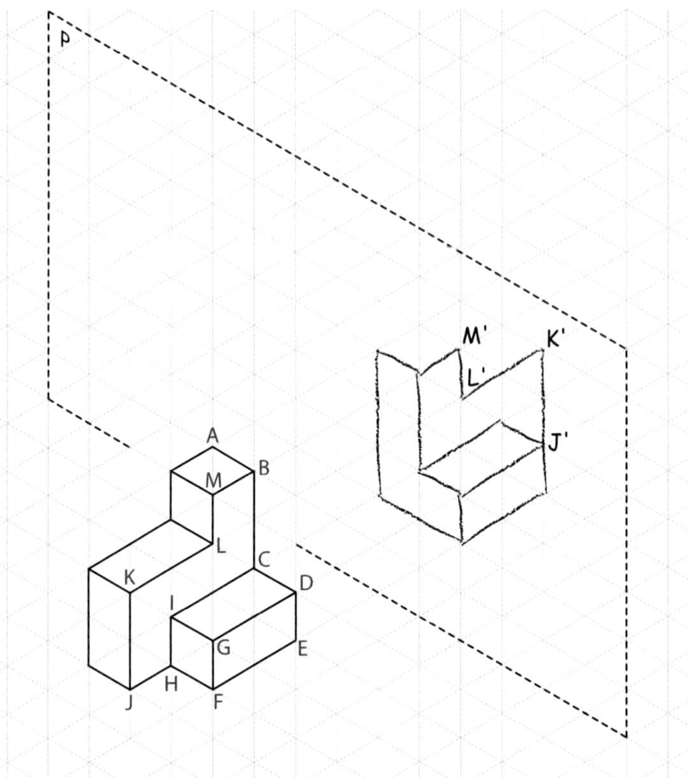

FIGURE 3.64. Reflection of Z-shaped surface.

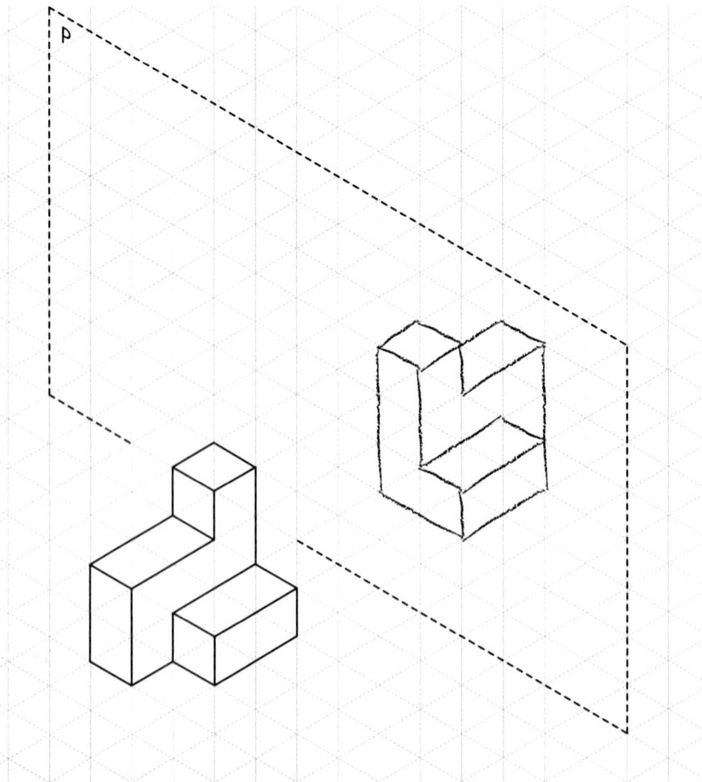

FIGURE 3.65. Completed object reflection.

3.12.05 Object Symmetry

To determine planes of symmetry for an object, you might want to sketch a dashed line on a surface of the object where you *think* a plane "splits" the object into two symmetrical parts. If you extend this line in your imagination, you can visualize an infinite plane. Look on each side of the dashed line you sketched on the object. Is one side a mirror image of the other side? If the sides are mirror images, then the dashed line you drew lies within a plane of symmetry; if the sides are not mirror images, then your dashed line does not lie within a plane of symmetry for the object.

You probably want to start with horizontal or vertical planes to identify an object's potential planes of symmetry. Figure 3.66 shows an object repeated three times with three potential planes of symmetry identified; two of the planes are vertical, and one is horizontal.

Notice that for the three planes identified, only planes 1 and 3, the vertical planes, are actually planes of symmetry. The part of the object on one side of plane 2 is not a mirror image of the part on the other side, so this plane cannot be a plane of symmetry. The object, then, has two planes of symmetry.

What about planes that are neither vertical nor horizontal? Those planes can be handled as was just shown for vertical and horizontal planes. First, sketch a line on the object that you think will lie within an imagined plane of symmetry; then examine each side of the object to see if the two halves sliced by the plane are mirror images of each other. This time the lines you sketch will be at an angle and not horizontal or vertical. Figure 3.67 shows an object with several planes of symmetry identified.

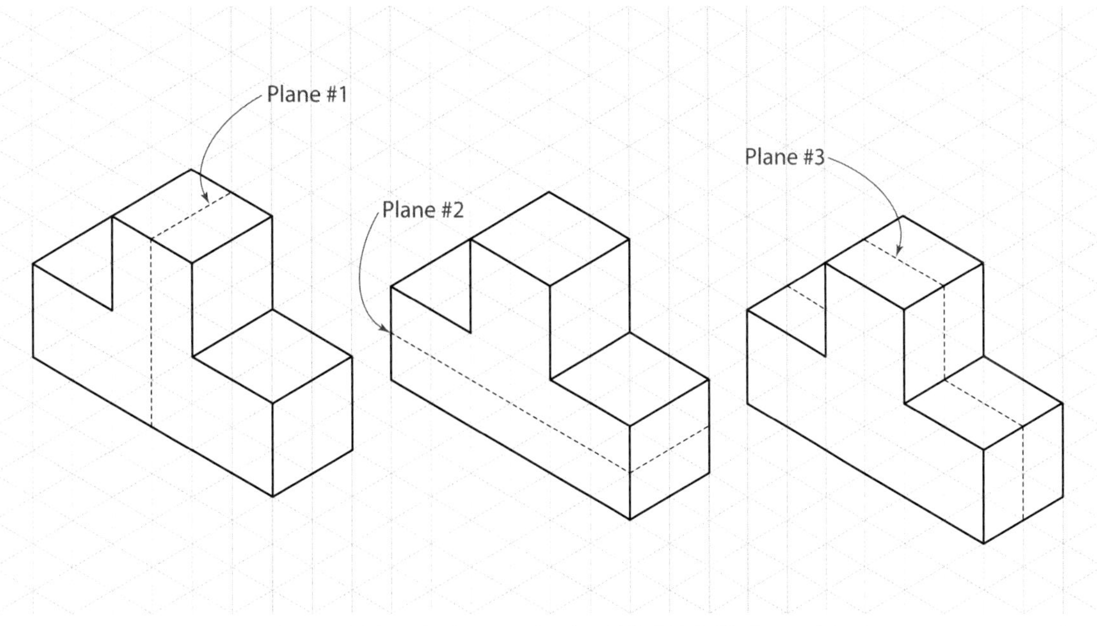

FIGURE 3.66. An object with three potential planes of symmetry identified.

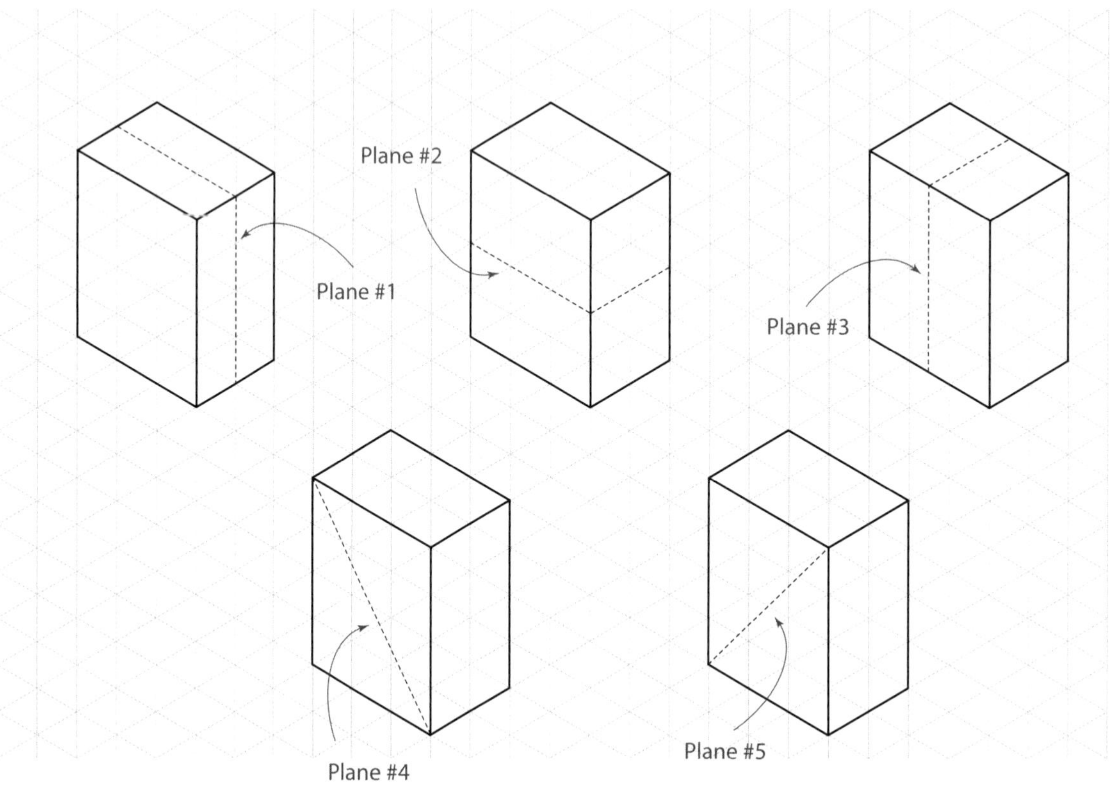

FIGURE 3.67. An object with five planes of symmetry identified.

3.12.06 Cross-Sections

When cutting planes are angled, the dimensions of the resulting cross-section "along" the plane becomes "stretched" compared with the dimensions of the cross section that results when the cutting plane is not angled. Figure 3.68 shows a simple object that has two cutting planes passing through it and the cross section corresponding to each cutting plane. For the vertical cutting plane, the shape of the cross section corresponds to the overall height and width of the object. When the cutting plane is angled, its height is stretched but its width remains the same. This point was illustrated previously in Figure 3.38. is, when the cylinder is cut along a plane perpendicular to its axis, a circular cross section results; however, when the cutting plane is angled, the diameter is stretched in one direction but remains the same in the other direction, resulting in the elliptical cross section.

To visualize the cross section that is obtained with an "angled" cutting-plane, think in terms of edges, either existing or imagined, on the object that are parallel to or perpendicular to the edges of the cutting plane. As the cutting-plane slices through the object, it intersects with the edges and surfaces of the object; the boundaries of

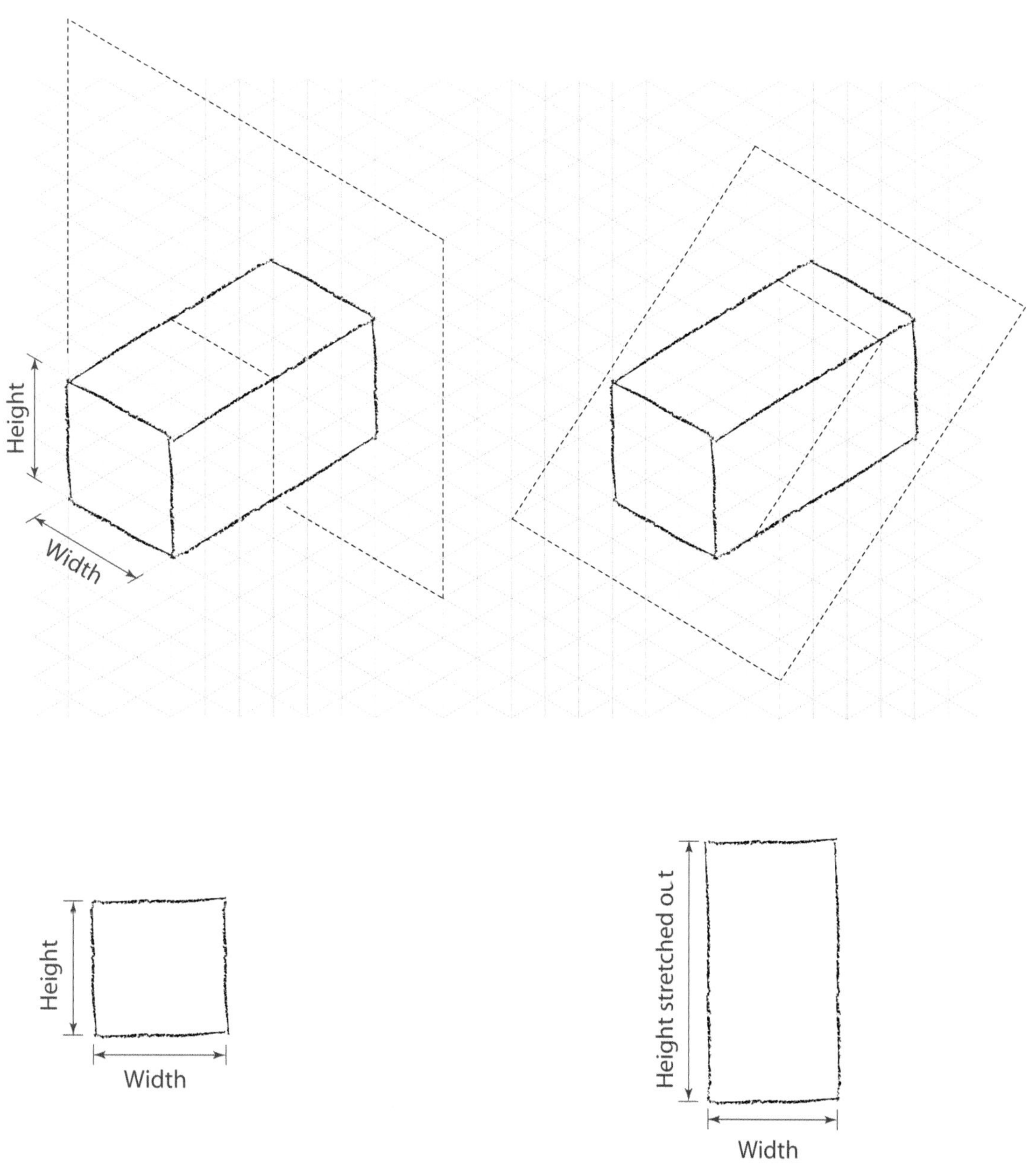

FIGURE 3.68. Effect of angling the cutting plane.

the cross-section will generally be parallel to the "edges" of the plane. (In reality, the cutting plane is infinite in size and, therefore, has no edges; but it is usually sketched as a finite size with edges so its orientation is shown.) After you have defined the edges of the cross section, you should mentally rotate the resulting planar cross section so it is perpendicular to your view direction. This allows you to "see" the cross section in its true shape and size. Obtaining the cross section for an object with an angled cutting plane is illustrated in Figure 3.69.

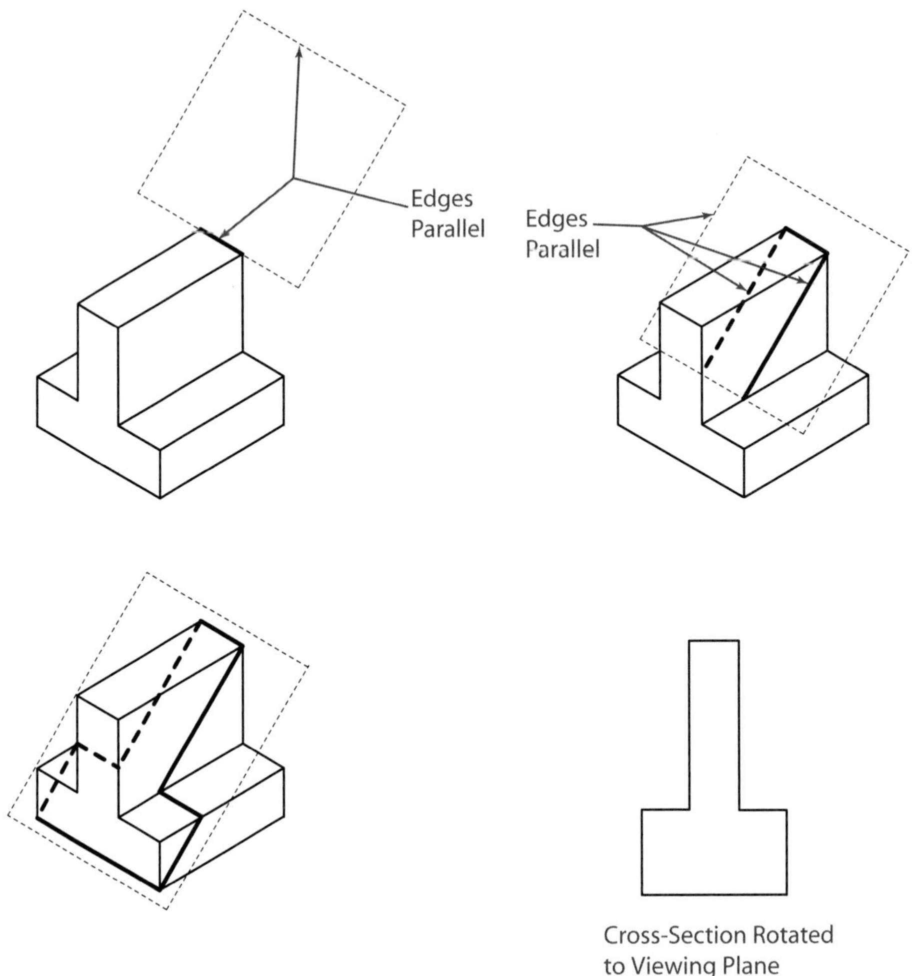

FIGURE 3.69. Cross section of an object with an angled cutting plane.

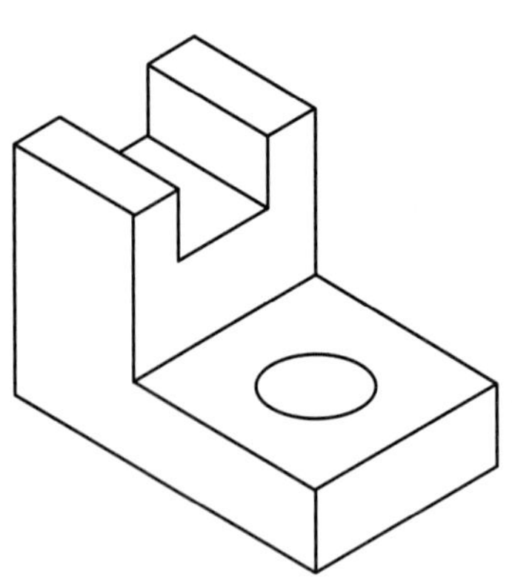

FIGURE 3.70. An object to be created through Boolean operations.

3.12.07 Combining Solids

The examples in the preceding three figures resulted from Boolean operations between simple objects combining to form a third, slightly more complicated object. The Cut, Join, and Intersect operations can be used in a series of steps to create a more complex object, which is often the case in 3-D modeling software applications. When creating a complex object using these methods, you first need to examine the final object you want to end up with; then visualize the steps needed to get to that final object. Figure 3.70 shows an object that you need to create using Boolean operations.

How would you create the object shown in Figure 3.70? One method you might employ is based on the concept of "material removal," which simply means cutting. Using this method, you need to create a block that is the overall size of the final object. Then you must create a smaller block that you will use to cut the larger block to form the basic staircase shape of the object. The first operation is illustrated in Figure 3.71.

Working with the newly formed object, now you need to create a cylinder and use it to cut the hole in the object, as shown in Figure 3.72.The final step is to create a block small enough to cut the "slot" in the top of the object, as illustrated in Figure 3.73. After completing all of these steps, you end up with the final object you set out to create.

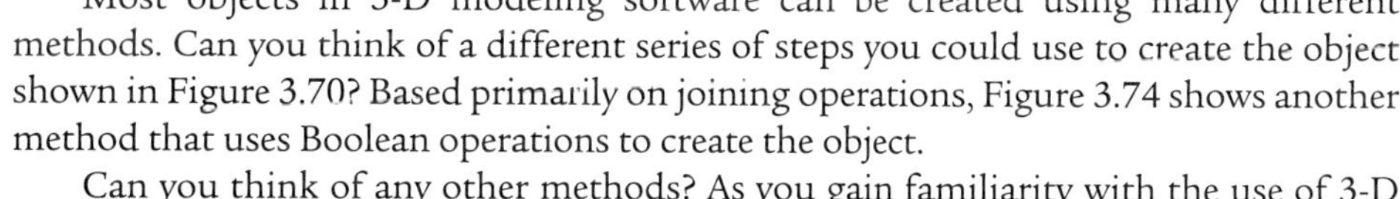

Most objects in 3-D modeling software can be created using many different methods. Can you think of a different series of steps you could use to create the object shown in Figure 3.70? Based primarily on joining operations, Figure 3.74 shows another method that uses Boolean operations to create the object.

Can you think of any other methods? As you gain familiarity with the use of 3-D modeling software, you will develop your own preferred methods for creating parts. Sometimes the method you use will depend on the object's final design characteristics. Other chapters in this text as well as texts devoted entirely to 3-D modeling software cover 3-D computer-aided modeling in greater detail and provide information about other types of operations that can be used to create solid models effectively and efficiently.

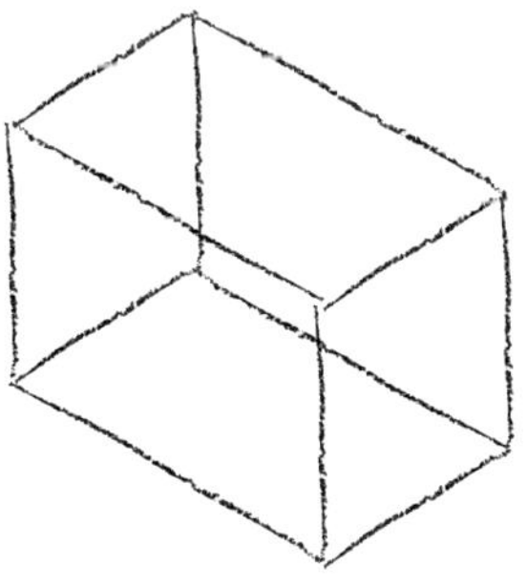

Create block

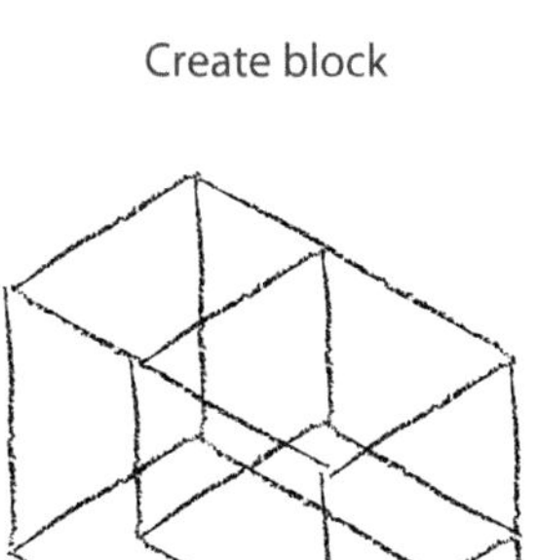

Create small block

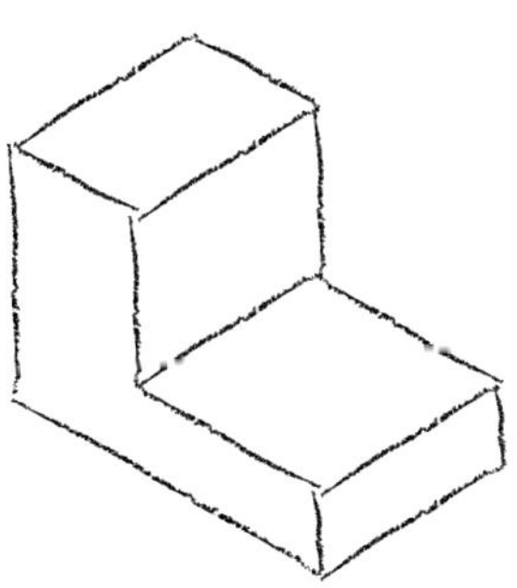

Small block cuts larger block

FIGURE 3.71. Creation of a stepped shape through cutting.

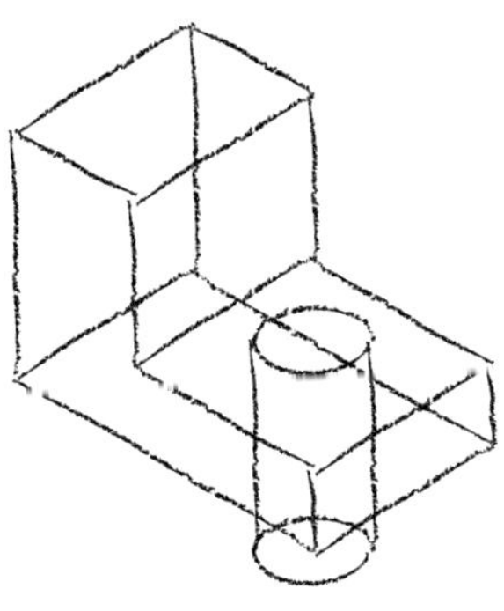

Create cylinder

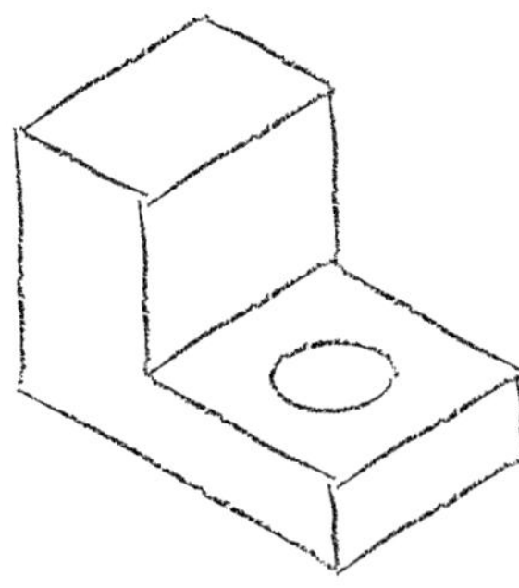

Cylinder cuts object

FIGURE 3.72. Cutting a hole in an object.

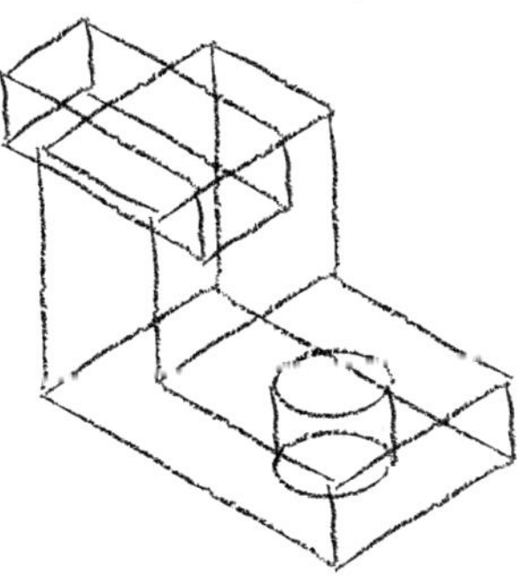

Create small block

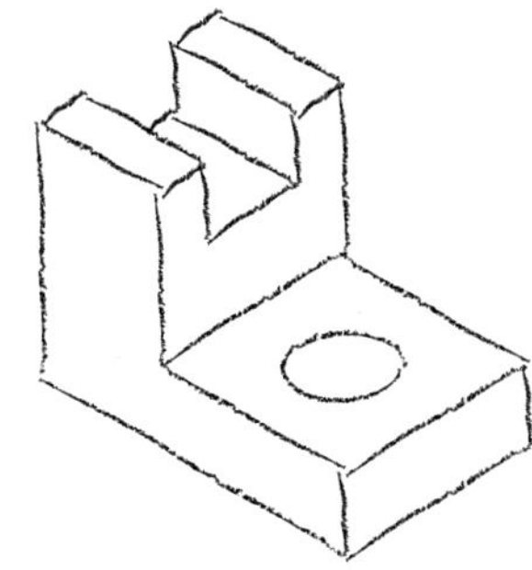

Small block cuts object -- desired result

FIGURE 3.73. Cutting an upper slot to achieve the desired result.

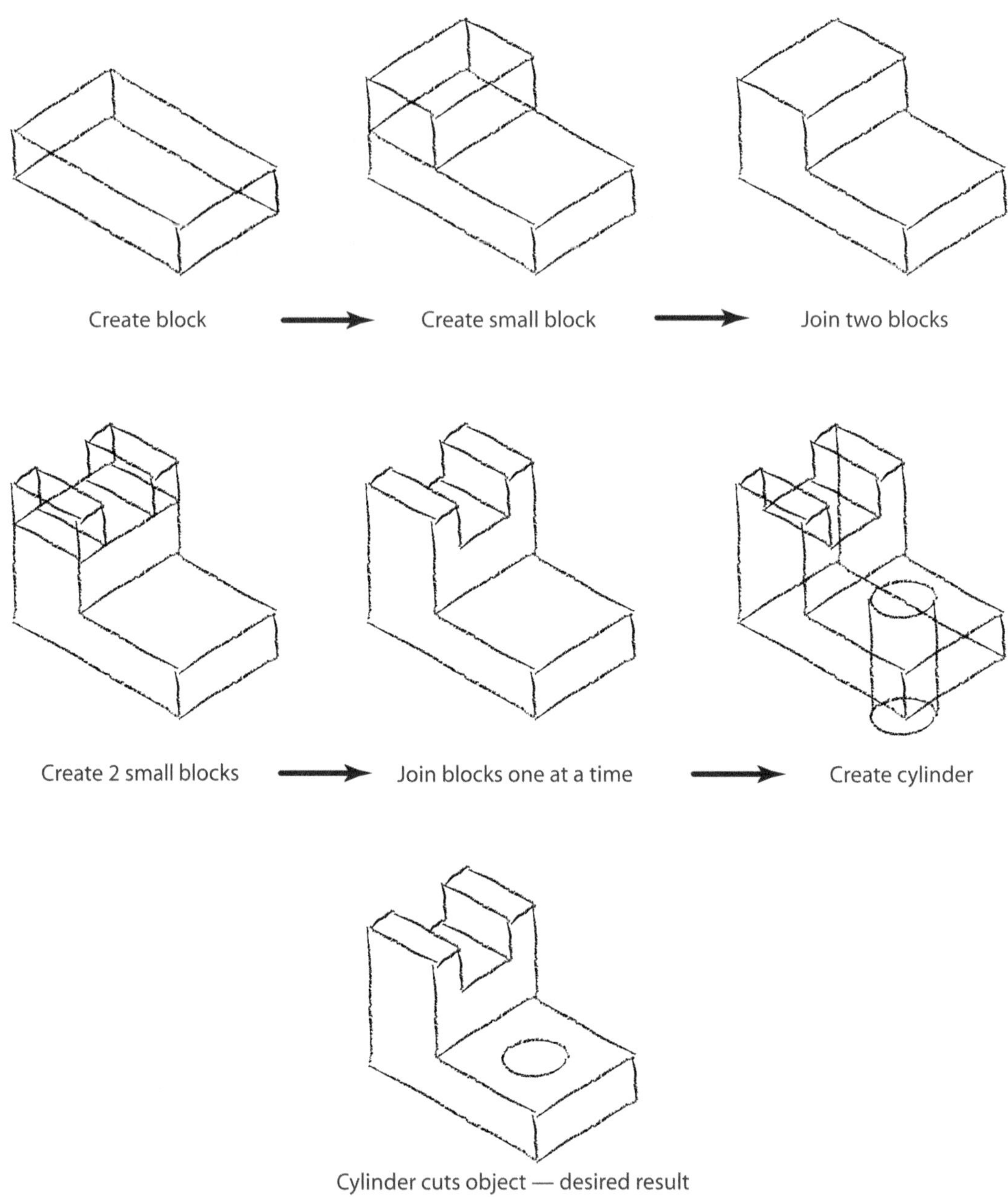

FIGURE 3.74. An object created by an alternative series of combinations.

CAUTION When creating corner views of simple objects, remember the general rules for creating simple isometric sketches described in an earlier chapter—that lines are included only at the intersection between surfaces, that no hidden lines are shown, and that only the visible portion of partially obscured surfaces are sketched. One common error novices often make is to include extra lines on a single surface of an object, especially when there are several changes in the height of the object. Figure 3.75 shows an improper sketch from a coded plan. Can you detect the "extra" unnecessary line?

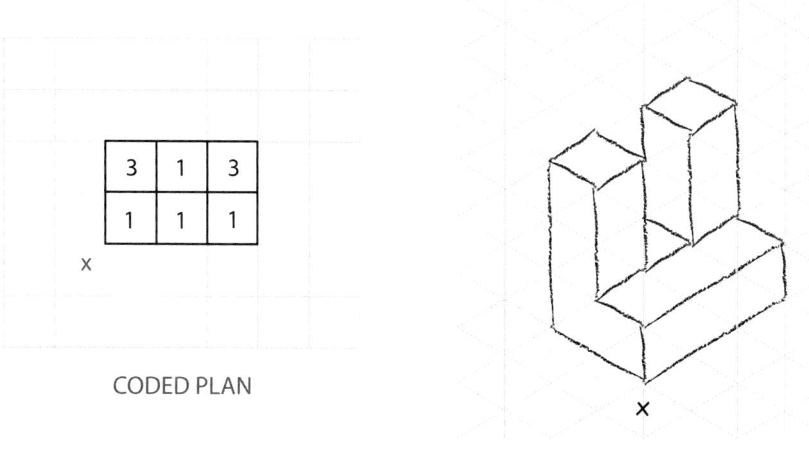

FIGURE 3.75. A common error: a sketch containing an extra line.

The extra line for the surface exists on the lower surface between the two "towers" on the object, as highlighted in Figure 3.76. Since a line defines the intersection of each of the towers with the lower surface, novices often extend the line through the gap even though there is no intersection between surfaces there. Figure 3.77 shows the correctly drawn object from the coded plan.

When rotating an object about an axis using the right-hand rule, right-handed people often forget that they must put down their pencil in order to rotate the object correctly in space. Right-handed people often use their left hand to define the rotation as they sketch with their right hand. If you forget to put down your pencil to define the direction of your rotation, you will end up with a rotation in the opposite direction of what you intended, as shown in Figure 3.78.

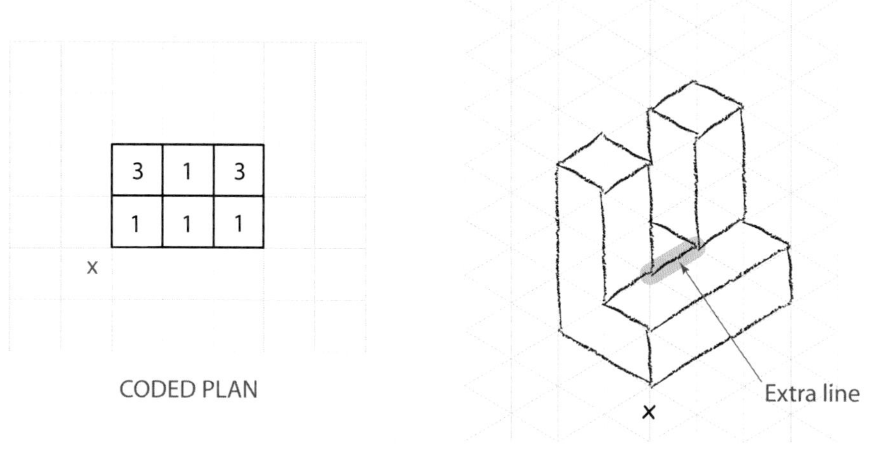

FIGURE 3.76. A sketch with the extra line highlighted.

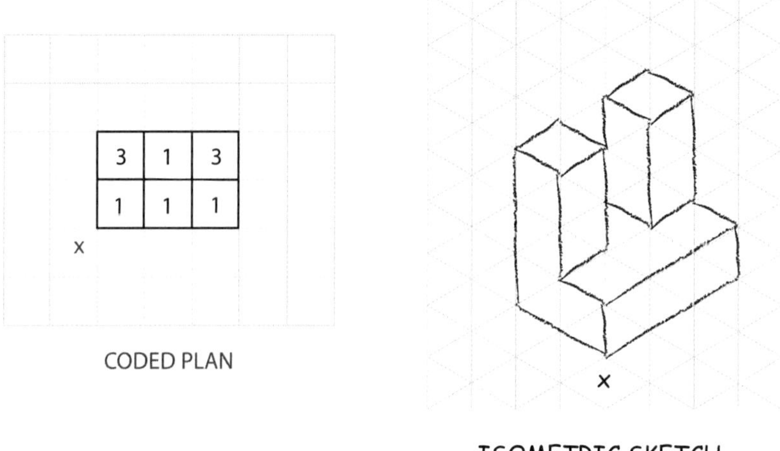

FIGURE 3.77. A correctly sketched object from a coded plan.

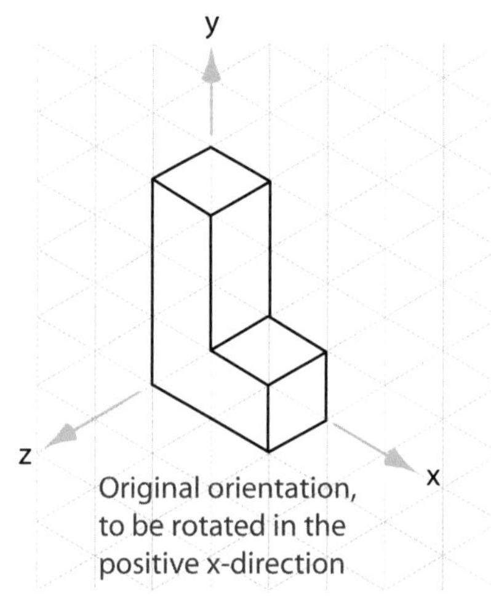

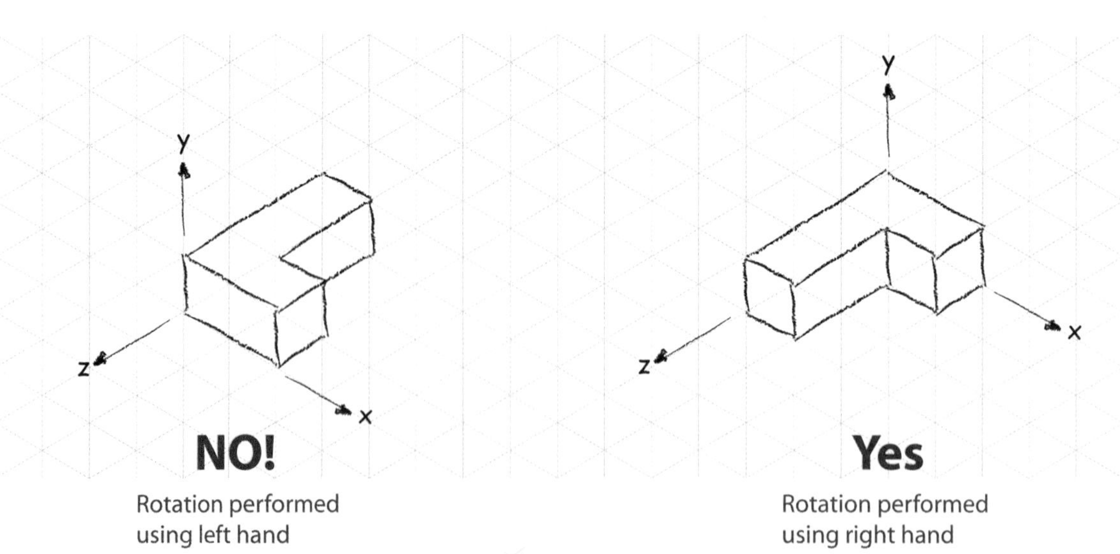

FIGURE 3.78. A common error: positive rotation (about the x-axis in this example) using the left hand instead of the right hand.

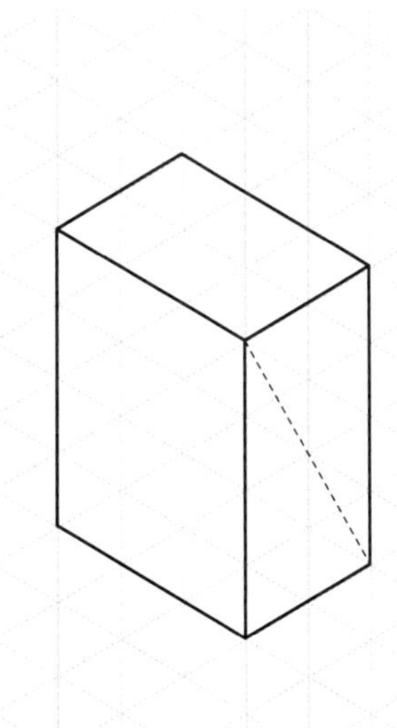

FIGURE 3.79. An object with a potential plane of symmetry identified.

When identifying angled planes of symmetry, be careful. Sometimes an angled plane produces two halves that look similar to each other but are not mirror images across the plane. For example, Figure 3.79 shows a potential plane of symmetry for an object. The two halves of the object appear to be identical halves; however, they are not mirror images across the plane. Therefore, the plane identified is not a plane of symmetry.

3.13 Chapter Summary

In this chapter, you learned about Gardner's definitions of basic human intelligences (including spatial intelligence) and the way spatial intelligence is developed and assessed. Spatial intelligence is important for engineering success, especially in engineering graphics and solid modeling courses. The chapter outlined several exercises that help develop spatial skills, including:

- Constructing isometric sketches from different corner views.
- Rotating 3-D objects about one or more axes.
- Reflecting objects across a plane and recognizing planes of symmetry.
- Defining cross sections obtained between cutting planes and objects.
- Combining two objects to form a third object by cutting, joining, or intersecting.

3.14 glossary of key terms

Boolean operations: In early versions of 3-D CAD software, commands used to combine solids.

combining solids: The process of cutting, joining, or intersecting two objects to form a third object.

corner views: An isometric view of an object created from the perspective at a given corner of the object.

cross-section: The intersection between a cutting plane and a 3-D object.

cut: To remove the volume of interference between two objects from one of the objects.

cutting-plane: An imaginary plane that intersects with an object to form a cross section.

3.14 glossary of key terms (continued)

intersect: To create a new object that consists of the volume of interference between two objects.

join: To absorb the volume of interference between two objects to form a third object.

mental rotations: The ability to mentally turn an object in space.

reflection: The process of obtaining a mirror image of an object from a plane of reflection.

spatial orientation: The ability of a person to mentally determine his own location and orientation within a given environment.

spatial perception: The ability to identify horizontal and vertical directions.

spatial relations: The ability to visualize the relationship between two objects in space, i.e., overlapping or nonoverlapping.

spatial visualization: The ability to mentally transform (rotate, translate, or mirror) or to mentally alter (twist, fold, or invert) 2-D figures and/or 3-D objects.

symmetry: The characteristic of an object in which one half of the object is a mirror image of the other half.

volume of interference: The volume that is common between two overlapping objects.

3.15 questions for review

1. What are some of the basic human intelligences as defined by Gardner?
2. What are the stages of development for spatial intelligence?
3. What are some of the basic spatial skill types?
4. What do the numbers on a coded plan represent?
5. What are some general rules to follow when creating isometric sketches from coded plans?
6. When a person is looking down a coordinate axis, are positive rotations CW or CCW?
7. Describe the right-hand rule in your own words.
8. Are object rotations about two or more axes commutative? Why or why not?
9. What is one difference between object reflection and object symmetry?
10. Are all objects symmetrical about at least one plane? Explain.
11. The shape of a cross section depends on two things. Name them.
12. What is the effect on the resulting cross section of a cutting plane that is tilted?
13. What are the three basic ways to combine solids?
14. In the cutting of two objects, does it matter which object is doing the cutting?

3.16 problems

1. For the following objects, sketch a coded plan, labeling the corner marked with an *x* properly.

2. Indicate the coded plan corner view that corresponds to the isometric sketch provided.

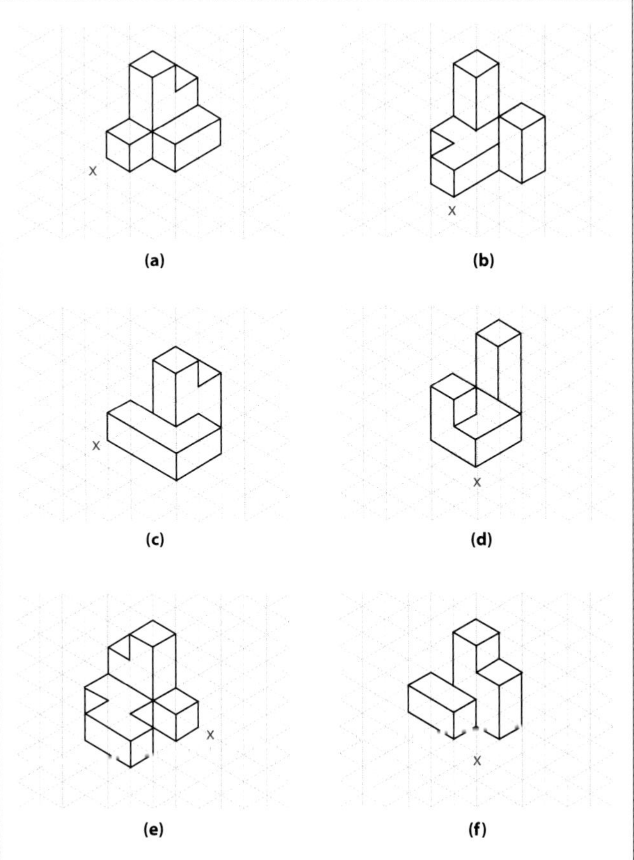

FIGURE P3.1.

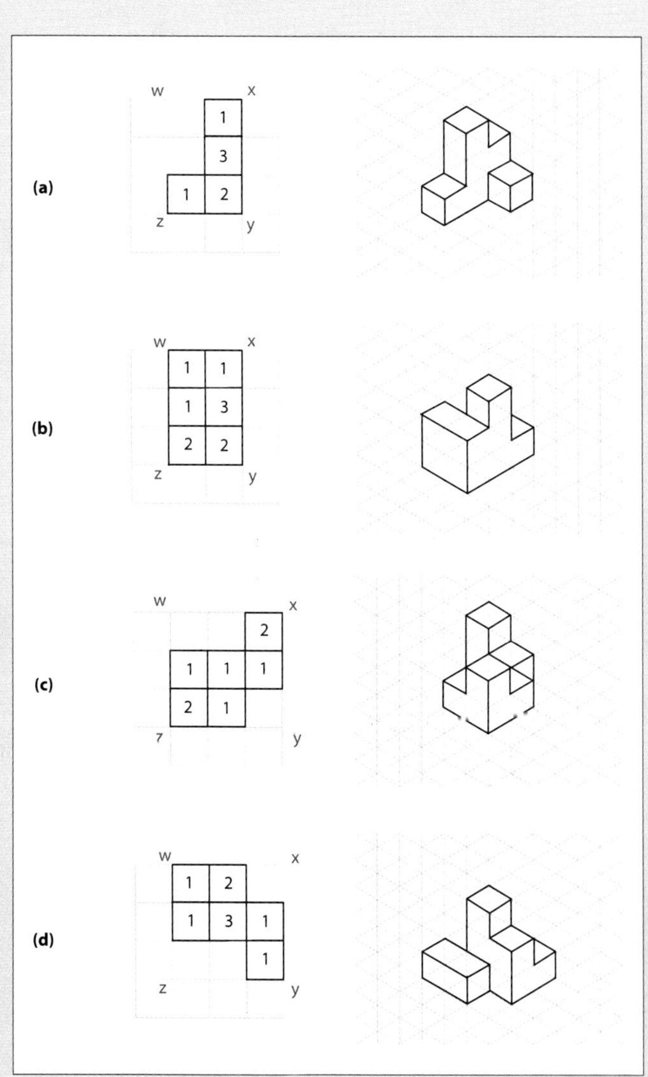

FIGURE P3.2.

3.16 problems (continued)

3. Use isometric grid paper to sketch the indicated corner view (marked with an *x*) for the coded plan.

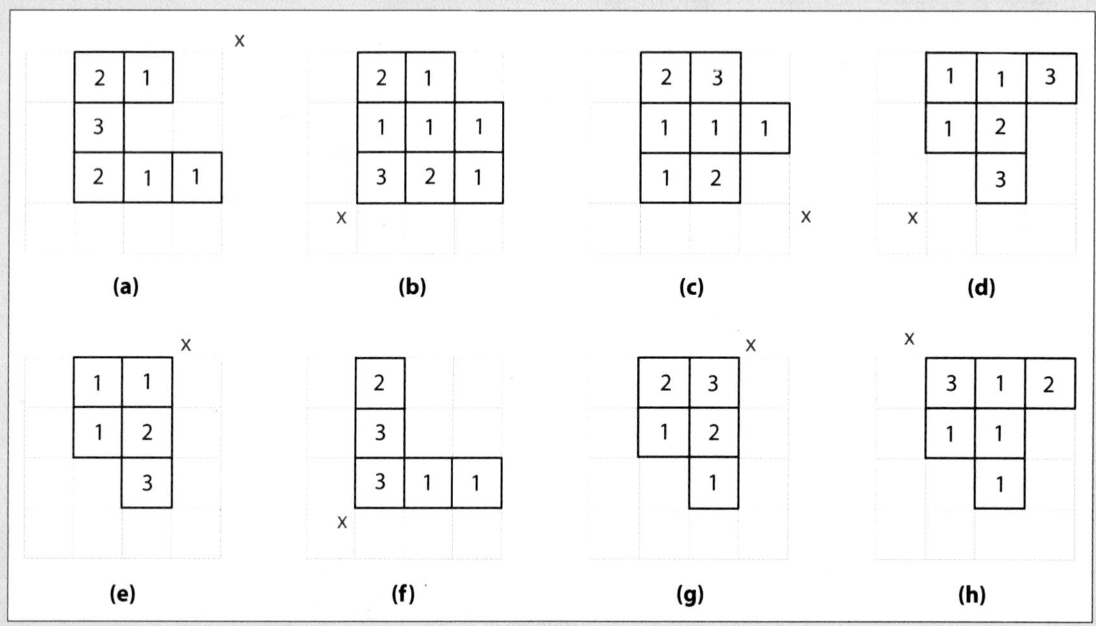

FIGURE P3.3.

3.16 problems (continued)

4. Using the notation developed in this chapter, indicate the rotation the following objects have experienced.

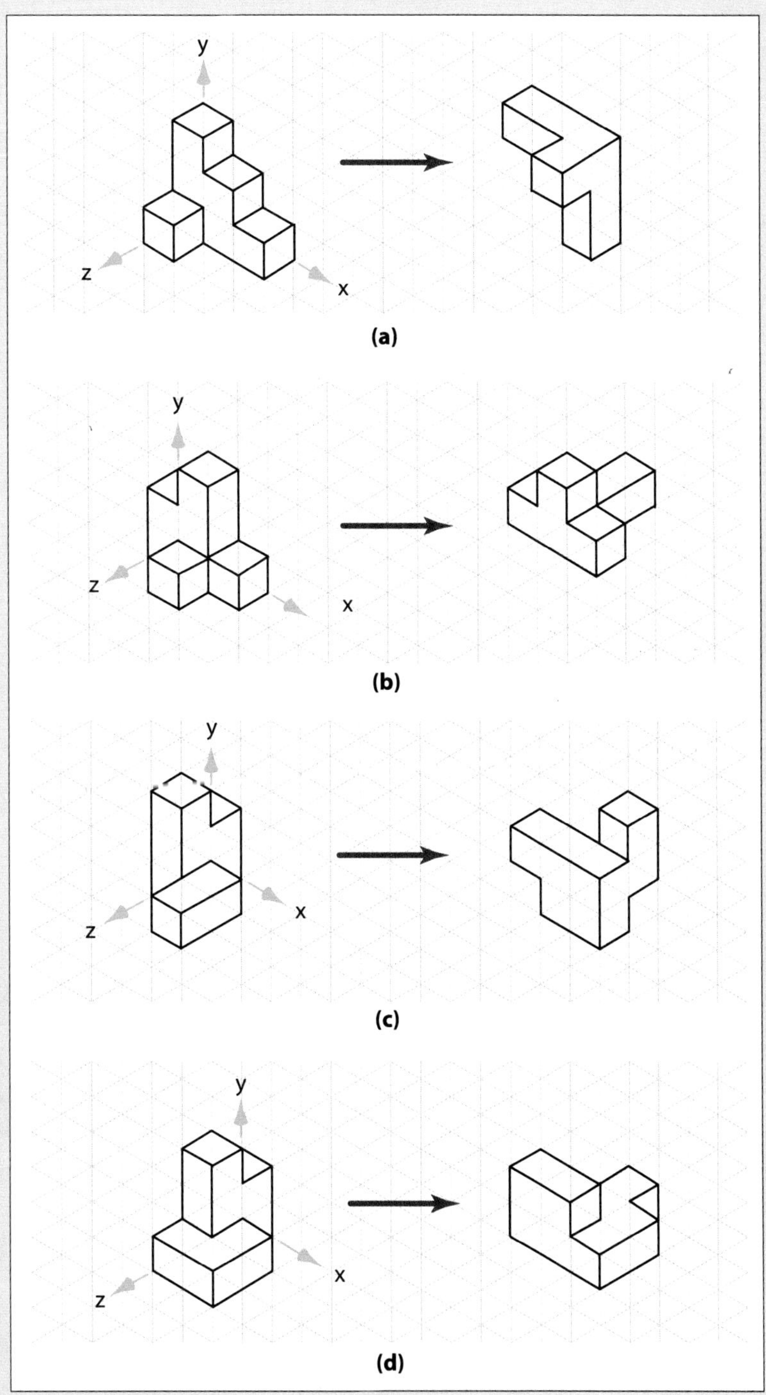

FIGURE P3.4.

3.16 problems (continued)

5. Rotate the following objects by the indicated amount and sketch the results on isometric grid paper.

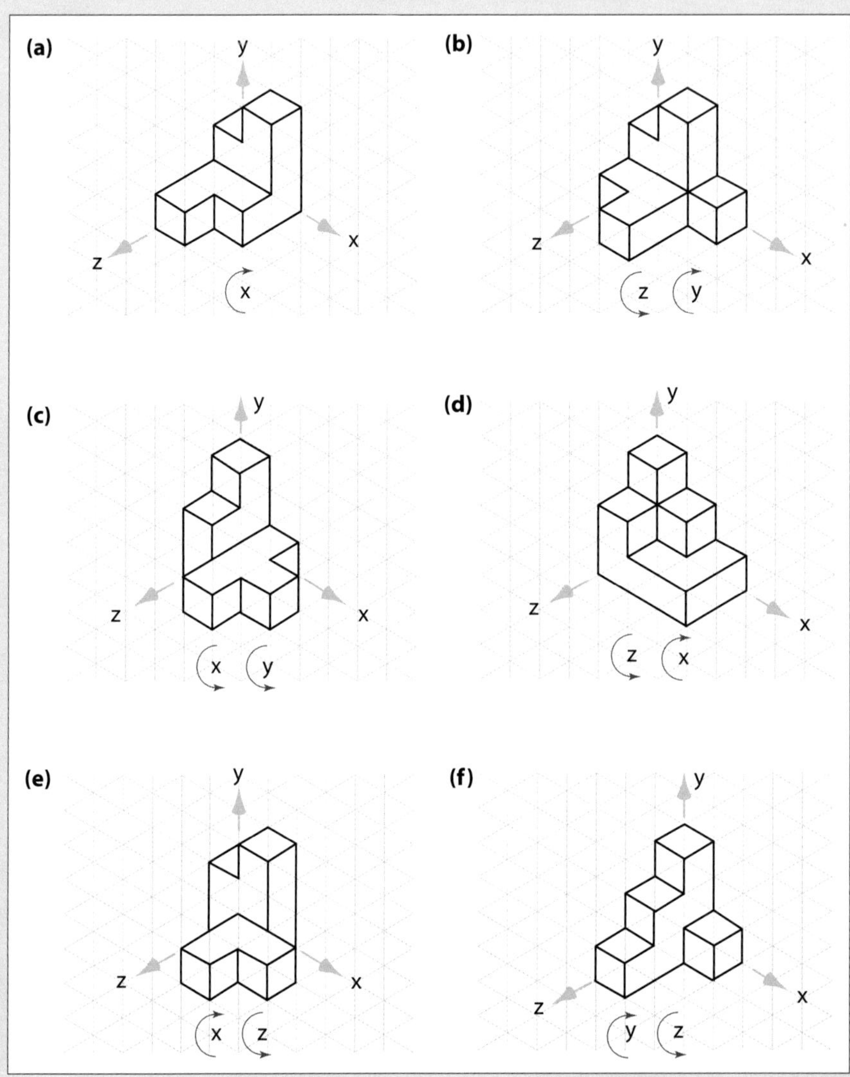

FIGURE P3.5.

3.16 problems (continued)

6. Copy the following object on isometric grid paper and sketch its reflection across the indicated plane. Note that the sketch of the reflection has been started for you.

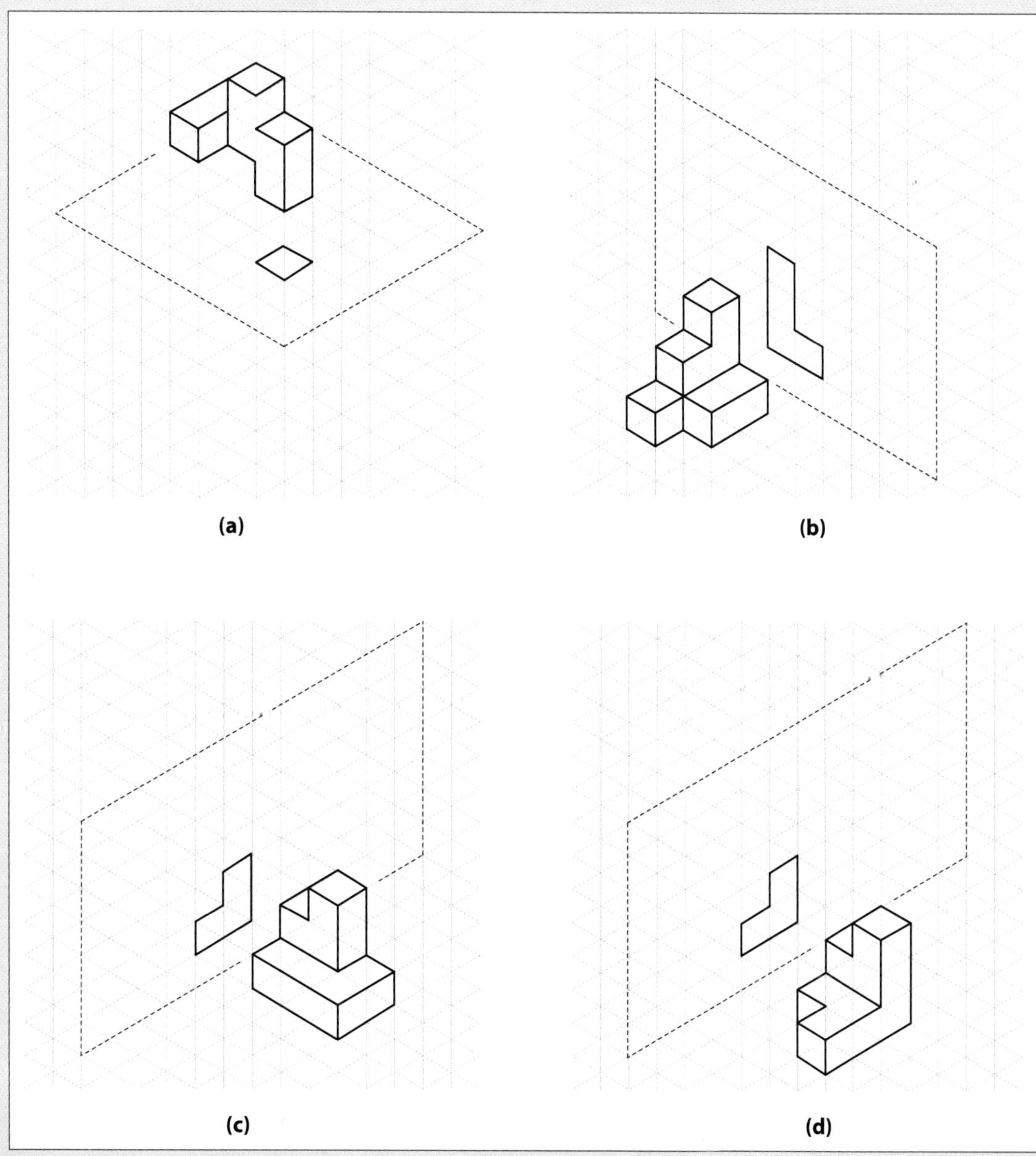

FIGURE P3.6.

3.16 problems (continued)

7. How many planes of symmetry do each of the following objects have?

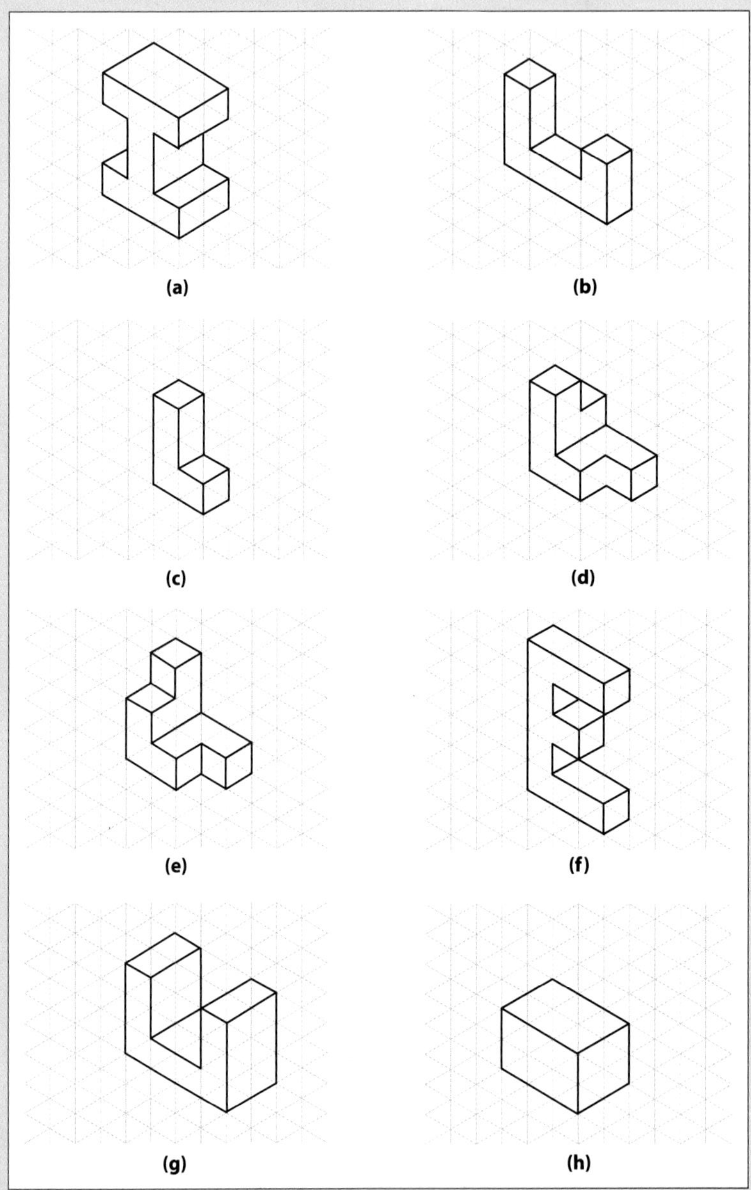

FIGURE P3.7.

3.16 problems (continued)

8. Sketch the cross section obtained between the intersection of the object and the cutting plane.

9. Sketch the result of combining the following objects by the indicated method.

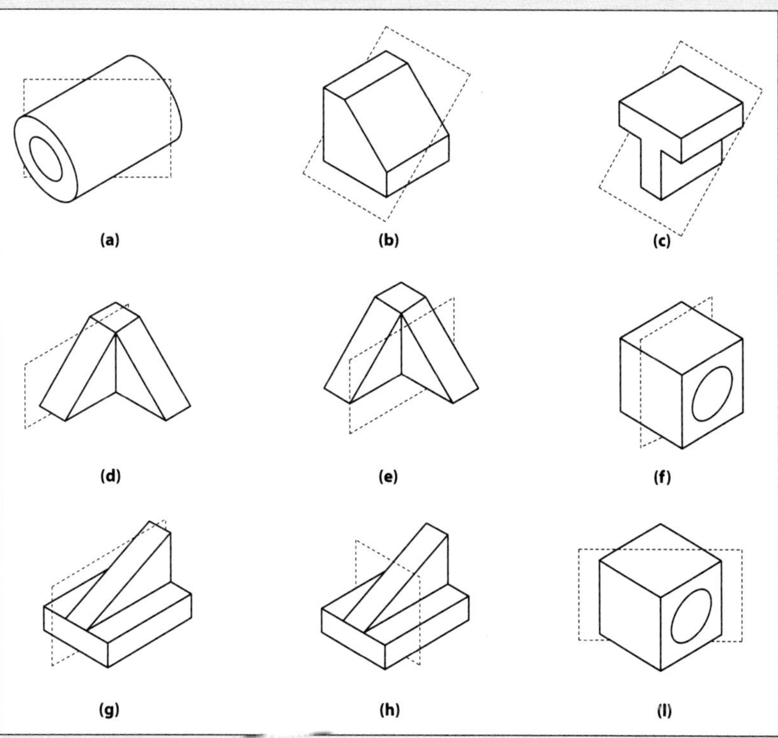

FIGURE P3.8.

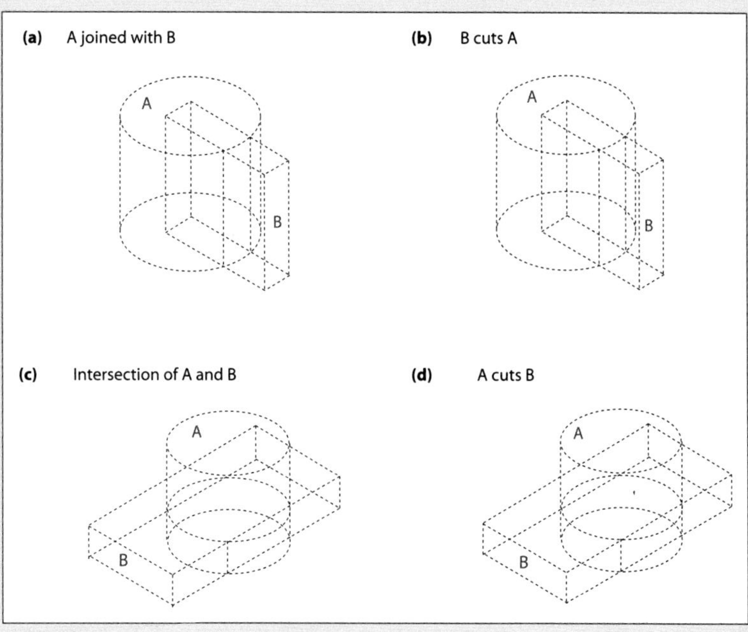

FIGURE P3.9.

3.16 problems (continued)

10. Describe by words and sketches how you would create the following objects by combining basic 3-D shapes.

11. Create isometric sketches from these coded plans using the corner view that is circled or the corner prescribed by your instructor.

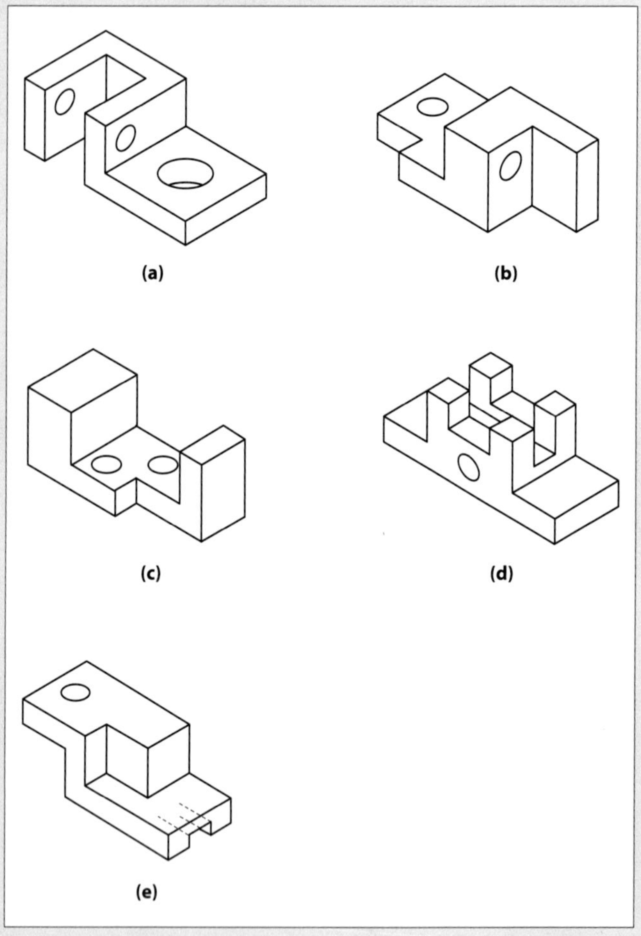

FIGURE P3.10.

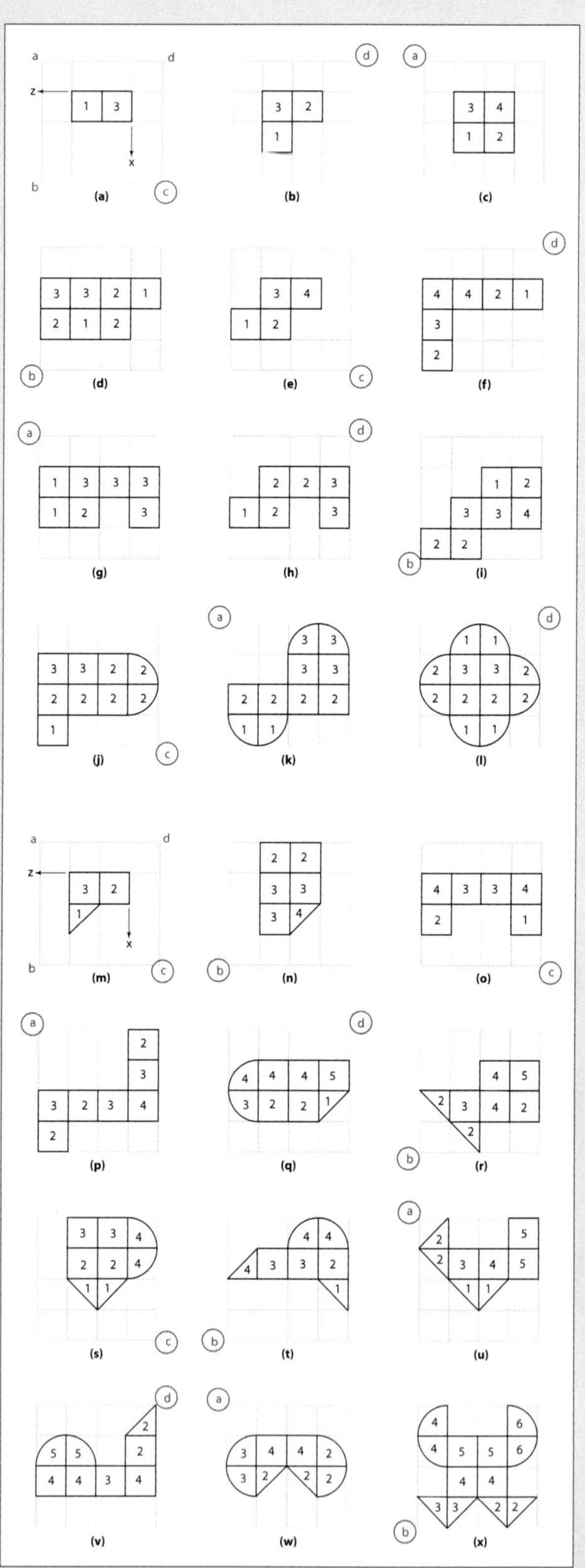

FIGURE P3.11.

3.16 problems (continued)

12. Add the reflected images or redraw these objects with symmetry using the xy, yz, or xz planes as indicated or the planes prescribed by your instructor.
13. The object shown in (a) is show again in (b) rotated by −90 degrees about the y-axis to reveal more detail. Rotate the object sequentially about the axes indicated or about the axes prescribed by your instructor.

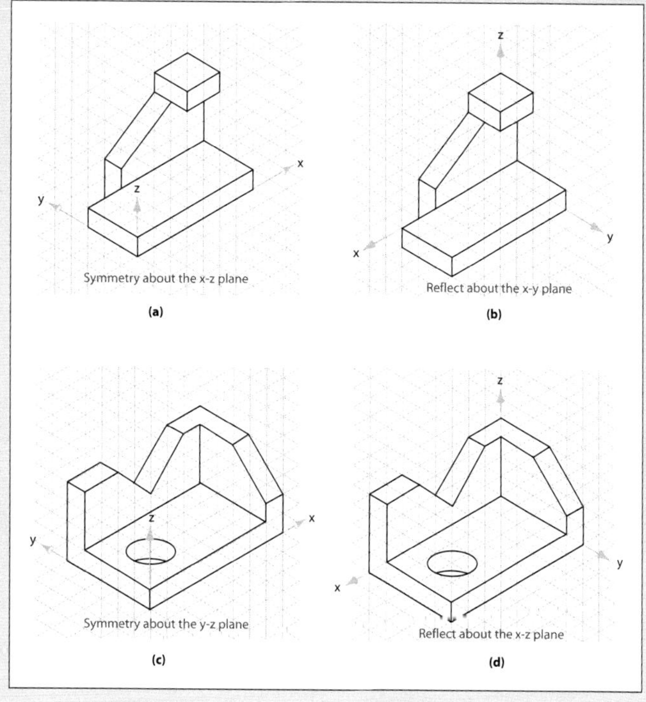

FIGURE P3.12.

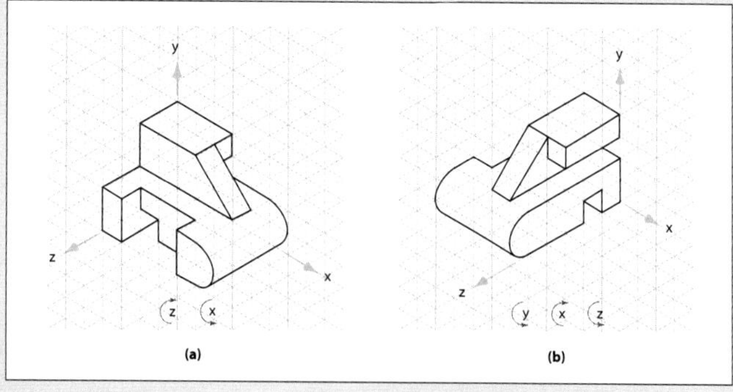

FIGURE P3.13.

3.16 problems (continued)

14. Create an isometric sketch of the objects created from coded plans A and B. Rotate object A sequentially about the axes indicated or about the axes prescribed by your instructor. Show the new object created by the indicated Boolean combination of object A and object B or the Boolean operation prescribed by your instructor when the coordinate axes of A and B are aligned.

15. Triangular volume A, triangular volume B, and rectangular volume C are shown intersecting in space. On the dashed outline drawings, darken and add edges to show all visible edges of the final volume created by the indicated Boolean operations.

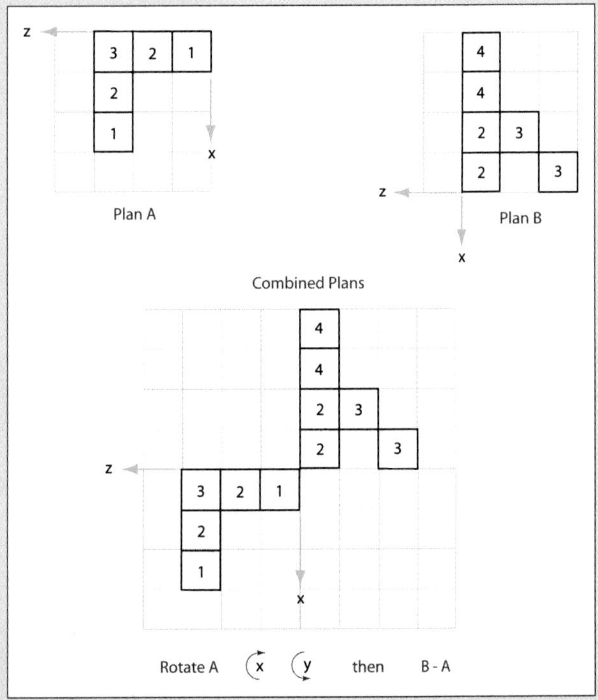

FIGURE P3.14.

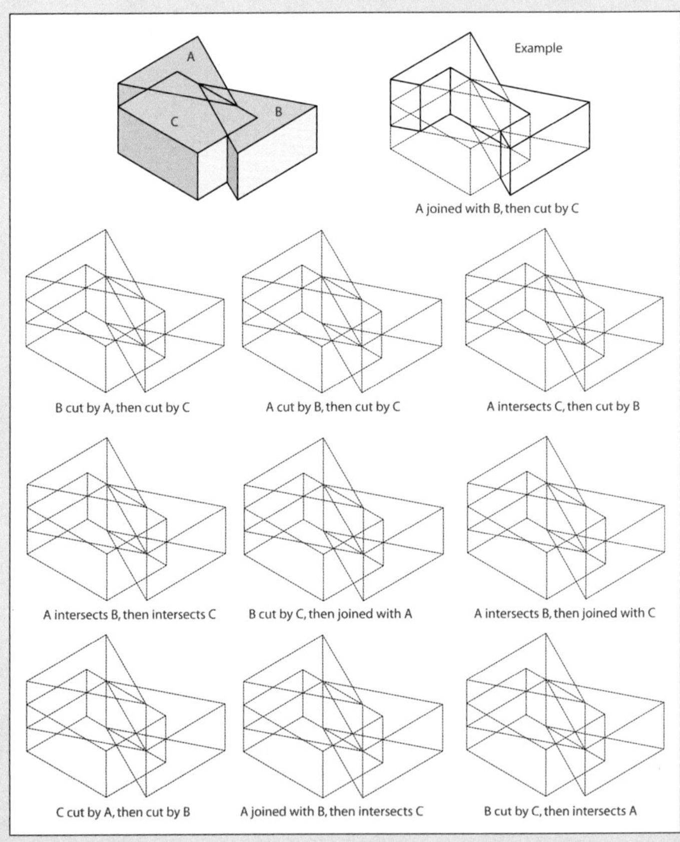

FIGURE P3.15.

4 Working in a Team Environment

objectives

After completing this chapter, you should be able to

- Understand the benefits of working in a team
- Organize team projects and team member responsibilities
- Communicate and work in a team
- Assess the strengths and weaknesses of your team
- Apply strategies for improving team performance
- Solve problems with team members
- Work in a team effectively

4.01 introduction

Assume your instructor has assigned a "design" project for the semester. Moreover, your instructor has assigned everyone in the class to work in teams. You probably have worked on team projects in the past. Perhaps your experiences were positive; however, you may have been frustrated at times because not everyone on the team put forth the same amount of effort in completing the project. You may be a bit skeptical about the viability and value of working on a team.

In real life, most engineering projects are accomplished by teams. Think of the space missions. The astronauts would not be able to travel into space without the effort of thousands of engineers, technicians, ground crew, and other support staff. As another example, think of the last movie you watched. The movie "team" included the director, producer, cast, camera operator, costumer, sound effects crew, and many other people who pulled together to develop the successful final product. Automobiles, computers, iPods, cell phones, and countless other everyday products were designed, produced, packaged, marketed, and distributed through a team effort. So if you really want to have a successful career as an engineer or a technologist, you have no choice but to learn to work in a team environment. In fact, you may even find that you enjoy the team atmosphere as you begin to appreciate the way individuals pull together and complement one another to create products that are innovative and timely.

While working on team projects in high school or college classes, you may have experienced less than satisfactory results. In those often dysfunctional teams, task assignments may not have been distributed evenly or team members may not have delivered on promises to complete their tasks. While being on a dysfunctional team is an experience you would like to avoid, being a part of a good team is a rewarding experience. This chapter will outline reasons for working on a team, ways to get started on the road to a successful team, strategies for making your team effective and efficient, and tools for dealing with problems that will inevitably confront you and your team. As with most activities, being prepared for the challenges and opportunities, and having a plan for how to deal with them will improve your chances of success and lead to an enjoyable and productive team experience. Further, learning how to develop successful team environments will prepare you for a future technological career. As you read this chapter, pay particular attention to the sections on organization, time management, and communication, because they are the keys to a successful project—as an individual or as a team.

4.02 Why Work in a Group?

In a later chapter, you will learn about creativity and the design process. Many of the projects suggested in that chapter are difficult, if not impossible, to do by yourself. Amost all of them can be enhanced by the interpersonal interaction that comes from being part of a good team and the environment that comes with being a member of a team. Obviously, more team members means more bodies at work (and the likelihood of a better final product); but a team is more than a group of people who divide up a task into manageable parts. The diversity of different life and professional experiences of team members leads to a larger group of ideas and a variety of approaches in solving problems. Team discussions can generate ideas, expand options, and improve the final product. Even questions from naysayers are helpful—both in clarifying ideas and identifying fatal flaws—before a great deal of time and effort is spent on an idea that ultimately leads to a dead end.

In addition, employers value employees (whether for summer jobs, internships, co-op positions, or full-time jobs after graduation) who can work as a member of a team and who are team players. An engineer rarely works alone. Most projects require a range of training and skills beyond what the most skilled engineer can be expected to do. Even if a project is small enough for a single person to design it from start to finish, engineers rely on information collected or generated by others and need others to fabricate or construct what they design; often engineers rely on others to make sure the project has been built correctly and to identify problems.

Working on a team does not involve personal relationships; a team is based on professional relationships that require you to respect and value the skill sets that other members of the team bring to the project. A team is a group of colleagues who work together to complete an assigned task.

4.03 What Does It Mean to Be a Team Player?

Being a good team member takes work. Most people are used to working on their own—making decisions, prioritizing tasks, and being accountable for their own work. Working with others requires a different approach than working alone. To be a successful part of a team, you need to consider several issues. You should be prepared *not* to be in charge of everything. For some people, this requires a great deal of effort; for other people, it is less taxing. At times, you will be the supervisor; other times you will be supervised. You need to be flexible and understand that a team consisting only of leaders (or only of followers) is not likely to perform well.

Also be prepared to have some interesting (and some frustrating) encounters with your new work mates. Be prepared to exchange points of view and to learn from those around you. Everyone on a team is responsible for success and is accountable for failure.

Most importantly, prepare to learn how to be a team member. Share your strengths with the team and be willing to contribute. Remember, the combined efforts of all team members should yield a better outcome than the efforts of one individual. Learn new team skills and be adaptable.

Many teams have problems when everyone tries to be in charge or when no one tries to be in charge. The result can be the same: uneven distribution of work, incomplete work, missed deadlines, subpar performance, and frustration. Even though a team is a united effort, each individual is accountable for the overall performance of the team.

Individuals generally react differently in groups than they do on their own. If you miss deadlines or produce inferior work as an individual, you can expect to be held accountable if your work habits are the same when you are part of a team. Conversely, if you produce high-quality work on your own and do the same as part of a team, you will be rewarded accordingly. Remember that team members are accountable first for their individual performance and second for the group's performance. Keep everyone informed of your progress.

4.04 Differences between Teaming in the Classroom and Teaming in the Real World

Selecting personnel and identifying skills are the most important tasks of assembling a team to work on a real-world project. While it may be advantageous to pick people who have worked together previously and who have established a good working relationship, you need to make sure that all of the skills required for the completion of the project are represented by at least one person on the team. For example, if the team is designing a building, the team must have a member who understands, among other things, the:

- Design of a foundation.
- Design of the structure.
- Design of elevator and/or escalator systems.
- Design of air-conditioning systems.

Additional skills are likely to be on the list if the building is to be made of reinforced concrete, or if it is to be constructed in Alaska or California or Louisiana.

In the real world and in the classroom, the goal is to complete a successful project on time and within budget. However, the skills and training of potential team members in the classroom are all virtually the same (unlike the real world). Furthermore, the *primary* goal in the classroom is for each member of the team to learn about each task required in the project. Whereas a mechanical engineer is not expected to teach other members of a building-design team how the air-conditioning system works or why the particular components were selected, each member of a classroom team is expected to explain her part of the team project. Unless the team members complete all of the tasks together, each member must teach the rest of the team what she did on her part of the project.

4.05 Team Roles

For your team to operate like a "well-oiled machine," you need to understand that the members must fill specific **team roles**, if effective collaborative work is to result. Typically, well-functioning teams have a leader, a timekeeper, and a note taker at a minimum. If there are additional team members, assigning someone to the role of devil's advocate is also a good idea. The **team leader** does just that—she leads. This does not mean the team leader dictates or makes all of the decisions for the group. The team leader sets the meeting time, sets the agenda for the meeting, and generally keeps the meeting moving. The team leader also makes sure the team stays on target and remains focused on the task at hand.

The **note taker** keeps a written record of the team's progress. He or she records what tasks have been assigned to whom and records the expected completion dates of the tasks. The note taker is responsible for sending the minutes of the meeting to all team members. The minutes are a written record of what transpired during the meeting and serve as a reminder of who is responsible for completing what task(s).

The **timekeeper** makes sure the schedule is maintained and that meetings do not run over the allotted time. If meetings routinely last longer than planned, team members may skip them or resent coming to them—either of which leads to less productive team encounters.

Finally, the role of the **devil's advocate** is to challenge ideas without being too overbearing or unpleasant. The devil's advocate makes sure that all options are considered and that ideas are sound. However, a devil's advocate should not challenge ideas just for the sake of the challenge; doing this can annoy teammates and detract from the overall effectiveness of the team's operation.

Depending on your personality, you might be naturally inclined toward one role over another. For example, you may naturally be a critic who performs the role of devil's advocate very well. In the classroom setting, you should try out other team roles, so you can develop additional team skills. You may need to hone your note taking skills, and filling that role on the team may help your personal development. In classroom projects, team members can rotate roles so everyone has a chance to experience each role. By performing roles that are unfamiliar to you, you learn to appreciate people who work in these roles. Developing an appreciation for and respecting the skills of the other members of your team are the first steps toward your becoming an effective team member.

4.06 Characteristics of an Effective Team

Most successful teams either knowingly or unknowingly operate by certain ground rules that contribute to overall team productivity. Some of these ground rules are described in subsequent sections of this chapter, giving you the opportunity to learn the rules and adapt them to your particular setting and project.

4.06.01 Decisions Made by Consensus

For a diverse team, it will be nearly impossible to get 100 percent agreement on all of the decisions. Trying to achieve this unreachable goal will lead to frustration and poor productivity for the team. **Consensus** means finding an option that all team members will support. It does not mean that all team members would select that option as their first choice, although some of them probably would. When making decisions by consensus, everyone on the team is invited to voice an opinion. Some people may be naturally shy and unwilling to speak up. The team leader should note when a team member has not voiced an opinion and invite the individual to speak up. Silence should not be interpreted as agreement—often it is not. Another important aspect in making team decisions is to consider the data carefully. Decisions made based on feelings, where data is ignored, are usually not optimal.

4.06.02 Everyone Participates

No member of a team should be allowed to sit back and watch others do the work. As mentioned previously, it is important for every member of the team to voice an opinion during meetings. It is just as important for every member of the team to participate in the work of the team. Tasks should be assigned to members based on talents/skills, and no one should be allowed to choose not to do something. The leader is responsible for making sure that every member participates equally in the work of the team. This does not mean that every task needs to be divided into equal parts, but it does mean the *overall* work should be divided evenly.

4.06.03 Professional Meetings

Team meetings should be productive and engaging. If they are ineffective and a waste of time, team members will likely skip them or not participate fully. This, in turn, will lead to poor-quality work from the team. Team meetings work best when a procedure has been established for the conduct of the meeting and an agenda has been created in advance. The agenda should be prepared by the team leader and e-mailed to participants in advance of team meetings. As a rule of thumb, the **agenda** should include (1) a review of progress to date, (2) a review and possible revision of the project schedule, and (3) new task assignments for team members as needed. This list is not exhaustive—your agenda will be dictated somewhat by the project you are working on. The timekeeper is responsible for making sure the team follows the agenda so that all items on the agenda are completed within the time allotted for the meeting. Punctuality at team meetings is a necessary ingredient for success. A person who shows up late for a meeting is not being fair to or respectful of her teammates. Attendees should do their best to be on time.

4.07 Project Organization—Defining Tasks and Deliverables

Your team was likely formed to work on a project, maybe even a design project. In a subsequent chapter, you will learn about the design process and its various stages. For now, understand that design is an iterative process that proceeds from stage to stage until completion. At some point in the process, you may have to return to an earlier stage to redo something the team thought was completed. Redoing earlier work is a normal part of the design process, especially in cases where you are trying something new and do not know if your idea or solution will actually work. During each stage, you meet as a team to review your progress and to determine what needs to be done next. In the early stages of the project, you probably made a list of all of the tasks that needed to be done. You should review this list at each meeting because the list will likely change as the project evolves.

Once you are sure you have a complete list of what needs to be done before the next meeting, it is time to make a list of tasks and assign responsibility to the person who will be completing each task. When reviewing the items on the list of tasks to accomplish, you need to determine which tasks depend on the outcome of other tasks. For example, if you are going to machine a part, you need to create a drawing of the part. To create the drawing, you may need to create a solid model of it in your CAD system. Thus, it would be unreasonable to expect someone to do the machining *before* the modeling work has been done—the task of machining depends on the outcome of the modeling task.

Another consideration when assigning tasks is to determine which tasks can be done by an individual and which require a group effort. If someone on your team has a difficult schedule to work around, you may want to assign individual tasks to that person most of the time (but not always) to accommodate their schedule. In addition to looking at the individual/group effort required of each task, you also should try to estimate the amount of time each task will require for completion. If one group member ends up completing a task that requires ten hours while two other members complete tasks that require one hour each, resentment is likely to build, thereby hindering group progress. However, as stated earlier in this chapter, division of labor for the project should be balanced overall. So if the person assigned the ten-hour task has been a slacker on previous assignments, perhaps that person should complete a significant task the next time one is assigned.

When assigning tasks, you should try to match the talents and abilities of each team member to the requirements of the task. However, you want to rotate duties so one person is not burdened with all of the writing or all of the modeling or all of the calculating (similar to the way the team roles are rotated so everyone has the opportunity to experience each role). As you are thinking about the assignment of tasks, ask yourself the following questions:

- Which team member is *best qualified* to do the task?
- Who is *able* to do the task in terms of either time or skill?
- Who is *willing* to do the task?

You need to make choices between assigning a task to a person who can accomplish it and assigning the task to the best person. The best outcome may not result from the best person being assigned to a task. If the person is overloaded as a result of the task assignment, she may not do a very good job or may not be able to complete the task. Balancing task assignments is key to producing the best possible project. A project may not be the best one a team can produce; rather, it is the best project a team can produce within the limits of available resources.

4.08 Time Management—Project Scheduling

Once you have organized your team and started work on your project, you should begin developing a plan for completing the project. Think of the plan as being dynamic, not static, since you are likely to be making changes to the plan as the project unfolds. Think of the initial plan as a flowchart of activities or a calendar of events that should include items such as who was assigned to work on each task and where each task fits in the overall project. When examining the various tasks that make up the final project, think about the interrelationships between tasks. What task precedes/follows each task? How does information flow from one task to another? In the previous example, modeling was the first task to be accomplished, which was followed by the creation of a drawing, culminating in a fabricated part. The task that precedes and follows each task is well-defined, but the method of communicating information between tasks may not be as straightforward. Ideally, this information flows seamlessly through the CAD software; however, you may need to run file translator routines to move information between tasks.

Perhaps the most important activity in project scheduling is to determine how each task fits within the overall project and when each task should be completed for your project to end successfully. Usually when you are organizing your project, you can begin at the beginning or you can begin at the end. If you start at the beginning, the organization of the project can be done in a cyclic manner. As a team, ask the following questions:

1. What needs to be done first?
2. What do we need to know before we can do WhatNeedsToBeDoneFirst (WNTBDF)?
3. Now that we have a new WNTBDF, repeat step 2.

Sometimes it is more efficient to begin by considering the *final* deliverable for the project (product, design, prototype, sketch, etc.). You then work backward through the process, identifying what needs to happen before a specific task can be accomplished. Another way of thinking about this is to consider all of the other tasks that must be completed before a specific task can be accomplished. When organizing your project from the beginning or the end, you also need to consider who will be completing each task—you cannot establish a timeline without considering the realities of everyone's schedule.

Sometimes the result of this activity is to discover that the timeline for your well-planned project does not match the deadlines established by your client (or instructor).

In this case, you must revisit the task list and eliminate tasks or compress the time to complete each task. In others words, you must determine how good of a project you can deliver in the time allowed. Even in the real world, you do not always have enough time to produce the best design (or the client does not have enough money to build the best product). The goal is to produce the best product within the constraints given. These constraints are usually time, money, materials, and talent.

By now, you may have realized that the schedule for your project and the organization of tasks in your project (presented in the previous section) are linked. The following sections include information on two tools you may find useful as you organize and plan your project: the Gantt chart and the Critical Path Method.

4.08.01 Gantt Charts

When working in teams, it is essential to establish a well-thought-out plan for completing the project. If you are working on a project as an individual, the planning stage is not nearly as critical, since no one else is depending on you to complete a task to an acceptable level of quality within a certain time frame. One useful tool to help you organize your project, assigning a timeline for the completion of various tasks, is a Gantt chart. A **Gantt chart** is a table that lists the tasks in the leftmost column that must be completed, and it identifies the dates across the top row by which each task must be completed. Shading indicates the times for working on each of the project tasks. Without a detailed plan that lays out due dates and establishes a timeline, most projects get bogged down in trivial details and important tasks are delayed. This will often put the success of the project in jeopardy of being completed on time. Figure 4.01 shows a simple Gantt chart for a student project in reverse engineering, a topic that will be discussed in more detail in a subsequent chapter.

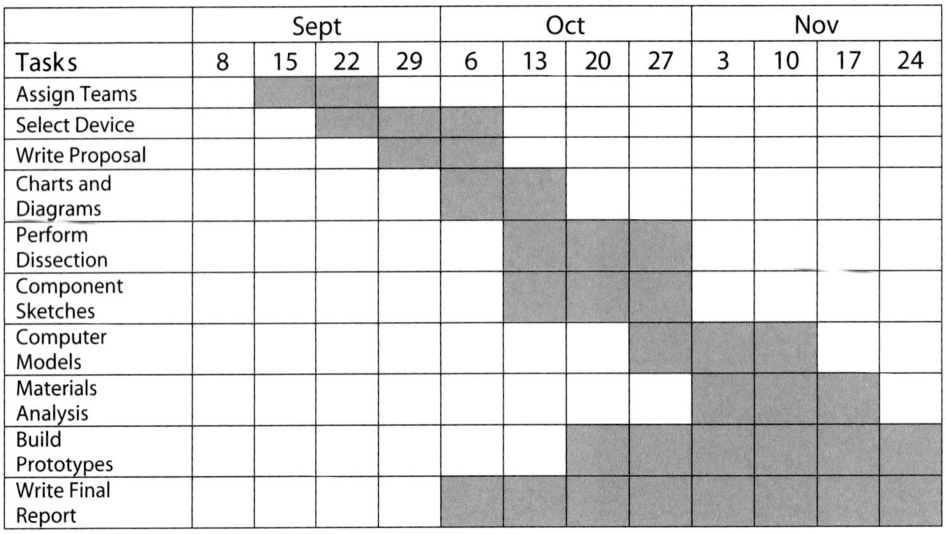

	Sept				Oct				Nov			
Tasks	8	15	22	29	6	13	20	27	3	10	17	24
Assign Teams												
Select Device												
Write Proposal												
Charts and Diagrams												
Perform Dissection												
Component Sketches												
Computer Models												
Materials Analysis												
Build Prototypes												
Write Final Report												

FIGURE 4.01. Gantt chart for reverse engineering project.

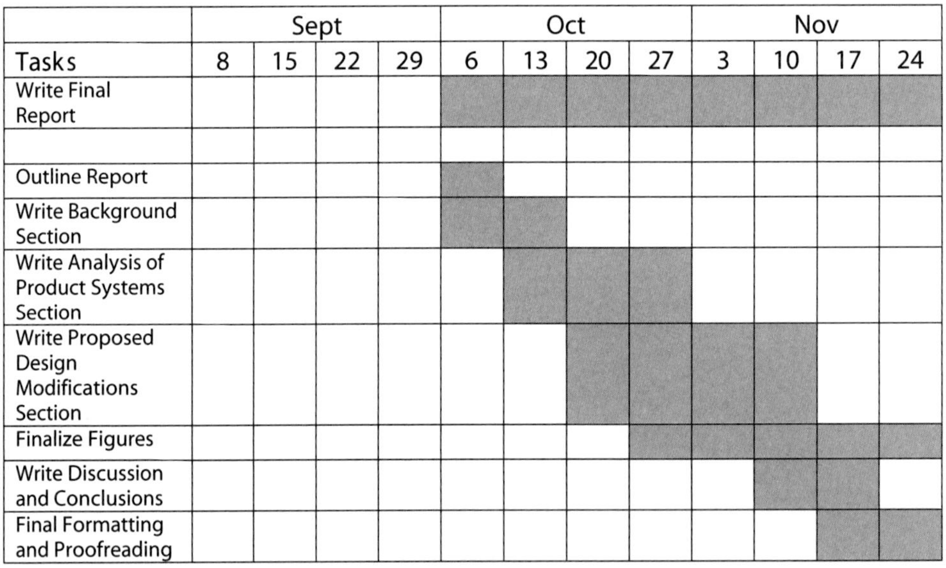

	Sept				Oct				Nov			
Tasks	8	15	22	29	6	13	20	27	3	10	17	24
Write Final Report												
Outline Report												
Write Background Section												
Write Analysis of Product Systems Section												
Write Proposed Design Modifications Section												
Finalize Figures												
Write Discussion and Conclusions												
Final Formatting and Proofreading												

FIGURE 4.02. Gantt chart for report writing task.

Note that each of the major task headings can be broken down further and a Gantt chart created for each major task. For example, Figure 4.02 shows a new Gantt chart created just for the last task listed in the previous Gantt chart.

4.08.02 Critical Path Method

The **critical path method (CPM)** is used in project scheduling to determine the least amount of time needed to complete a given project. CPM is also used to determine which activities are most critical to the on-time completion of the project (hence, the name) and which activities are not as critical to the overall project schedule. The **critical path** includes the sequence of the activities that have the longest duration. When these activities are strung together on the critical path, you can determine the shortest possible duration of the project. Activities that are not on the critical path can be allowed to "float" with regard to schedule, which will not impact completion of the overall project.

The CPM is like a flowchart for the project. It helps everyone visualize what has been accomplished, in what stage of the project the team is (what percentage is complete, whether the project is ahead or behind schedule, etc.), and what needs to be done next at each step of the project. If names are associated with each task, the CPM can also serve as a reminder of who is waiting for a finished task before that person can begin the next task. The information required to construct a CPM diagram is:

- A list of all activities required to complete the project.
- The amount of time each activity will take to complete.
- The dependencies between tasks (i.e., what task relies on the completion of another task before it can be started).

As an example of a CPM, consider a project broken down into six major activities. The task breakdown is characterized as follows:

Activity	Duration	Depends on
1	2 days	–
2	4 days	1
3	6 days	2
4	5 days	1
5	2 days	2, 4
6	8 days	3, 5

The critical path diagram for this project can be constructed as shown in Figure 4.03. Note that arrows are used to show forward progress through the project, and dependencies between tasks are shown graphically. For this example, there are three possible "paths" through the project from start to finish, as shown in Figure 4.04. The first path includes Activities 1, 2, 3, and 6. This path has a total duration of 20 days. The second path includes Activities 1, 2, 5, and 6 with a duration of 16 days. The final path includes Activities 1, 4, 5, and 6 with a duration of 17 days. The critical path is the first one (with Activities 1, 2, 3, and 6) because this path has the longest duration. Based on the critical path, the *shortest* possible completion schedule for the project is 20 days; anything less than this is impossible.

Each activity on the critical path is now a critical activity; that is, if any of the activities are not completed on time, the overall time needed to complete the project will be lengthened. This means that Activity 4 and Activity 5 (the only activities not on the critical path) have some float time–if they are not completed on time, the overall project will still be on schedule. If you are a project manager, this information shows you where to concentrate your efforts. If it looks as though Activity 4 is beginning to interfere with Activity 3, you can suspend work on Activity 4 for a short while to make sure Activity 3 proceeds unhindered. Or if Activity 2 starts to flounder, you can shift resources away from Activity 4 to make sure the critical activity (Activity 2) is completed on schedule.

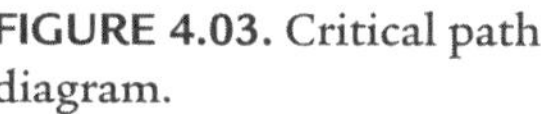

FIGURE 4.03. Critical path diagram.

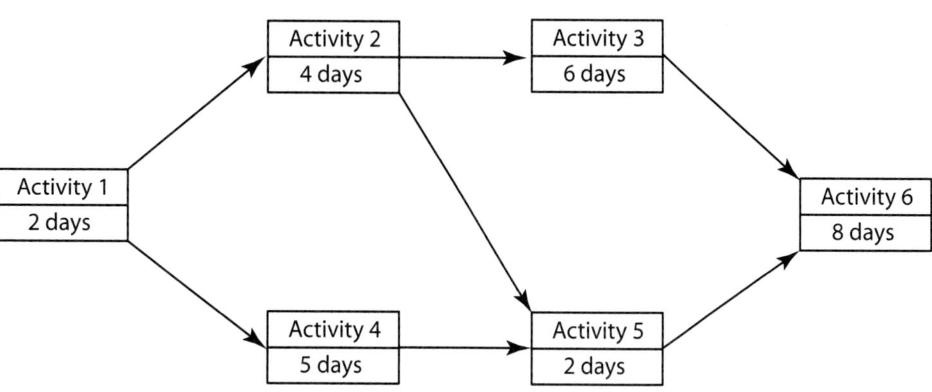

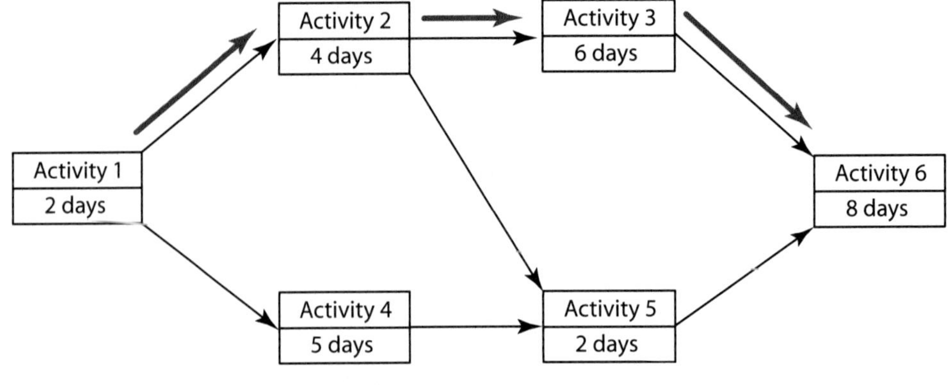

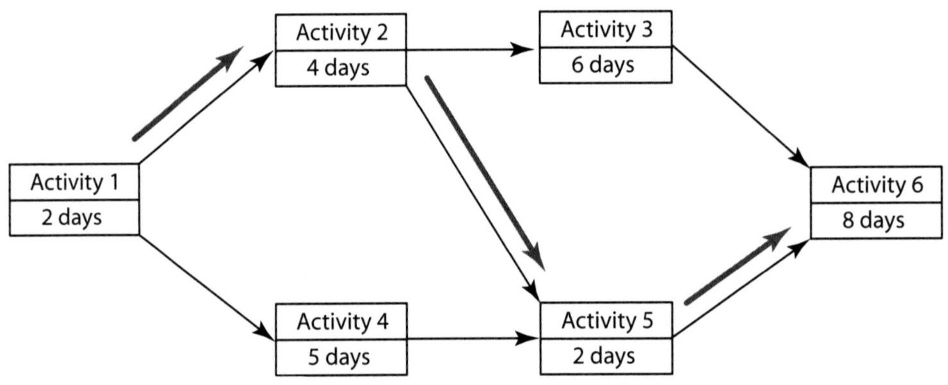

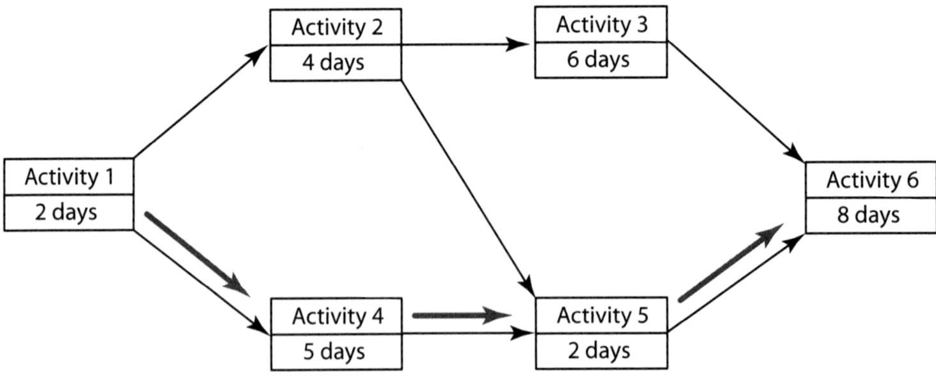

FIGURE 4.04. Three possible paths through project completion.

Because completion of any project is dynamic rather than static, as with Gantt charts, you should review your critical path diagram periodically to ensure that it still accurately reflects the realities of your project.

4.09 Communication

Following is a discussion about communication between the team and the outside world and communication among members of the team. Each member of the team must communicate openly and honestly during task assignments (Do I agree with the plan? Do I have other commitments that will interfere with the timeline? Can I commit enough time to the task? Do I have the skills required to complete the task alone?). A team member's silence is usually interpreted as agreement with the plan, or at least acceptance of the plan. Most problems between team members result from a lack of open and honest communication.

4.09.01 Agreeing How to Communicate

E-mailing, chatting, and text messaging are useful modes of communication, and you have probably used all of them; however, regular face-to-face meetings are essential for a team's successful communication. Electronic modes of communication can be used, but nothing takes the place of face-to-face meetings. Because no one has time for an unnecessary or poorly conducted meeting, when team members do meet, they want the meetings to be productive. Decide early in the project how often the team needs to convene to conduct its business. Team members have other obligations, and the team project will get only a portion of their time. Having a regularly scheduled meeting time enables all team members to plan their activities around project time.

Not all information is shared in meetings; and the options of e-mail, notes, memos, and voice mail are appropriate under the right circumstances. All team members need to agree on how to communicate.

In your initial team meeting, find out how the team wants to handle communication. How much time do members have for meetings? Does everyone have e-mail? How often do they check it? Is voice mail reliable? The value of written records of team meetings and decisions cannot be overemphasized. By investing whatever time is needed to agree on how to communicate, team members will save time during the course of the project.

Documentation is essential after meetings. The note taker's responsibility is to summarize the conduct of the meeting and communicate his or her notes to the rest of the group. Sometimes the team keeps a bound notebook of team-meeting notes, with the notebook passed from one note taker to the next (when team roles are being rotated). Usually the note taker will convert meeting notes to an electronic document and e-mail them to all team members, summarizing the following:

- Did we say what needed to be said?
- Did we begin and end on time?
- What was decided?
- What tasks were assigned?
- Who is responsible for each task?
- What questions still need to be answered?
- What must be done before the next meeting?

After the note taker e-mails the document out for review, you should look over the notes carefully, making comments if any points conflict with how you remember the meeting. If you do not take the time to review the notes or if you do not bother to raise objections with the notes as written, the notes will become the permanent record of what transpired at the meeting. Once again, silence is interpreted as agreement. You cannot come back later to say that you did not know you were responsible for completing a task when that assignment was included in the meeting notes that you agreed (either actively or passively) were complete and accurate.

4.09.02 Communicating Outside Meetings

As the team's level of trust increases, the need for face-to-face meetings may decline. Alternative modes of communication can be used to keep members informed of progress, changes, and the need for team meetings. The team should also decide how members unable to meet with the entire group will communicate with the team, as well as how the team will communicate meeting results to absent members. Progress reports should include information about what is or is not happening. Sometimes what is not happening is as important as what is actually taking place. Other points to include in a progress report include pointing out tasks that need immediate attention or changes that need to be made. One of the most important parts of a progress report is recommendations for what needs to be done to get a task or an activity back on track.

A **process check**, another way of communicating in face-to-face meetings or in electronic forms of communication, can help move the team (or individuals) toward improved performance. Usually, a process check is a reflection of:

- What the team members did well that they want to continue doing.
- What did not go as well as the team would have liked.
- What the team can do to improve the things they want to improve and not detract from the things they think are going well.

The team should periodically conduct a process check at the beginning or end of a team meeting; doing so will help resolve people's differences and reinforce good feelings. Either way, a process check facilitates a professional, functioning team.

4.09.03 Communicating with the Outside World

A key to effective teamwork is effective communication. In the previous sections, you learned how to communicate with other team members. In reality, teams also need to be able to communicate with the outside world. In a university setting, the outside world may include instructors and classmates. In the working world, the outside world may consist of bosses, coworkers, clients, and/or the general public. The modes with which your team might need to communicate include progress reports, final reports, final design documentation, and project presentations.

When preparing a progress report, focus on the status of the project and any obstacles the team has encountered. Typically, a progress report is prepared for an instructor or boss, so the points must be clear and direct. Progress reports are meant to give an overview of the progress since the last report, so shorter is probably better as long as all necessary details are included. If your team has identified additional resources that are necessary for the successful completion of the project, these resources should be identified in the progress report as well. In final reports to the client or boss, you should provide the details of your design. Final reports are typically several pages in length (depending on the project) and outline the choices you made, the analysis you performed, and any test results you obtained. Design documentation usually includes drawings, specifications, and the details of any analysis you performed. Your design documentation should show how you arrived at the answer(s) you did. Presentations should show the highlights of your project. Usually, you will be given separate instructions from your professor about her expectations for a group presentation. Make sure you adhere to any guidelines you are given. If you are going to make a group presentation, you should practice at least once as a team to make sure you stay within the time limit and cover all necessary points.

For classroom projects, you will probably need to convey to your instructor how each person on the team contributed to the overall project. Specifically, you may be asked to address the following:

- Who contributed?
- What did each person contribute?
- How much credit does each person deserve?

Be honest and fair in your appraisal of your teammates' work on the project, as well as in your appraisal of your own work. Giving someone a pass that does not deserve one is not fair to the rest of the team, nor is judging someone too harshly.

4.10 Tools for Dealing with Personnel Issues

The following sections will give you tools for avoiding problems usually attributed to team members who become difficult to work with because of a personality trait that does not adapt well to team work or because the team member is not properly committed to completing the assignment. Keeping everyone motivated is best done proactively rather than reactively.

4.10.01 Team Contract

A **team contract** (also known as a code of conduct, an agreement to cooperate, or rules of engagement) is a formal written document, which should be readily available during all team meetings. The document should be established only after careful, thorough, and honest discussion by all team members. The contract lists the rules the team agrees to live by. Often team members need to revise a contract once they spend enough time working together that they (and others on the team) discover pet peeves. A good method for establishing a contract is to ask each member to bring to a team meeting a list of two or three of the biggest problems they encountered while members of previous teams. At the meeting, the team should consider each item in a round-robin fashion, until all members' lists have been exhausted. The resulting contract is a list of rules/agreements that, if followed, will incorporate the items on each person's list. This means that if everyone follows the rules in the contract, no one will violate any pet peeve or cause any of the previously experienced team-related problems.

Changes to the contract may be required as the team progresses, because well-meaning members, in spite of their best intentions, revert to inappropriate behavior. However, this does not have to mean an end to the team. Although revised contracts often include rewards and penalties to assist members in bringing about the desired behavior, you want to avoid coercing a member into accepting a contract. It is imperative that everyone on the team be treated with respect and that all disagreements are viewed as legitimate. Benjamin Franklin, upon the signing of the Declaration of Independence, said, "We must . . . all hang together or, most assuredly, we shall all hang separately." This quote applies to teams as well.

4.10.02 Publication of the Rules

Once you have determined how to operate as a team, write down the agreed-upon rules. Figure 4.05 shows a sample team contract established by a student project group.

Once you have created the contract, make sure every team member receives a copy and use the ground rules as a tool for effective communication and teamwork. Establishing, revisiting, and revising (if necessary) a contract can start the team out on the right foot or keep the team on track as crunch time materializes. Make sure all members support the rules. If you find that the rules are not working, change them, making sure everyone agrees to the new set of rules.

4.10.03 Signature Sheet and Task Credit Matrix

You should include a signature sheet with every assignment, whether it is requested by your instructor or not. On the sheet, explain that the individuals' signatures mean that the individuals participated in the assignment, have a general understanding of the entire submission, and are deserving of the credit indicated on the task credit matrix.

The **task credit matrix** is a table that lists each member on the team, each task, and indicates how the credit should be shared among the members of the team. If a team member does not deserve credit for the project, he or she should not be allowed to

FIGURE 4.05. Sample team contract.

Sample Contract

1. All members will attend meetings or notify the team by e-mail or phone in advance of anticipated absences.
2. All members will be fully engaged in team meetings and will not work on other assignments during meetings.
3. All members will complete assigned tasks by agreed-upon deadlines.
4. Major decisions will be subject to group discussion and consensus or majority vote.
5. The roles of recorder and timekeeper will rotate on a weekly rotational basis (all members will take their turn — NO EXCEPTIONS).
6. The team meetings will occur only at the regularly scheduled (weekly) time or with at least a two-day notice.

sign the signature sheet and a zero should be entered in the task credit matrix. It is not expected that each member will have contributed to each task. What is expected is that the team split the work fairly and that each individual deserves credit. Note that the credit does not have to be equal, although ideally (and usually) it is. Your team may decide to weight different tasks, based on the amount of effort or contribution required. You also may decide to split the tables, one for the task breakdown (to aid the instructor in knowing to whom to direct questions) and a second expressing the team's desire for distribution of points or credit. Often instructors will allow the team this privilege.

4.11 Chapter Summary

In this chapter, you learned about the importance of teams in engineering and in working on student projects. You learned about organizing a team and assigning roles for efficient, effective team meetings. The need to rotate roles among members was also discussed. You learned about the keys to successful team meetings and about organizing projects for optimal teamwork. The critical path method and Gantt charts were introduced as tools to help keep a project moving forward and for maintaining a reasonable project schedule. Finally, you learned about the importance of communication when working on a team. Two types of communication were discussed: internal communication among team members and external communication for working with other people, such as bosses and instructors.

4.12 glossary of key terms

agenda: The list of topics for discussion/action at a team meeting.

consensus: A process of decision making where an option is chosen that everyone supports.

critical path: The sequence of activities in a project that have the longest duration.

critical path method (CPM): A tool for determining the least amount of time in which a project can be completed.

devil's advocate: The team member who challenges ideas to ensure that all options are considered by the group.

Gantt chart: A tool for scheduling a project timeline.

note taker: The person who records the actions discussed and taken at team meetings and then prepares the formal written notes for the meeting.

process check: A method for resolving differences and making adjustments in team performance.

task credit matrix: A table that lists all team members and their efforts on project tasks.

team contract: The rules under which a team agrees to operate (also known as a code of conduct, an agreement to cooperate, or rules of engagement).

team leader: The person who calls the meetings, sets the agenda, and maintains the focus of team meetings.

team roles: The roles that team members fill to ensure maximum effectiveness for a team.

timekeeper: The person who keeps track of the meeting agenda, keeping the team on track to complete all necessary items within the alotted time frame.

4.13 questions for review

1. What are some of the advantages of working in teams?
2. What are some disadvantages of working in teams?
3. What are some important member roles that all teams should include, and what are the responsibilites of each role?
4. How should tasks be assigned in a team project?
5. In what ways is the Critical Path Method useful to the successful completion of a project?
6. What can be done to ensure that the workload in a project is fairly distributed among the team members?
7. What are some methods that can be used to ensure that all team members contribute equally to the team effort?
8. What can be done to solve the problem of a non-performing team member?
9. What can be done to ensure effective communication within a team?

4.14 problems

1. In a memo to your instructor, describe the team roles outlined in this chapter.
2. In a memo to your instructor, describe an unsatisfactory team experience you encountered previously.
3. Meet with your team and develop a code of conduct for future team meetings.
4. Create a Gantt chart for the steps (and timeline) you would take to secure a summer job. Some of the tasks might include writing a resume, looking online for opportunities, and preparing letters.
5. If you have been assigned a project by your instrutor, create a Gantt chart for its completion.
6. Create a critical path diagram for planning and hosting a surprise birthday party for your mother or another family member.

5 Creativity and the Design Process

objectives

After completing this chapter, you should be able to

- Describe the steps in the design process
- Identify engineering tasks at the various stages of the design process
- Explain the role of creativity in the engineering design process
- Use several techniques that aid in the development of creative thinking

5.01 introduction

What distinguishes engineers from scientists is that engineers design and create solutions to the many technical problems of the world, whereas scientists study problems and report their findings in literature. For example, environmental scientists studied Earth's atmosphere several years ago and found that, as a society, the world is creating too many air pollutants—especially from automobile emissions. Certain engineers acted on the results of those studies and designed cars that were more fuel-efficient and that produced less pollution. Other engineers worked on changing the composition of fuel so that fewer pollutants were produced during combustion. Engineers are now working on solar-powered and electrically powered vehicles and on hydrogen fuel cells to further reduce the quantity of air pollutants emitted into the atmosphere. Other engineers are working on ways to use biofuels for cars.

Engineering is a design-oriented profession. Engineers design devices or systems and figure out how to mass-produce them, how to package them, and how to ship them to their intended destination. As an engineer, the type of device or system you design may depend on your discipline, but the design procedure you use will be essentially the same regardless of the industry in which you work.

As described in a previous chapter, engineers function in many capacities in business enterprises. Some engineers are involved in manufacturing, some in marketing, some in management, and some in testing. However, design is the central function of engineers. Typically, as you move through an engineering career, you will be responsible for different aspects of the enterprise at different times; but at some point, you will probably play a primary role in the design of new products. At other times, you may oversee the design of new products as a supervisor of a team of engineers. You also may be responsible for making sure the entire enterprise—including designers, manufacturers, and marketers—is running smoothly and the various team members are communicating with one another and working together to achieve the optimal end product. Because design is such a central portion of the engineering endeavor, you need to understand the design process thoroughly. The remainder of this chapter describes the **engineering design** process and the role creativity plays in the process.

5.02 What Is Design?

Design is a goal-oriented, problem-solving activity that typically takes many iterations—teams rarely come up with the "optimal" design the first time around. As an example, think of the minivan. When Chrysler developed the first minivan, it was considered a revolutionary concept in design and other car manufacturers quickly developed competing models. With each model, improvements were made to the original design such that the minivans of today are much improved compared to the initial product. The key activity in the design process is the development and testing of a descriptive model of the finished product before the product is finally manufactured or constructed. In the case of a manufactured product, the descriptive model usually includes solid 3-D computer models, engineering sketches and drawings, and possibly rapid prototypes. For a civil engineering project, the descriptive model includes drawings, specifications, and sometimes a scale model made of wood or plastic. Three-dimensional CAD models also are becoming more prevalent in the civil/construction industry. Engineering design includes a systematic approach to product definition, conceptualization, development, testing, documentation, and production/construction. Design is usually accomplished in a group environment with many people contributing ideas and skills to complete the finished design. Hence, creativity and interpersonal skills are important attributes of the design engineer.

When designing a device or a system, the engineer must keep certain factors in mind. These factors are usually related to the function and cost of the resulting system. For **sustainable design**, life cycle analysis and environmental impact are especially important factors. In most cases, engineers must make a number of choices during the design process. For example, consider the automobile. Engineers may choose from metals, composites, or plastics for the materials that make up the car's body. Each type of material has advantages and disadvantages. Although steel is strong and ductile, it is prone to corrosion (rust) and is relatively heavy, reducing the fuel efficiency of the car, which, in turn, leads to increased pollution. Composites are strong and can absorb a great deal of energy during crashes, but can be brittle and may be more expensive than steel. Plastics are readily formed in almost any shape and are resistant to corrosion, but are relatively weak, making safety a significant concern. Although plastics are widely used in car bumper systems, they typically are not used in the car's body. These are just a few of the factors that engineers must consider as they design an automobile body.

Engineers make choices by weighing the often competing factors associated with function, cost, and environmental impact. A car could be built that causes virtually no pollution and that is perfectly safe; however, the average person may not be able to afford it. So engineers make trade-offs between cost, safety, and environmental impact and design a car that is reasonably safe, is relatively inexpensive, and has minimal emissions.

Design is an aspect in virtually every discipline of engineering; however, chemical engineers typically view design differently than do mechanical engineers. For a chemical engineer, design includes determining the correct chemicals/materials to combine in the correct quantities and in the correct order to achieve the desired final product. Chemical engineers determine when to stir the mixture, heat it, or cool it. Electrical engineers may design computer chips or wiring for a building or the antenna system for a car or a satellite receiver. Civil engineers, like most chemical engineers, typically design one-off systems with features and/or specifications that are unique to a single application or location. They may design a single bridge or roadway, or a water distribution system or sewage system. Because civil and chemical engineering designs usually are not mass-produced, it is often impossible to create a **prototype** for testing before construction begins. Imagine the cost of building a "practice" bridge for every bridge that a civil engineering team designs and constructs.

Mechanical engineers typically design products that will be mass-produced for consumer use—cars, bicycles, washing machines, etc. Therefore, prototyping is an important part of the design process for mechanical engineers. Prototypes are the initial design concepts that are often created so that further design analysis can be performed before machines are retooled to produce, say, 10 million copies of a product. The process of creating and testing prototypes often saves a company money because engineers can work out the kinks or discover flaws early in the design stage. In the past, the design process included the production of several prototypes for testing and analysis. Today much of the testing can be accomplished using computer software tools, greatly reducing the need for prototypes. However, most manufacturing companies still produce at least one prototype before going into the production of a new product. The foundation for many computer-based testing and analyses in the design process is a 3-D solid model, which is a focal topic of this text.

Although modifications exist in the design process for engineers of different disciplines, there are similarities too. Almost all designs require drawings, sketches, models, and analysis (calculations). The remainder of this chapter will focus on design in a manufacturing arena—the type of design most familiar to mechanical engineers. This type of design results in products that are mass-produced. Where appropriate, variations to the design process for one-off designs (as in civil and chemical engineering) will be described.

5.02.01 Computers in Design

Computers have been used in engineering design for several decades. In the early years, computers were used primarily for their number-crunching capabilities. In other words, computers were employed to perform the tedious calculations involved in engineering design. Over the years, the role of computers in the design process changed significantly. Graphical computer workstations evolved and with them the ability of engineers to see their designs before building them. Engineers also can do much of the testing of design iterations on-screen, eliminating the need for numerous prototypes. Numerical methods such as **finite element analysis (FEA)**, modal analysis, and thermal analysis have enabled engineers to design systems in a fraction of the time that traditional design methods require. Modern design software often is easily incorporated into the manufacturing process, enabling the designer to establish cutting tool paths on-screen for the efficient manufacture of computer-generated models. Even other manufacturing capabilities, such as rapid prototyping, in which physical prototypes can be created within a matter of hours rather than days, have been a direct result of computer-aided design capabilities.

Today **computer-aided design (CAD)** is an efficient design method. The basis for CAD is the construction of a graphical 3-D model on the computer. This model can then be tested by any of the available numerical methods. Design modifications can be accomplished on-screen and the modified 3-D models tested again. When the engineer is satisfied that the design will meet or exceed all of the design criteria, a 2-D drawing can be created using the 3-D model as its basis. From this drawing, a physical prototype can be created and tested by traditional means to ensure compliance with the design criteria. Then the drawing is usually handed over to the manufacturing division for mass production. Alternatively, when a **computer-aided manufacturing (CAM)** system is available, the part or parts are produced directly from the 3-D computer models.

5.02.02 Classification of Engineering Designers

Engineering design is a broad concept with many integrated stages, competing alternatives, and diverse requirements for success. As such, many design teams in industry have specialty engineers who are responsible for certain aspects of the design process. Most design projects have a team leader, or **chief designer**, who oversees the work of the individual team members. In the early stages of the process for mass-produced designs, **industrial designers** lend their creative skills to develop the product concept and style. For civil engineering projects, **architects** may be employed for their creative talents in **conceptual design**. Specialists in CAD, **CAD designers**, develop the computer geometry for the new design. **Design analysts**, specialists in FEA and other software tools, check the new design for stress and load distribution, fluid flow, heat transfer, and a host of other simulated mechanical properties. **Model builders** are engineers who make physical mock-ups of the design using modern rapid prototyping and CAM equipment. **Detail designers** complete the final design requirements by making engineering drawings and other forms of **design documentation**. Before you begin reading about the **design process**, you should consider the role that creativity plays in the process.

5.03 Creativity in Design

Creativity is an important feature of the design process. Often engineers get hung up on the "rules" and "constraints" of design and forget to think creatively. Historically, there is a strong link between engineering and art. One of the earliest recognized engineers is Leonardo da Vinci. In fact, some people say that da Vinci was an engineer who sometimes sold a painting so he could make a living. If you examine the sketches and drawings of da Vinci, you will see that he was interested in the development of products

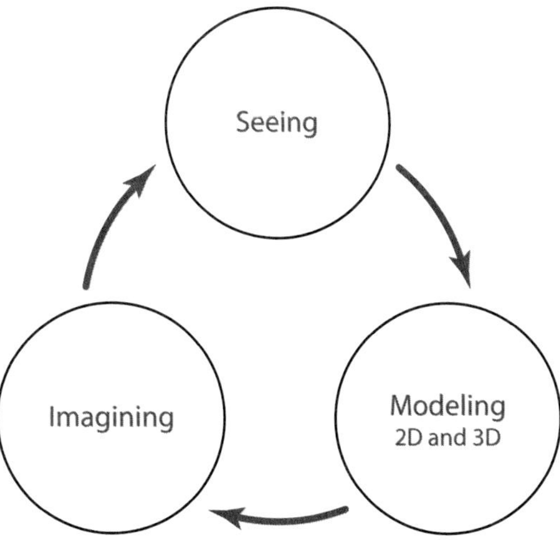

FIGURE 5.01. Visual thinking model.

to help improve people's lives, including some of the first recorded conceptual designs of airplanes and helicopters. Creativity is at the heart of innovation, and it can be enhanced with both individual and group activities. Some of the more common activities used to facilitate creative thinking are described subsequently. However, psychologists have found that the brain works most creatively when the hands are engaged in completing a mindless task. You probably have experienced occasions when you try to remember something but cannot, then find that the thought pops into your head when you begin another task. So to free your mind for creative thoughts, it may be best to take a break and wash the dishes, do some yard work, lift weights, or do laundry. Performing any of these mindless tasks will help free your mind for creative thoughts.

5.03.01 Visual Thinking

Visual thinking is the process of expanding one's creative ideas using visual cues and feedback. The visual cues can take the form of sketches or computer models; however, in the *initial* stages of design, sketching is often viewed as a necessary ingredient for creative thought. Visual thinking can be thought of as a circular feedback loop, as illustrated in Figure 5.01. The visual thinking process can start at any place in the feedback loop; but for the sake of simplicity, start with the step labeled "Imagining." You first imagine an idea for a new design or product and then sketch the idea in some graphical mode (2-D sketch, 3-D sketch, or computer image). Seeing the idea adds to your understanding of it, which can be extrapolated more deeply. You get a better mental image with a visual cue, which allows you to take the preliminary idea and refine it, sketch it again, etc. You can continue the process until you have a well-defined sketch or idea of the product for formal analysis and design.

5.03.02 Brainstorming

Brainstorming is the most common form of group ideation and concept generation and is a process used for generating as many ideas as possible. Brainstorming is typically done as early in the design process as possible (i.e., before you start solving the problem, before breaking the project into tasks, and before deciding who will do each task). Once individuals begin to focus on specific tasks, it may be too late to consider alternatives. Too often, teams will tackle a problem by taking the first idea presented and pursuing that idea without considering any alternatives. Teams also can use brainstorming to generate a list of the tasks that need to be done before the project can be completed. Five simple steps define a brainstorming session:

Step 1: Assemble your project team and make sure you allow ample time for the session. Diversity in the group will enhance the quality and breadth of the ideas generated. Select a group leader to run the session and select a group recorder to take notes.

Step 2: Define the idea of the design project to be discussed. Write down the idea and make sure everyone in the group understands it.

Step 3: Discuss the rules about brainstorming and make sure everyone in the group agrees to abide by them. If necessary, keep the rules on display as a reminder to members who may stray. Rules for brainstorming are intended to help the team generate more ideas. Comments about an idea, whether positive or negative, can stifle the brainstorming process. Although the following rules are simple, you may find them difficult to follow:

1. Everyone participates.
2. Every idea is recorded.
3. Judgment is suspended—there is no such thing as a bad idea.
4. No criticism is permitted.
5. No commentary is permitted.
6. No one dominates the process.

Step 4: Start the brainstorming session by asking everyone in the group to offer an idea. If possible, the recorder writes down all responses so everyone can see them. Alternatively, each member of the team can keep a list of his own ideas. These lists will prevent ideas from being lost and will help restart the process if there is a pause in the flow of ideas.

Step 5: At the end of the session, spend time going through all of the ideas. Combine, categorize, and eliminate the ideas to narrow the list. Once team members are sure that all possible ideas have been included, the team should discuss the advantages and disadvantages of each idea. After talking about the pros and cons, each member should rank the best three ideas on the list. The team should keep the ideas with the highest rank and decide as a group which approach to use.

Instead of conducting the entire brainstorming session by the process outlined previously, you may consider using the following steps for a brainstorming session:

1. Individually spend ten minutes writing down ideas for tackling the assigned project.
2. Combine the individual lists onto a flip-chart, blackboard, or whiteboard. Team members may ask questions to clarify an idea, but no other comments are allowed during this process.
3. Continue brainstorming as a group until all ideas have been exhausted.

5.03.03 Brainwriting (6-3-5 Method)

Brainwriting is an alternative to brainstorming. In brainwriting, each member of the group focuses on sketching his ideas rather than verbalizing them. With brainwriting, you typically start with a team of six people. Each person sketches three ideas on a sheet of paper, leaving ample room for additional graphics and annotations. The idea sketches are then passed around the table so fellow members can add their own comments and ideas, as shown in Figure 5.02. Usually the idea sketches are passed around the group five times. The expectation is that by the fifth time around, a favorable design idea will have emerged. Brainwriting also is called the *6-3-5 method* (six people, three ideas each, five times around the table).

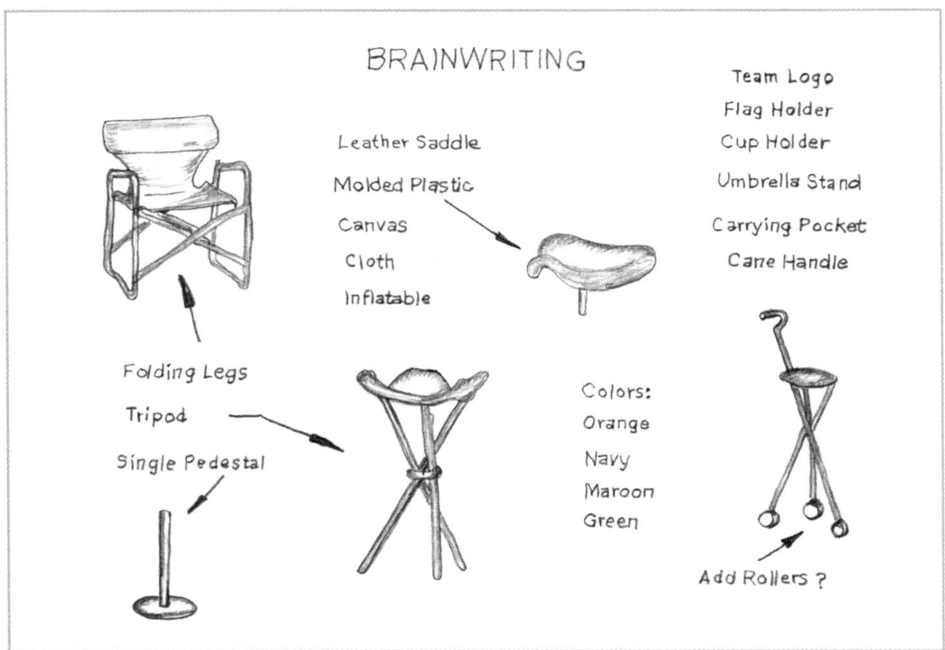

FIGURE 5.02. Brainwriting.

5.03.04 Morphological Charts

Morphology refers to the study of form and structure. A **morphological chart** can be used to generate ideas about a new design concept. The chart has a leading column that lists the various desirable functions of the proposed design. Along each row, various options for each function are listed, as shown in Figure 5.03. You can use brainstorming techniques to list as many options as possible for each function. The group then reviews and decides on a priority pathway through the options to address each desired function.

5.03.05 Concept Mapping

Concept mapping is a technique used to network various ideas together, as shown in Figure 5.04. During idea generation, the main design concept is placed in the center of the map with the various options linked outward in a brainstorming-like session. Each option then serves as a node for other choices. In that manner, the team can explore a big picture of all ideas and see a strong visual image of the connectivity of the different ideas.

FIGURE 5.03. A morphological chart.

MORPHOLOGICAL CHART					
Function	**Options**				
Seat Style	Saddle	Molded Dish	Strap	Inflatable	
Seat Materials	Leather	Plastic	Canvas	Rubber	Cloth
Number of Legs	One	Three	Four		
Leg Material	Wood	Aluminum	Plastic	Wrought Iron	
Leg Assembly	Pin	Hinge	Force Fit	Folding	
Accessories	Cup Holder	Beverage Cooler	Umbrella Holder	Flag Post	
Aesthetic Offerings	Team Logo	Choice of Colors			
Carrying Style	Carrying Handle	Handle Like Cane	Strap On Back	Roller Wheels	

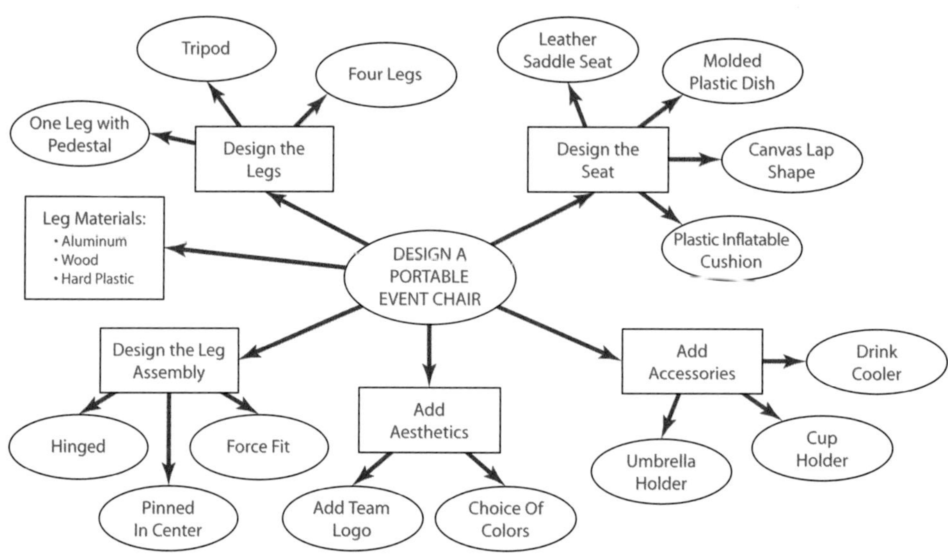

FIGURE 5.04. A concept map.

5.04 The Engineering Design Process

Design is a multistep process. However, there is considerable disagreement on exactly how many steps are involved in the process. Figure 5.05 shows one example of the sequence of steps in the design process. The design team often starts with stage 1 (Identify) and continues to stage 7 (Produce) and then begins again. Often the process does not proceed sequentially from stage 1 to stage 7; the design team might discover a serious problem in stage 4 (Analyze) and then return to stage 2 (Conceptualize) before moving on to stage 5 (Prototype).

Textbooks and writings on the design process include many different versions of and names for the stages in the design process. The stages presented previously are just one way to look at the design process; therefore, you should not think of them as the definitive word on the subject. However, the stages described in the remainder of this chapter are related to the graphical tools you will study in this textbook. For this reason, they have been adopted here. Knowing the number of stages and the labels for each stage is not nearly as important as understanding the overall process.

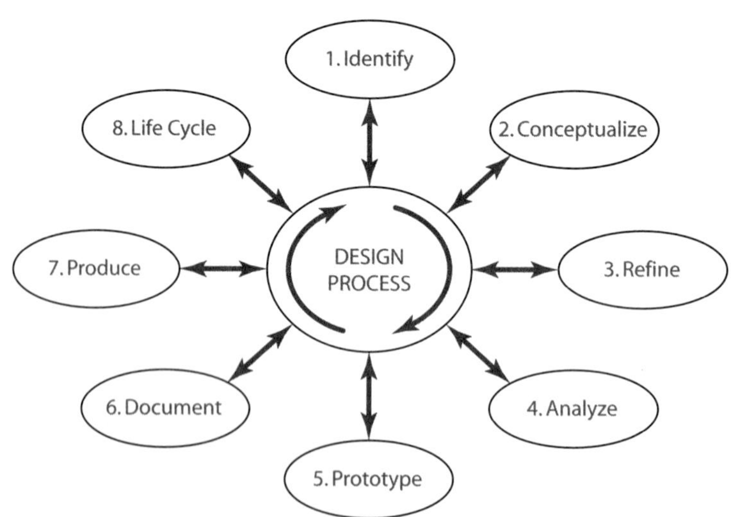

FIGURE 5.05. An engineering design process.

5.04.01 Stage 1: Problem Identification

Good design practice starts with a clearly defined need for a new product or system. Alternatively, a revised, improved design for an existing product may be required. A market survey may demonstrate that the new product or system is useful, has market appeal, is producible with today's technology, and will make a profit for the company supporting the design effort. Sometimes a new design idea is simply an alternate solution to a problem answered by an existing competing product. Indeed, many of today's highly successful products are the evolutionary result of free market enterprise. In the civil engineering world, a design project is typically the result of a client requesting a specific structure or system. For example, a governmental agency such as a county may request that a civil engineering firm design a new water distribution system to serve the needs of the county's residents. In this case, the client may have already defined the problem.

In the **problem identification** stage, the design engineer must address questions and answers from the customer's/client's perspective and from the engineer's perspective. For example, Table 5.01 shows two different perspectives for a new urban bicycle design.

When a product is designed, one of the design considerations is how long will it be used before it is no longer effective, which is called the **life cycle** of the product. Some products are designed for replacement on a regular basis. Thus, environmental considerations and disposal of a product must be considered throughout the design process. In the design process called **green engineering**, environmental concerns are considered throughout the process, not just at the end, because an engineer's choices in the problem definition stage often influence the overall environmental impact and life cycle of the product. Once the functional requirements have been identified, the design team can start the design process by generating some concepts.

5.04.02 Stage 2: Concept Generation

Concept generation is the most creative phase of the design process. You learned about some methods for creative thinking and concept generation earlier in this chapter. Typically, concept generation starts with brainstorming, brainwriting, or a similar team meeting where ideas are tossed around and discussed. Criticism is usually limited

TABLE 5.01. Functional Requirements for a New Urban Bicycle.

CUSTOMER'S PERSPECTIVE	ENGINEER'S PERSPECTIVE
MODERATE STREET SPEED	SUSTAINABLE SPEED OF XX MPH MAXIMUM SPEED OF YY MPH
COMPACT SIZE	DIMENSIONS NOT TO EXCEED A X B X C
SAFE	STRUCTURAL STRENGTH CONTROLLABILITY BRAKING CAPABILITY TIRE PUNCTURE RESISTANCE
COMFORTABLE	ERGONOMICS OF SEAT EFFICIENCY OF POWER TRAIN POSITION(S) OF HANDLEBAR
ATTRACTIVE	CHOICES OF PAINT COLOR LIGHTS AND REFLECTORS
AFFORDABLE	SELECTION OF MATERIALS MANUFACTURING PROCESS NUMBER OF PARTS SALES VOLUME

FIGURE 5.06. Concept sketches.

in the concept generation stage, since maximizing the number of good ideas is desirable. At the end of the concept generation stage, the team should have selected a few main ideas that it will focus on for future refinement and analysis. Sketching is an integral way to develop concepts for a new design. Figure 5.06 shows some examples of concept sketches.

5.04.03 Stage 3: Concept Selection and Refinement

Once a few quality concepts have been identified, the design team must converge on one or two final concepts to further explore in the design process. A common technique for selecting the final concept(s) is to use a **weighted decision table**, as shown in Figure 5.07. With a weighted decision table, all of the common attributes and desirable features of each concept are list in the first column. A weighting factor for each feature/attribute is then established (e.g., using a scale of 0 to 10). The various design options are listed in subsequent columns in a parallel fashion to the listed features/attributes. The team then conscientiously scores each option for every feature/attribute, each time applying the weighting factor to the score, as illustrated in Figure 5.07. Adding all of the scores for individual attributes yields a final "bottom-line" number that can be used to select the highest-ranked option.

Sometimes the initial concepts may need to be refined before a final decision can be made. Refinement will likely include the development of 3-D computer models for defining geometry not accurately expressed in the concept sketches. For example, as shown in Figure 5.08, different computer models of a new product can assist the members of the design team in visualizing the specific model that has the marketing appeal they are seeking and in making the final decision.

			DESIGN 1		DESIGN 2		DESIGN 3		DESIGN 4	
FEATURE ATTRIBUTE		WEIGHT 0–10	**OPEN BOTTOM** SCORE -- TOTAL 0–10	**S * W**	**THUMB GRIP** SCORE -- TOTAL 0–10	**S * W**	**DOUBLE LOOP** SCORE -- TOTAL 0–10	**S * W**	**SINGLE LOOP W/FINGER SUPPORT** SCORE -- TOTAL 0–10	**S * W**
AESTHETICS										
	Color	5	3	**15**	3	**15**	3	**15**	3	**15**
	Form	8	7	**56**	7	**56**	7	**56**	7	**56**
ERGONOMICS										
	Grip ability	8	7	**56**	6	**48**	2	**16**	9	**72**
	Drinking Ease	6	3	**18**	5	**30**	4	**24**	8	**48**
FUNCTIONALITY										
	Adapts to Hand	8	5	**40**	7	**56**	2	**16**	8	**56**
STABILITY										
	Base size	6	7	**42**	9	**54**	3	**18**	9	**54**
	Height	8	7	**56**	8	**64**	3	**24**	8	**64**
MANUFACTURABILITY										
	Injection Molding	5	3	**15**	3	**15**	3	**15**	3	**15**
	Slip Molding	5	3	**15**	3	**15**	3	**15**	3	**15**
			Weighted Total	**313**	**Weighted Total**	**353**	**Weighted Total**	**199**	**Weighted Total**	**395**

FIGURE 5.07. A weighted decision table for concept selection.

FIGURE 5.08. Computer models for concept selection.

FIGURE 5.09. Object mass properties.

MASS PROPERTIES REPORT (Concept 4)

Mass Properties of ergonomic cup (Material – ABS Plastic)

Output coordinate System: -- default --

Density = 0.037 pounds per cubic inch

Mass = 0.948 pounds

Volume = 25.722 cubic inches

Surface area = 134.693 inches^2

Center of mass: (inches)
X = 0.071
Y = 1.648
Z = 0.000

Principal axes of inertia and principal moments of inertia: (pounds * square inches)
Taken at the center of mass.

Ix = (0.033, 0.999, 0.000)	Px = 2.401
Iy = (-0.999, 0.033, 0.000)	Py = 3.770
Iz = (0.000, 0.000, 1.000)	Pz = 3.921

Moments of inertia: (pounds * square inches)
Taken at the center of mass and aligned with the output coordinate system.

Lxx = 3.768	Lxy = 0.046	Lxz = -0.000
Lyx = 0.046	Lyy = 2.402	Lyz = -0.000
Lzx = -0.000	Lzy = -0.000	Lzz = 3.921

Moments of inertia: (pounds * square inches)
Taken at the output coordinate system.

Ixx = 6.341	Ixy = 0.157	Ixz = -0.000
Iyx = 0.157	Iyy = 2.407	Iyz = -0.000
Izx = -0.000	Izy = -0.000	Izz = 6.498

5.04.04 Stage 4: Design Evaluation and Analysis

In this stage, the selected concept is further analyzed by any number of numerical methods. Before the advent of CAD and analysis tools, this stage of the design process involved building and testing physical models. Now the building and testing can be done on the computer, saving companies a great deal of time and money. The tests are conducted to determine mechanical properties of objects or systems and their performance during simulated conditions.

One of the simpler types of analysis that can be performed with a computer model is the computation of the object's mechanical properties, such as mass, center of gravity, and moment of inertia. All of these properties may not be meaningful to you right now; however, they are key quantities used in performing most types of static and dynamic analyses. A **mass properties analysis** report, shown in Figure 5.09, is a useful document for evaluating and presenting the static mechanical conditions of the design.

Further analysis of the design might include an FEA of stress contours and deformation. Heat transfer and aerodynamic flow also can be simulated using modern computational software (Figure 5.10). These numerical methods will be discussed in more detail in a later chapter and are themselves topics of entire texts.

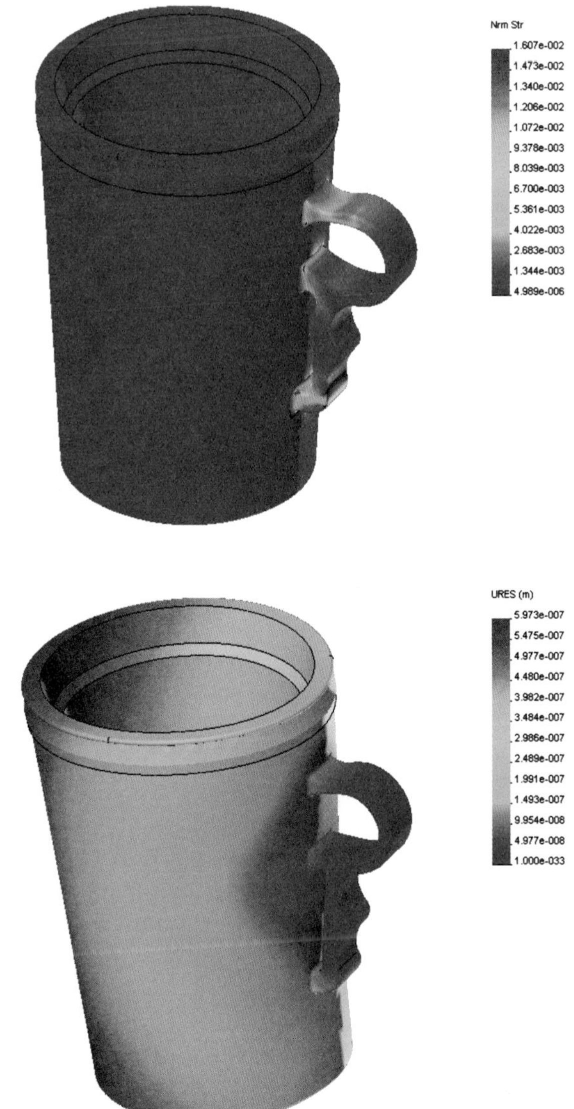

FIGURE 5.10. A numerical analysis model.

5.04.05 Stage 5: Physical Prototyping

Most designers and clients would like to see a physical model of the design—they want to look at it, hold it in their hand, and show it to other interested parties. Several different types of physical models can be developed during this stage of the design process. Engineers can have the shop people build a *scale model*, an actual *true-size model*, or just a simple *mock-up concept model* that shows the general physical appearance of the design.

In recent years, modern technology has accelerated the production of prototype models in the design process. CAM systems can take data from a 3-D solid model and cut the pattern using computer numerical control (CNC) machines. Figure 5.11 shows a part that was created through CNC machining. CNC machining will be covered in more detail in a later chapter of this text.

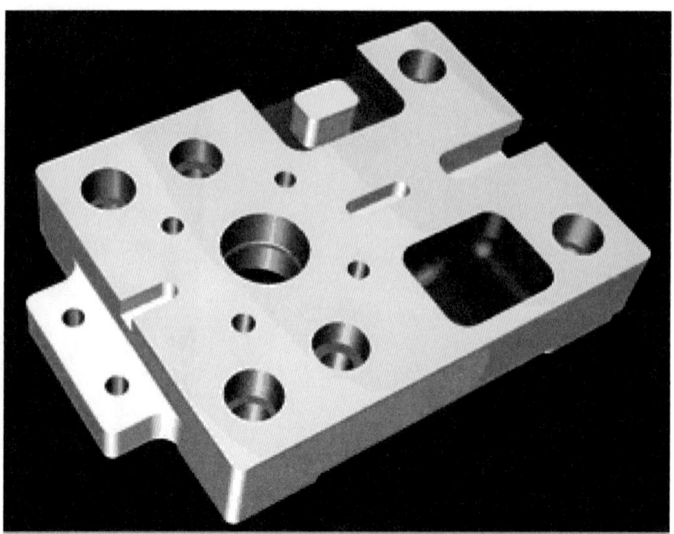

FIGURE 5.11. A part created through CNC machining.

Today rapid prototyping systems can perform some of the same functions of traditional machining tools, except that they require far less time (hence, the term *rapid*) and fewer resources than traditional methods. Some of the modern rapid prototyping methods include stereolithography (SLA), Selective Laser Sintering (SLS), Laminated Object Manufacturing (LOM), Fused Deposition Modeling (FDM), Solid Ground Curing (SGC), and inkjet printing techniques. Most recently, 3-D printers have become affordable prototyping alternatives for the office environment. Reasonable 3-D models can be printed using 3-D printers, as shown in Figure 5.12.

5.04.06 Stage 6: Design Documentation

There are many forms of design documentation, but the most common form is a finished detailed drawing, as shown in Figure 5.13. A detailed drawing shows the information needed to manufacture the final part. A good portion of the rest of this text discusses detailed engineering drawings. You will learn how to create drawings and how

FIGURE 5.12. Models created with a 3-D printer.

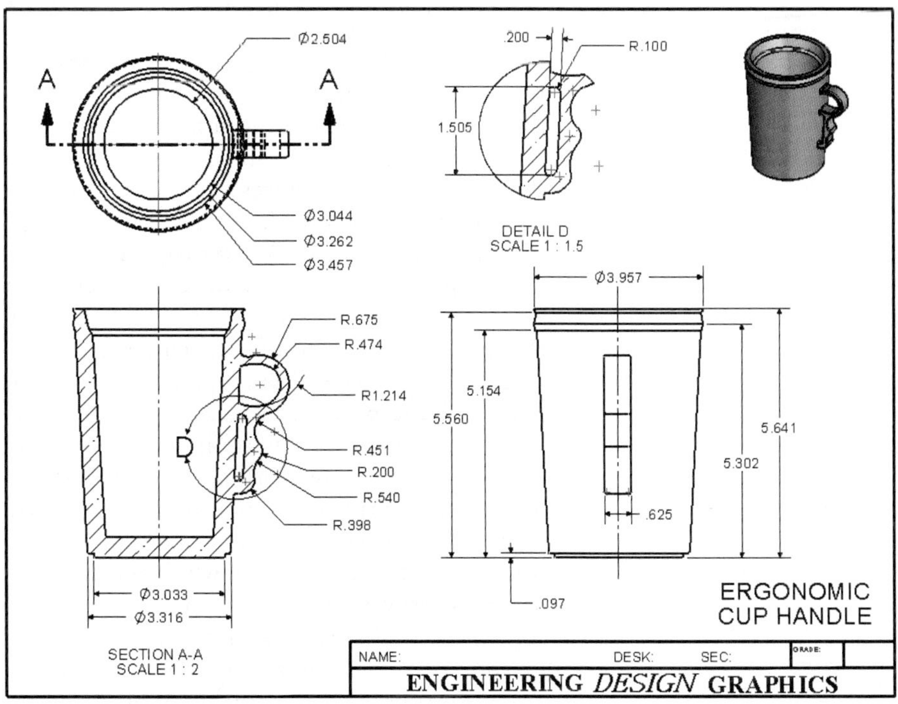

FIGURE 5.13. A detailed design drawing.

to interpret them correctly. You will learn about dimensioning and tolerancing for annotation of the drawing. You will learn about conventions developed over the years that provide everyone with the same understanding of what is on the drawing. You also will learn how drawings are created from 3-D computer models of parts and systems.

5.04.07 Stage 7: Production

Once the design documentation is complete, it is time to begin the production stage. For a civil engineering design, production is called the construction stage; for mechanical engineering, production is the manufacturing stage. Many engineers claim that the design process ends with the documentation stage. However, the way the product is designed may impact its production and distribution processes later on. For mechanical engineering projects in this stage, the goal is usually (but not always) to produce the product in large quantities, to meet performance standards, and to keep manufacturing costs low. Many different methods for manufacturing parts have been developed and are widely used, including machining, casting, rolling, and sheet metal cutting processes. These methods will be covered in more detail in a later chapter.

For civil engineering projects, because no prototypes have been built, design modifications are common in the production stage. Contractors may find, for example, that ductwork does not fit within the space provided on the drawings, or they may find that piping needs to be rerouted around an obstruction. For this reason, in civil engineering projects, it is important to continue to document the design by making notations on the drawings where changes are made. These drawings are called **as-built drawings**, and they reflect the way the project was actually constructed—not just the way it was designed. As-built drawings are an important part of the design process because if piping is rerouted through a building, for example, someone will need to know exactly where the plumbing is in case leaks or other problems arise.

5.05 The Concurrent Engineering Design Process

The new paradigm of **concurrent engineering** is sometimes referred to as "design for manufacturability." In traditional engineering, the part being designed progresses through each stage, moving from one team to the next. At each new stage, the team takes the design from the previous team and applies its own expertise. The first time the manufacturing engineer sees the part for production is when the design and analysis teams finish their work. With concurrent engineering, designers, analysts, and manufacturing engineers work together from the initial stages of the design process. In this way, each person can apply his own expertise to the problem at hand *from the start.* Thus, early in the design process, the manufacturing engineer might say, "If you made this minor modification to the part geometry, we would save $100,000 in retooling costs." The design change in question could be easily implemented during the initial phases of the process; whereas, without concurrent engineering, the change (and related cost savings) might be impossible in the final stage.

Modern computer workstations have enabled concurrent engineering to become a reality in the workplace. With local area networks, wide area networks, and the Internet, data can be moved from one desktop to another almost effortlessly. Members of the concurrent engineering team who work in different countries can share design ideas nearly as easily as engineers who work in the same building. Using the principles of concurrent engineering, manufacturers can save thousands, even millions, of dollars. In addition, computer-aided concurrent engineering design is more efficient and results are often of higher quality compared with designs produced in the past.

5.06 Patents

Patents are a way to protect the intellectual property of a new design. In the United States, a patent generally gives the inventor sole claim to intellectual property rights for twenty years. Application for a patent is made to the director of the United States Patent and Trademark Office and typically includes three components:

1. A written document made up of a specification (description and claims) and an oath or a declaration.
2. A patent drawing (Figure 5.14) in those cases in which a drawing is necessary.
3. Filing, search, and examination fees.

Frequently, inventors hire a lawyer to do the legal work and to make sure the idea has not already been patented. In some cases, the individual inventor conducts a patent search using a for-hire Internet site.

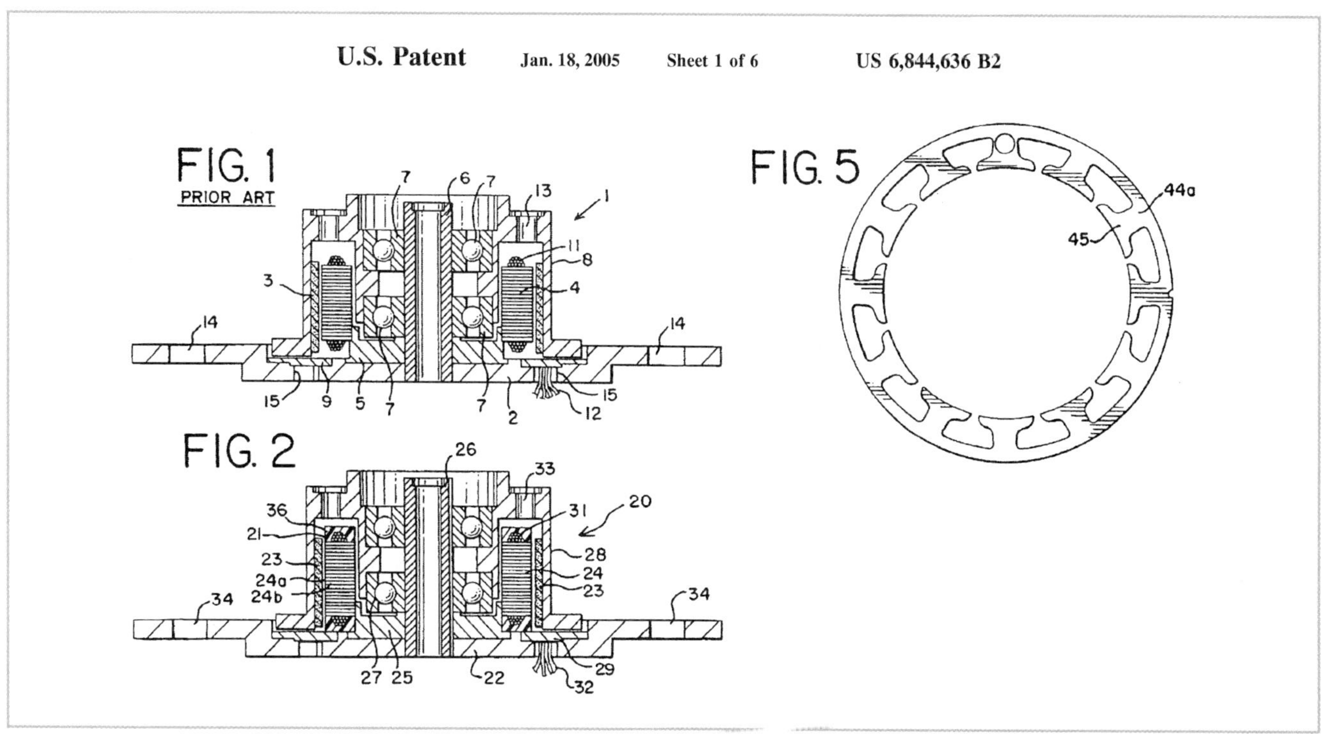

FIGURE 5.14. A patent drawing.

5.07 Chapter Summary

In this chapter, you learned that design is an iterative process and that the process has several stages; however, there is no general agreement about the exact number of stages. You learned that engineers must weigh competing factors such as cost, function, and environmental impact when making design decisions. You learned about design in the information age and ways computers are used throughout the process, greatly reducing costs. You learned about the importance of creativity in engineering design and about several techniques you can use, either as an individual or as part of a team, to foster creative thinking. Finally, you learned about concurrent engineering, which is enabled by computer technologies and can be used to reduce product costs and improve product quality.

casestudy
Integrated Project: Conception of the Hoyt AeroTec Target Bow

Hoyt USA was founded in 1931 by sportsman and bow maker Earl Hoyt. The company is located in Salt Lake City, Utah, where it has both engineering and fabrication facilities. Hoyt USA has a long-standing reputation as a high-quality maker of bows for sports, recreation, and competition. In 1972, the company revolutionized competition archery when it introduced its first metal-handled collapsible recurve bow at the 1972 Olympic games. Given only to the U.S. team, the metal-handle design offered significant advantages over other bows, which were mostly made of wood at that time. A metal handle was relatively immune to the effects of changing temperature, humidity, and time, which affected the geometry, stiffness, and vibration properties of the bow. These variations made it difficult to use a wooden bow to land arrows shot after shot in the same place on a target. In addition, lighter arrows produce higher stresses in a wooden bow, sometimes causing the bow to break. By contrast, the strength of the handle's metal enabled an archer to use lighter arrows, which reached more distant targets in shorter times. Soon after Hoyt USA introduced the metal-handled bow, other bow manufacturers followed suit.

In the early 1990s, Hoyt USA wanted to improve the design of its metal bow handle to improve its share of the target archery equipment market. This market was very competitive, with a typical recurve bow lasting only three or four years before it needed to be replaced. In the development of a new design, Hoyt USA had to consider several things. For product performance, the main considerations were strength, weight, and vibration. A new product had to be stronger than previous ones. Super-light carbon arrows, light composite bow limbs, and synthetic strings produced increasing levels of stress in bows, to the point where even metal handles were breaking. But additional strength could not be gained at the expense of weight. Many archers already complained about the excessive weight of metal-handled bows. Any new product also needed to be less flexible and have less vibration than existing bows. Target archers considered excessive flexibility, noise, and vibration to be undesirable characteristics.

The new product needed to be developed with analysis and production in mind. Detailed stress analysis was necessary to ensure that there would be no breakage problems, as had been the case with other products. Also, the new product had to be designed so it could be easily produced in state-of-the-art fabrication facilities consisting mostly of computer-controlled four-axis milling machines, which Hoyt USA was expanding at the time.

A conventional metal-handled recurve target bow.

The new concept that Hoyt USA developed was a structural support member located behind the grip of the bow handle. The support member was designed with an outward bend so it would not touch the arm of the user. Touching the bow at any location except the grip was a violation of the rules in target archery. The new design allowed the forces in the bow handle to be more widely distributed, resulting in reduced stresses without a significant increase in weight.

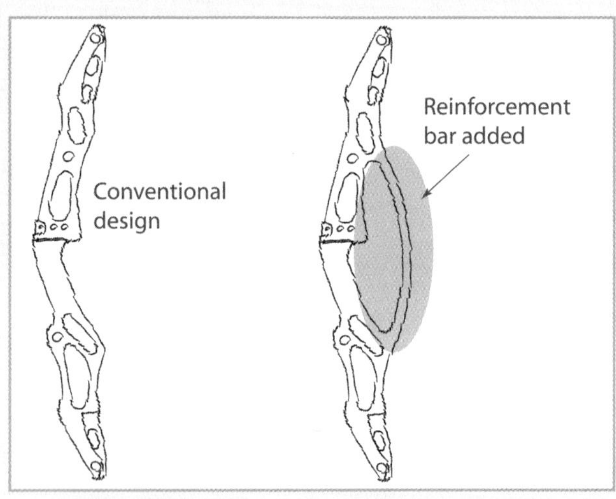

Concept sketches showing the addition of a structural reinforcement bar to a conventional bow handle to reduce its flexibility and vibration.

Conceptual sketches for the new design were developed first. The sketches enabled Hoyt USA engineers to communicate ideas among themselves, as well as to engineers outside their group, managers, production specialists, and potential customers. About twenty design variations were examined on the drawing board, after which three or four were selected for further development. A solid model was built using a computer, because the model could be used for stress analysis with finite element methods and because the model could be exported directly to the fabrication machines on Hoyt USA's production floor.

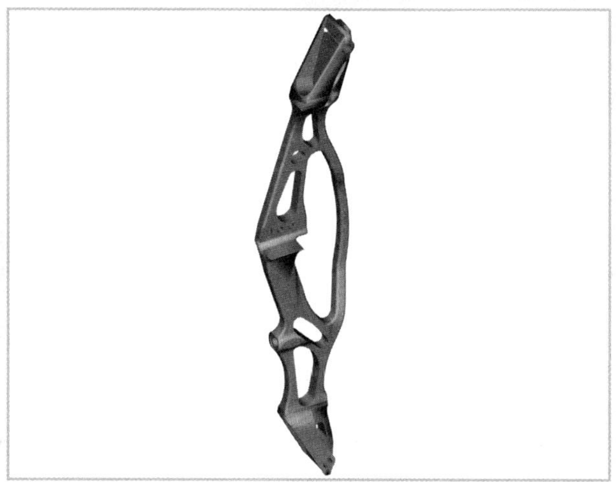

A solid model of the new design used for analysis and fabrication.

After final selection and refinement of the design as a result of the stress analyses and field testing, Hoyt USA began full production of the new product less than one year after the concept was first discussed. To protect the innovative design from being copied by competitors, Hoyt USA applied for and was granted a U.S. patent. Patents also were secured in foreign countries. The design concept was trademarked TEC for "Total Engineering Concept." Because of the relatively radical appearance of the product (and the rather conservative nature of archers), at first, the product was slowly and cautiously adopted by the market. But after a few years (and some outstanding performances at the Olympics), TEC products by Hoyt USA were eagerly embraced by the market as a superior technology.

The production version of the structurally reinforced handle.

DISCUSSION QUESTIONS/ACTIVITIES

1. Explain the design process that the developers at Hoyt used to engineer the TEC bow.
2. Create a weighted decision table based on the questions and answers that might have been generated during the design process of the TEC bow.
3. In what ways were concurrent engineering techniques used throughout the development of the TEC bow?

5.08 glossary of key terms

architects: Professionals who complete conceptual designs for civil engineering projects.

as-built drawings: The marked-up drawings from a civil engineering project that show any modifications implemented in the field during construction.

brainstorming: The process of group creative thinking used to generate as many ideas as possible for consideration.

brainwriting: A process of group creative thinking where sketching is the primary mode of communication between team members.

CAD designers: Designers who create 3-D computer models for analysis and detailing.

chief designer: The individual who oversees other members of the design team and manages the overall project.

computer-aided design (CAD): The process by which computers are used to model and analyze designed products.

computer-aided manufacturing (CAM): The process by which parts are manufactured directly from 3-D computer models.

concept mapping: The creative process by which the central idea is placed in the middle of a page and related concepts radiate out from that central idea.

conceptual design: The initial idea for a design before analysis has been performed.

concurrent engineering: The process by which designers, analysts, and manufacturers work together from the start to design a product.

design analysts: Individuals who analyze design concepts by computer methods to determine their structural, thermal, or vibration characteristics.

design documentation: The set of drawings and specifications that illustrate and thoroughly describe a designed product.

design process: The multistep, iterative process by which products are conceived and produced.

detail designers: The individuals who create engineering drawings, complete with annotation, from 3-D computer models or from engineering sketches.

engineering design: The process by which many competing factors of a product are weighed to select the best alternative in terms of cost, sustainability, and function.

finite element analysis: A numerical method used to analyze a product in terms of it structural, thermal, and vibrational performance.

green engineering: The process by which environmental and life cycle considerations are examined from the outset in design.

industrial designers: The individuals who use their creative abilities to develop conceptual designs of potential products.

life cycle: The amount of time a product will be used before it is no longer effective.

mass properties analysis: A computer-generated document that gives the mechanical properties of a 3-D solid model.

model builders: Engineers who make physical mock-ups of designs using modern rapid prototyping and CAM equipment.

morphological chart: A chart used to generate ideas about the desirable qualities of a product and all of the possible options for achieving them.

patents: A formal way to protect intellectual property rights for a new product.

problem identification: The first stage in the design process where the need for a product or a product modification is clearly defined.

prototype: The initial creation of a product for testing and analysis before it is mass-produced.

sustainable design: A paradigm for making design decisions based on environmental considerations and life cycle analysis.

visual thinking: A method for creative thinking, usually through sketching, where visual feedback assists in the development of creative ideas.

weighted decision table: A matrix used to weigh design options to determine the best possible design characteristics.

5.09 questions for review

1. What are the main stages in the design process?
2. Why is creativity important in the engineering design process?
3. How does engineering design differ from the type of design artists perform?
4. What is meant by concurrent engineering?
5. How is a computer used in the modern-day design process?
6. What are some of the differences in design for a civil engineering project versus a mechanical engineering project?

5.10 design projects

The following sections will outline specifications for design projects. These projects were tested with students at the University of California at Berkeley over the years and are suitable for use in a first- or second-year design course at a university. The projects are designed for completion by a team of four or five students.

5.10.01 STANDARD PROJECT MATERIALS

Use the following standard list of materials, in addition to any special items listed in the specific design rules for the project, to construct the device assigned by your instructor. No other materials are permitted.

- Paper, 30# (maximum): 2 square meters maximum; 2 layers maximum
- Poster board, single-ply, medium weight: 1 square meter maximum
- Foam core modeling board, 3/16″ nominal thickness: 1 square meter maximum
- Twine, 60 lbs. (maximum) labeled breaking strength: 3 meter length maximum
- Wood dowel, 1/4″ nominal diameter: 1 meter length maximum
- Mailing tube, 2″ nominal diameter, medium-weight cardboard: 1 meter length maximum; no endcaps
- Rubber bands (sample to be supplied), #62 or #64: 10 maximum
- Elmer's Glue-All glue: 30 cc maximum
- Hot melt adhesive (polyolefin): 30 cc maximum
- Scotch brand transparent cellophane tape: 1 meter maximum

All of the materials can be purchased at local art supply or convenience stores. Equivalent material may be substituted only with the instructor's permission. Paints, markers, flags, and other decorative items not on the list may be used as long as they are purely decorative; for example, paint cannot be used as weight or ballast.

5.10.02 STANDARD PROJECT DELIVERABLES

The following list provides the standard deliverables for your project. Your instructor may assign additional deliverables and will let you know the due date for each deliverable. When you are organizing your team effort, you can use these deliverables as the milestones to produce a Gantt chart or critical path diagram to help you stay on track and complete the project on time.

Required Drawing Deliverables:

1. Conceptual sketches—alternative and final designs
2. Outline assembly drawings
    - Multiview of assembled project
    - Isometric or pictorial view of assembled project
    - Cutaway views as required for clarity
    - Sectional views as required for clarity
    - Overall dimensions only

   Balloons to identify subassembly or part numbers and names
3. Detail drawings
    - One multiview drawing per part (isometric or pictorial)
    - All dimensions, datums, and tolerances
    - Quantity
    - Material
    - Sectional views as required for clarity
    - Isometric views as required for clarity

5.10 design projects (continued)

4. Exploded assembly drawings
   - Blow-apart pictorial view of all assemblies
   - Blow-apart pictorial view of all subassemblies
   - Balloons to identify part numbers and part and subassembly names
   - Subassemblies as required (highly recommended)
5. Bill of materials
   - List of all parts by PN, showing name, quantity, and material
   - List of all materials needed for assembly (e.g., tape and glue)

Use millimeter dimensions and proper title blocks and borders for your engineering drawings. It is recommended that all drawings be cross-checked by different people. Alternate the functions of drafter, designer, and checker. The team leader must give final approval.

Final Demonstration and Oral Presentation:

Each team is expected to give an informative final presentation of its design, as well as a demonstration during the distance contest. Use descriptive graphics slides to complement the presentation. Keep the presentation short and direct.

Written Report:

In your written final report detailing the project results, describe the alternatives your team considered, describe which ones you selected, and explain why you selected them. Use drawings to illustrate key points in the design process. Include the results of your product testing and include a section on what you would have done differently.

DESIGN PROJECT #1: ESCAPE!

NASA is once again looking for a few good engineers. This time the agency is seeking conceptual ideas for an escape device that would allow launchpad crews and astronauts to leave the area quickly in case of a potentially explosive, toxic, or otherwise harmful situation. The device is to be remotely launched and should be designed to place personnel as far away from the launch point as possible. The device must land safely, leaving the personnel unharmed. However, for this project, you will demonstrate the concept of your device using a hard-boiled egg.

The Mission:

Your mission is to design and build a device that will launch a hard-boiled egg (USDA Grade A Large, which your team must provide) into the air and have it land as far from the launch point as possible. The device must land the egg totally intact (no cracks in the shell). The design, for example, may be composed of a mechanism for launching the egg and a device attached to the egg for lowering it slowly (like a parachute). You may surround the egg with a protective covering. However, the covering cannot penetrate the shell or be bonded to it. The function of your device will be graded on the distance from the point of launch relative to that of the other teams in the class. A stiff distance penalty will be assessed if the egg is damaged.

Design Rules:

The device must be constructed out of the standard project materials listed previously, in addition to *only* the following item: one hard-boiled egg (USDA Grade A Large).

Equivalent material may be substituted only with the permission of the instructor. Any design deemed by the instructor as unsafe will not be acceptable (e.g., no sharp flying objects, no explosive devices, and no raw or rotten eggs).

More Contest Rules:

- The device, including the launcher, must initially fit within a 1.0 m x 1.0 m x 1.0 m volume without external support, except for the triggering means.
- Once the egg has been launched, the device may expand to any size.
- The device must be freestanding and may not be taped, glued, or in any other way affixed to the ground.
- The device must be remotely triggered (e.g., by a string or rod). Team members (all parts of the body) must remain a minimum of 1 meter away from the device when the egg is launched.
- The device must be set up within 3 minutes; otherwise, a 3-meter distance penalty on the total distance will be assessed for every 10 seconds of overtime.
- Human power may be used to trigger the device, but not to impart motion to the egg. However, human power may be used to store energy (e.g., into the rubber bands) for any use.
- Distance is measured from the point the egg is completely clear of the launch structure (e.g., completely airborne with no attachment to the launch structure) to the point stops where any part of the device containing the egg touches a solid object connected to the ground.
- The egg must attain a distance of at least 3 meters from the launch position.
- The egg must survive the landing without cracking or sustaining any other visible signs of damage.
- Surviving eggs will be peeled and eaten by the team leader to ensure that they have not been altered in any way. If the team leader does not survive, the entire team will fail the project.
- The egg must be removed from its protective covering within 30 seconds.

5.10 design projects (continued)

- If the egg does not survive, the total distance will be recorded as zero.
- If the egg is damaged, it must be replaced in time for the next launch.
- The maximum distance from three trials will be recorded. A misfire will count as one trial.
- Spare parts are recommended and do not count in the materials inventory of the final assembly.

DESIGN PROJECT #2: FAST FOOD!

Food Service in the dormitories is experimenting with a new method of feeding students in the morning. Instead of going to the dining commons for breakfast, students will open a window before a prescribed time. Food Service will then deliver breakfast by launching it into the dorm room. That way, students do not need to wake up early just to get breakfast and can sleep late if they so choose. All students need to do is leave the window open in the evening to ensure that breakfast will be delivered in the morning. The dining service has asked your team to demonstrate a conceptual prototype of a device that will perform this function.

The Mission:

Your mission is to design and build a device that will launch a bagel with cream cheese into the air and have it land as far from the launch point as possible. The device must land the bagel totally intact and unsoiled. The design, for example, may be composed of a mechanism for launching the bagel and a box around the bagel to help protect it. However, the covering cannot pierce the bagel and cannot be bonded to it. The function of your device will be partially graded on the distance from the point of launch to the point of landing relative to that of the other teams in the class. A stiff distance penalty will be assessed if the bagel is damaged or soiled: you will have to eat it.

Design Rules:

The device must be constructed out of the standard project materials listed previously, in addition to *only* the following item: one plain bagel sliced horizontally and smeared with plain soft cream cheese to 1/4" average thickness. The bagel cannot be more than 24-hours old at the time of launch.

All of the materials (except the bagel with cream cheese) can be purchased at local art supply stores. Any design deemed by the instructor as unsafe will not be acceptable (e.g., no sharp flying objects, no explosive devices, and no spoiled food).

More Contest Rules:

- Once the bagel has been launched, the device may expand to any size.
- The device, including the launcher, must initially fit within a 1.0 m x 1.0 m x 1.0 m volume without external support, except for the triggering means.
- The device must be freestanding and may not be taped, glued, or in any other way affixed to the ground.
- The device must be remotely triggered (e.g., by a string or rod). Team members (all parts of the body) must remain a minimum of 1 meter away from the device when the bagel is launched.
- The device must be set up within 3 minutes; otherwise, a 3-meter distance penalty on the total distance will be assessed for every 10 seconds of overtime.
- Human power may be used to trigger the device, but not to impart motion to the bagel. However, human power may be used to store energy (e.g., into the rubber bands) for any use.
- Distance is measured from the point the bagel is completely clear of the launch structure (e.g., completely airborne with no attachment to the launch structure) to the point stops where any part of the device containing the bagel touches a solid object connected to the ground.
- The bagel must survive the landing without cracking, opening, soiling, or sustaining any other visible signs of damage.
- The surviving bagel with the longest distance will be eaten by the team leader to ensure that it has not been altered in any way. If the team leader does not survive, the entire team will fail the project.
- The bagel must be removed from its protective covering within 15 seconds.
- If the bagel does not survive, the total distance will be recorded as zero.
- If the bagel is damaged, it must be replaced in time for the next launch.
- The average of the three longest distances from as many launches as can be accomplished within a single 60-second period will be recorded. Thus, you should have multiple bagels and containers ready to launch. A misfire will be considered as zero distance.
- Spare parts are recommended and do not count in the materials inventory of the final assembly.

DESIGN PROJECT #3: REWARD!

Several problems are on the horizon for engineering graphics classes of the future. First, the classes are getting larger, requiring that lectures be held in larger rooms. This trend makes it difficult for the instructor to toss candy rewards to specific students in the class, because the instructor's throwing range is limited. Second, the course CD is apparently a flop in the market and customers from

5.10 design projects (continued)

all over the country are returning their disks to the publisher. To solve both problems at the same time, someone recommended that candies be strapped to CDs and thrown together. The aerodynamic properties of the CD can be used to increase the range of the candy. This idea was immediately adopted, so here is your project.

The Mission:

Your mission is to design and build a device that will launch a Hershey's chocolate Nugget (with almonds) taped to a CD so it passes through an 8′ x 8′ target frame placed as far from the launch point as possible. The target frame will be placed such that the opening faces the launcher, with the bottom of the target frame on the ground. The launching field will be relatively flat. Each team will have 3 minutes to hit the target at least once. The target distance from the launcher will be specified by the team. The single longest distance at which the target is hit will be recorded for each team. If the target is not hit on any of the tries, the final recorded distance will be recorded as zero.

Design Rules:

The device must be constructed out of the standard project materials listed previously, in addition to *only* the following items:

- One genuine Hershey's chocolate Nugget (with almonds) at room temperature, still in the wrapper
- One standard 120 mm diameter optical CD

Any design deemed by the instructor as unsafe will not be acceptable (e.g., no sharp flying objects, no explosive devices, and no spoiled food).

More Contest Rules:

- The device, including the launcher, must initially fit within a 1.0 m x 1.0 m x 1.0 m volume without external support, except for the triggering means.
- The device must be freestanding and may not be taped, glued, or in any other way affixed to the ground.
- The device must be remotely triggered (e.g., by a string or rod). Team members (all parts of the body) must remain a minimum of 1 meter away from the device when the Nugget is launched.
- When the launcher is armed and ready to be triggered, it must be entirely self-supporting and stable (e.g., it does not require any external support from team members).
- The device must be set up within 3 minutes; otherwise, a 1-meter distance penalty on the total distance will be assessed for every 10 seconds of overtime.
- Human power may be used to trigger the device, but not to impart motion to the Nugget. However, human power may be used to store energy (e.g., into the rubber bands) for any use.
- Distance is measured from the point the CD is completely clear of the launch structure (e.g., completely airborne with no attachment to the launch structure) to the 8′ x 8′ target frame.
- The Nugget with the longest distance will be eaten by the team leader to ensure that it has not been altered in any way. If the team leader does not survive, the entire team will fail the project.
- The single longest distance at which you hit the target from as many launches as can be accomplished within a single 3-minute period will be recorded. Thus, you should have multiple Nuggets on CDs ready to launch. A misfire will be considered as zero distance.
- Spare parts are recommended and do not count in the materials inventory of the final assembly.

DESIGN PROJECT #4: DEPLOY IT!

NASA is once again looking for a few good engineers. This time the agency is seeking conceptual ideas for the deployment of structures such as antennas and solar panels in spacecraft. The device is to be self-deploying and should be designed to extend as far from the base point as possible while still remaining connected to the base point. It is possible that deployment will occur in various environments—from gravity-free space to high gravity on planets and moons. However, you will demonstrate the concept for your device in earth gravity.

The Mission:

Your mission is to design and build a device that will deploy a structure to reach as far from the origin point as possible. The device must remain physically connected from the origin point to the point of furthest extension. Deployment must be automatic upon activation of a trigger mechanism, and the base structure is to be fixed to the ground. The design, for example, may be composed of a mechanism for extending a boom, cantilevered structure, or suspended structure from the origin point.

Design Rules:

The device must be constructed out of the standard project materials listed previously, in addition to *only* the following item: 4 meters of 3M #2090 Long-Mask masking tape, 2″ wide, to fix the base to the ground.

Any design deemed by the instructor as unsafe will not be acceptable (e.g., no sharp flying objects and no explosive devices).

5.10 design projects (continued)

More Contest Rules:

- The device, without external support, must initially fit within a 1.0 m x 1.0 m base area on the ground, except for the triggering means.
- The device must initially be less than 0.5 m in height, except for the triggering means.
- Once triggered, the device must deploy automatically to its final state without further assistance and may expand to any size.
- The device must be remotely triggered (e.g., by a string or rod). The trigger may be used only to release energy from the system. The trigger cannot add energy to the system. Team members (all parts of the body) must remain a minimum of 1 meter away from the device when the device is deployed. The means of triggering must be contained on the allowable materials list but will not be counted in the final materials inventory.
- The device may be fixed to the ground only in the original base area, using only the 3M tape specified for this purpose.
- The device must be set up within 3 minutes; otherwise, a 0.1-meter distance penalty on the total distance will be assessed for every 10 seconds of overtime.
- Human power may be used to trigger the device, but not to impart motion to the structure. However, human power may be used during setup to store energy (e.g., into the rubber bands) for any use.
- Distance is measured from the forward-most point of the base prior to deployment to the forward-most connected point of the structure after deployment (in a predefined direction).
- Except within the original base area, no part of the structure may touch the ground in the final deployed position. Incidental (accidental) contact with the floor is permitted during deployment. However, prolonged contact (e.g., using the ground for support, using a wheeled carriage, or bouncing along the ground) is not permitted. No external structures (e.g., wall, ceiling, or pipes) can be used for guidance or support at any time.
- The maximum distance from three trials will be recorded. A misfire will count as one trial.
- Spare parts are recommended and do not count in the materials inventory of the final assembly.

DESIGN PROJECT #5: VERTICAL LIMIT

Your local fire department is looking for conceptual ideas for rescuing people in high-rise buildings. The firefighters have asked you to develop and build a test model for their review. The device is to be self-deploying and is to be designed to extend as high as possible while still remaining connected to the ground. The structure is to be freestanding in its original and deployed states.

The Mission:

Your mission is to design and build a device that will deploy from a prescribed initial size to a freestanding structure that reaches as high as possible. Deployment must be automatic upon activation of a trigger mechanism. The base structure is to be fixed to the ground. The design, for example, may be composed of a mechanism for extending a boom or truss structure.

Design Rules:

The device must be constructed out of the standard project materials listed previously, in addition to *only* the following item: 4 meters of duct tape, 2″ wide, to fix the base to the ground.

Any design deemed by the instructor as unsafe will not be acceptable (e.g., no sharp flying objects and no explosive devices).

More Contest Rules:

- The device, without external support, must initially fit within a 1.0 m x 1.0 m base area on the ground, except for the triggering means.
- The device must initially be less than 0.5 m in height, except for the triggering means.
- Once triggered, the device must deploy automatically to its final state without further assistance and may expand to any size.
- The device must be remotely triggered (e.g., by a string or rod). The trigger may be used only to release energy from the system. The trigger cannot add energy to the system. Team members (all parts of the body) must remain a minimum of 1 meter away from the device when the device is deployed. The means of triggering must be contained on the allowable materials list but will not be counted in the final materials inventory.
- The device may be fixed to the ground only in the original base area, using only duct tape.
- The device must be set up within 3 minutes; otherwise, a 0.1-meter distance penalty on the total height will be assessed for every 10 seconds of overtime.
- Human power may be used to trigger the device, but not to impart motion to the structure. However, human power may be used during setup to store energy (e.g., into the rubber bands) for any use.
- Distance is measured from the ground to the highest point of the structure when it is fully deployed.
- No external structures (e.g., wall, ceiling, or pipes) can be used for guidance or support at any time.

5.10 design projects (continued)

- The maximum height from three trials will be recorded. The structure height must be maintained for the time it takes to measure the height (approximately 2 minutes). A misfire will count as one trial.
- Spare parts are recommended and do not count in the materials inventory of the final assembly.

DESIGN PROJECT #6: THERE 'N BACK

The problem of air pollution caused by automobiles has plagued cities worldwide for decades. Several solutions have been proposed over the years, including public mass transportation systems, electric vehicles, hybrid vehicles, low-emission fuels, human-powered vehicles, and solar- or wind-powered vehicles. None of these options have been very successful to date. Consequently, it is time to develop new concepts in powered vehicles. Recently, your instructor received an anonymous e-mail stating that energy in a vehicle might be stored in elastic elements. This idea was immediately adopted, and a study was commissioned to investigate the possibility of using a large number of surplus rubber bands to power a commuter vehicle.

The Mission:

Your mission is to design and build a small-scale concept vehicle that travels in a linear trajectory as far as possible and then automatically returns along the same trajectory. The device is to be powered by two rubber bands–either #62 or #64. On the day of testing, a travel line will be taped on the floor. Travel distances will be measured in the direction of the line only. Each team will have three launches of their vehicle. *The travel distance to be recorded will be the distance the vehicle travels backward along the trajectory line after the vehicle stops its forward travel.* The backward travel distance cannot exceed the forward travel distance. The single longest distance the vehicle travels backward in three attempts will be recorded for each team. If the vehicle has no forward or backward travel, the final distance will be recorded as zero.

Design Rules:

The device must be constructed out of the standard project materials listed previously (and *only* the materials listed).

Any design deemed by the instructor as unsafe will not be acceptable (e.g., no sharp flying objects, no explosive devices, and no burning or combustible materials).

More Contest Rules:

- The vehicle must be entirely self-contained (e.g., no external launching or guidance devices).
- The entire vehicle must initially fit within a 1.0 m x 1.0 m x 1.0 m volume without external support. After launching, the vehicle can expand to any size.
- The vehicle can be released by hand or remotely triggered. Any number of team members can be involved with the release. Once released, the vehicle cannot be touched.
- The device must be set up within 3 minutes for each launch; otherwise, a 0.5-meter distance penalty on the total distance will be assessed for every 10 seconds of overtime.
- The vehicle or a part of the vehicle must remain in contact with the ground at all times.
- Human power may be used to trigger the vehicle, but not to impart motion to the vehicle (i.e., no pushing or pulling the vehicle) However, human power may be used to store energy in the rubber bands for any use.
- Gravity cannot be used to produce motion (e.g., no launching from a ramp).
- Travel distance is measured in the direction parallel to the length of the path. If the vehicle hits the side wall of the hallway and stops, all vehicle motion is considered finished.
- The final recorded travel distance will be the distance from the closest point of forward travel (from the starting line) on the vehicle to the closest point of return travel (from the forward mark) on the vehicle.
- Objects expelled from the vehicle are still considered a part of the vehicle for measurement of travel distance.
- Spare parts are recommended and do not count in the materials inventory of the final assembly.

section two

Modern Design Practice and Tools

The widespread availability of computers has made three-dimensional modeling the preferred tool for engineering design in nearly all disciplines. Solid modeling allows engineers to easily create mathematical models parts and assemblies, visualize and manipulate these models in real time, inspect how they mate with other parts, and calculate their physical properties. The geometry of a part, as well as allowable errors, is determined to a large degree on the fabrication method used to make it. Thus, a basic understanding of fabrication processes is important for creating three-dimensional designs. There are many available manufacturing processes, and all have advantages and disadvantages. Three-dimensional modeling also is used as the foundation for many sophisticated computational analysis techniques such as mass properties, interference checking, and finite-element analysis. The ease with which this can be done has moved what was formerly a complicated analytical process performed by specialists into the realm of standard practice by many design engineers as part of the design process.

6 Solid Modeling

objectives

After completing this chapter, you should be able to

- Introduce solid modeling as an engineering design graphics tool
- Explain how solid models are created
- Show how parts and models can be decomposed into features
- Develop strategies for creating a solid model
- Explain how solid models support the entire product life cycle

6.01 introduction

Solid modeling is a computer-based simulation that produces a visual display of an object as if it existed in three dimensions. **Solid models** aid in forming a foundation for the product development process by providing an accurate description of a product's geometry and are used in many phases of the design process and life cycle of the product. This chapter will focus on methods for creating robust solid models of mechanical parts; however, these methods can be applied to other domains as well.

Solid models are created with specialized software that generates files for individual as well as assembled parts. These models are then used in a variety of applications throughout the design and manufacturing processes, as shown schematically in Figure 6.01. During the product concept stage, solid models are used to visualize the design. As the product is refined, engineers use solid models to determine physical properties such as the strength of the parts, to study how mechanisms move, and to evaluate how various parts fit together. Manufacturing engineers use solid models to create manufacturing process plans and any special tools or machines needed to fabricate or assemble parts. Solid models also can be used to generate formal engineering drawings to document the design and communicate details of the design to others. People responsible for the product life cycle may depend on solid models to help create images for service manuals and disposal documentation. Even sales and marketing functions use graphics generated from solid models for business presentations and advertising. Thus, it is very important not only to learn how to create solid models but also to understand how others will use the models. Solid models must be built with sound modeling practices if they are to be useful in downstream applications. In this chapter, you will learn how to create robust solid models that not only look like the real thing but also support the entire product life cycle. You also will learn about the history of CAD tools and the importance of solid modeling as part of an engineering design graphics system.

6.02 Tools for Developing Your Idea

Many tools have been developed for creating accurate images of an object as an aid in analyzing its function, recording its history, or visualizing its appearance. One of the simplest tools is a pencil, which is used to make sketches of an object on paper. More formal tools include rulers, protractors, compasses, and various types of manually operated drafting machines. These tools are used to make more accurate, standardized drawings according to precise rules and conventions, as discussed in a previous chapter.

CAD systems are among the most sophisticated graphics and design tools available to engineers and designers. Many types of CAD systems are on the market today. The simplest systems are general purpose drawing or drafting packages that can be used to create 2-D images, similar to the way pencil images are created on paper (except faster and easier). More complex packages allow you to create simulations of 3-D models that can be used not only to generate conventional 2-D drawings of a design but also to create 3-D images for visualization. The core of a CAD model is a geometric **database**. The database includes information about the geometry and other engineering properties of an object. The CAD software uses the database to display the model and to conduct further engineering analysis. A short discussion of CAD history will demonstrate how these systems evolved and provide some insight into the modeling processes used by designers with various CAD systems.

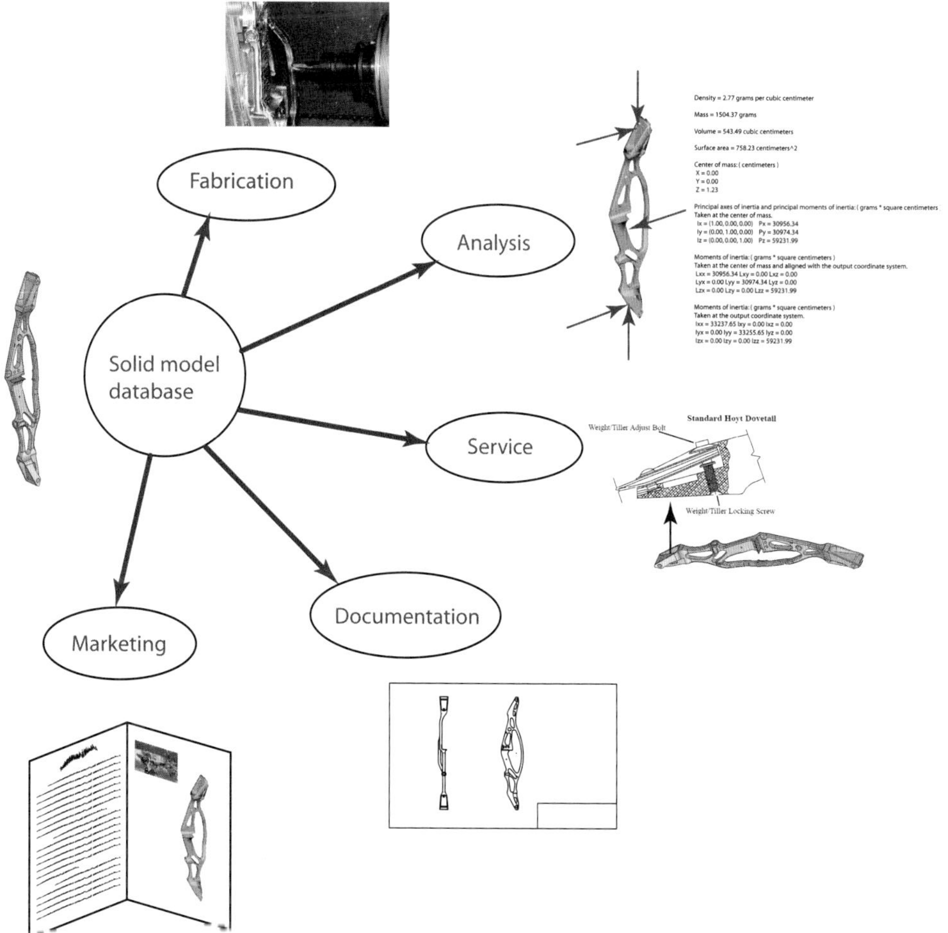

FIGURE 6.01. Uses for a solid model database.

6.02.01 Two-Dimensional CAD

The first CAD systems were developed in the late 1960s at a time when computational resources were very limited. Graphics displays had refresh times measured in seconds, and the data storage capabilities were limited to fractions of a kilobyte. As a result, only very simple models could be created. Those models were basically electronic versions of conventional pencil-and-paper drawings. The user had to specify the location of each vertex in the model for the particular view desired. If the user wanted another view, she had to start from scratch, just as you would to do if you were creating a drawing on paper.

Since CAD models are used to define the geometry or shape and size of objects, the models are composed of geometric entities. In the earliest CAD models, those entities represented the edges of the object, just as you would draw the edges of an object with a pencil. In fact, at that time there was very little distinction between a 2-D drawing of an object and a 2-D CAD model of the object. The 2-D CAD model was simply a database that contains the edges of the object, dimensions, text, and other information that you would find on the drawing, but in electronic form instead of on paper.

The simplest geometric entity is a point. Points in two dimensions are defined according to their location in a coordinate system, usually Cartesian coordinates (x,y). In a CAD system, the coordinate system represents locations on the "paper," or computer screen. Points are generally used to locate or define more complex entities, such as the endpoints of a line segment or the center of a circle. A point on an entity that marks a particular position, such as the endpoint of a line segment or the intersection of two entities, is referred to as a **vertex**.

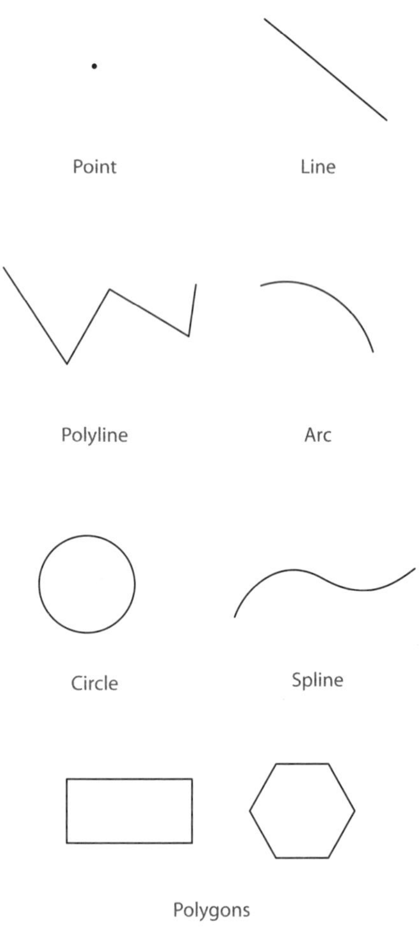

FIGURE 6.02. Some entities used for 2-D CAD.

FIGURE 6.03. With 2-D models, visualization of a 3-D object must be done mentally.

Two-dimensional geometric entities are those that can be created as a path or curve on a plane. Those entities include lines, circles, **splines**, arcs, polygons, and conic sections, which are shown in Figure 6.02. The entities can be assembled to create images of a desired object as it would be seen from different viewing directions, as shown in Figure 6.03.

One weakness of 2-D CAD systems is that to visualize and manipulate a 3-D model of the object, you must mentally assemble and reform the 2-D views. Another weakness of 2-D CAD (and pencil drawings) is that it is possible to create images of objects that are physically impossible to build, such as the three-pronged fork and the triangle shown in Figure 6.04.

6.02.02 Wireframe Modeling

In the early 1980s, 2-D CAD drafting packages evolved into 3-D modeling systems. In these newer systems, 3-D information could be included for the model. The computer could then perform the calculations needed to create views of an object as if it was seen from different directions. These systems were still limited to using entities such as

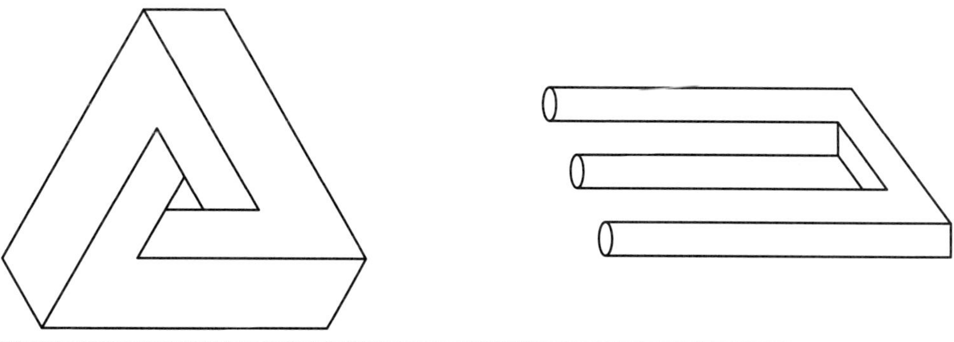

FIGURE 6.04. Impossible 3-D objects can be drawn with 2-D elements.

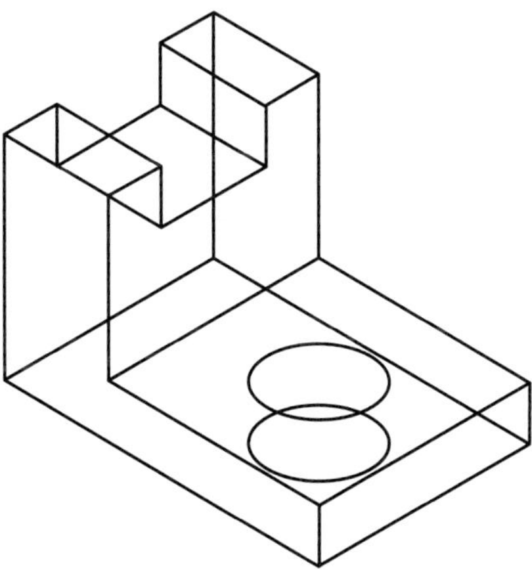

FIGURE 6.05. A wireframe model of a 3-D object.

lines, circles, and arcs; but the assemblage of the entities was no longer restricted to being on a single plane. The geometric entities were represented in a 3-D database within a 3-D coordinate system with x-, y-, and z-coordinates. Since simple curve or path entities were used to define the edges of an object, such models were called **wireframe models**. Think of a wireframe as being similar to a box kite. A wireframe model of a bracket is shown in Figure 6.05.

Wireframe models are still very limited in their representation of parts. The same wireframe model can represent an object from two different viewing directions, as demonstrated in Figure 6.06. Thus, the models were sometimes difficult to visualize as solids. Some models were ambiguous, being interpreted by viewers as different objects. Look at Figure 6.07 and try to imagine the solid object represented by the wireframe model in (a). Can you visualize the shape of the object? Does this figure represent more than one object? When the hidden edges are removed and the surfaces shaded, as in (b) and (c), it is much easier to see the desired shape.

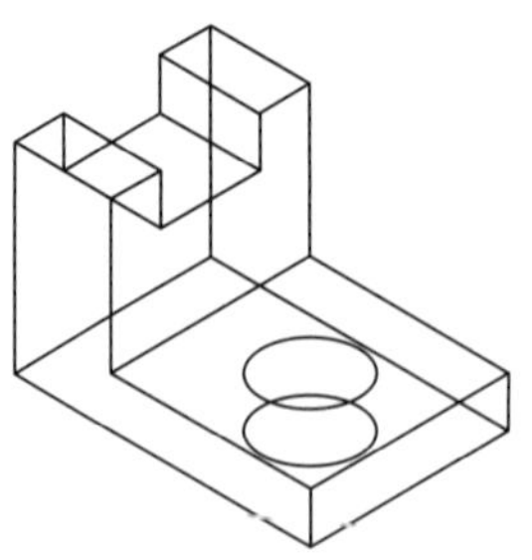

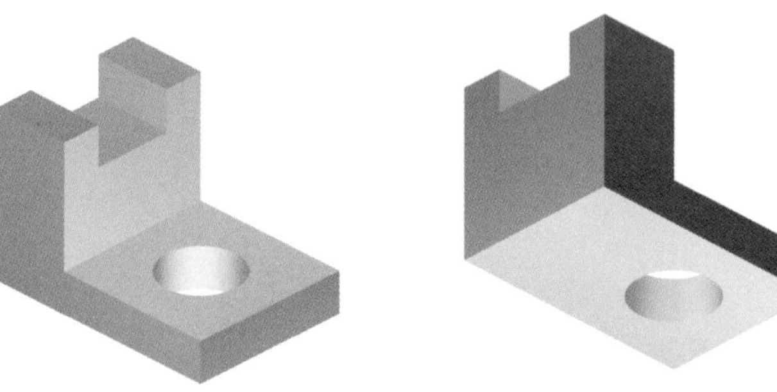

FIGURE 6.06. Two possible view interpretations of the same wireframe model.

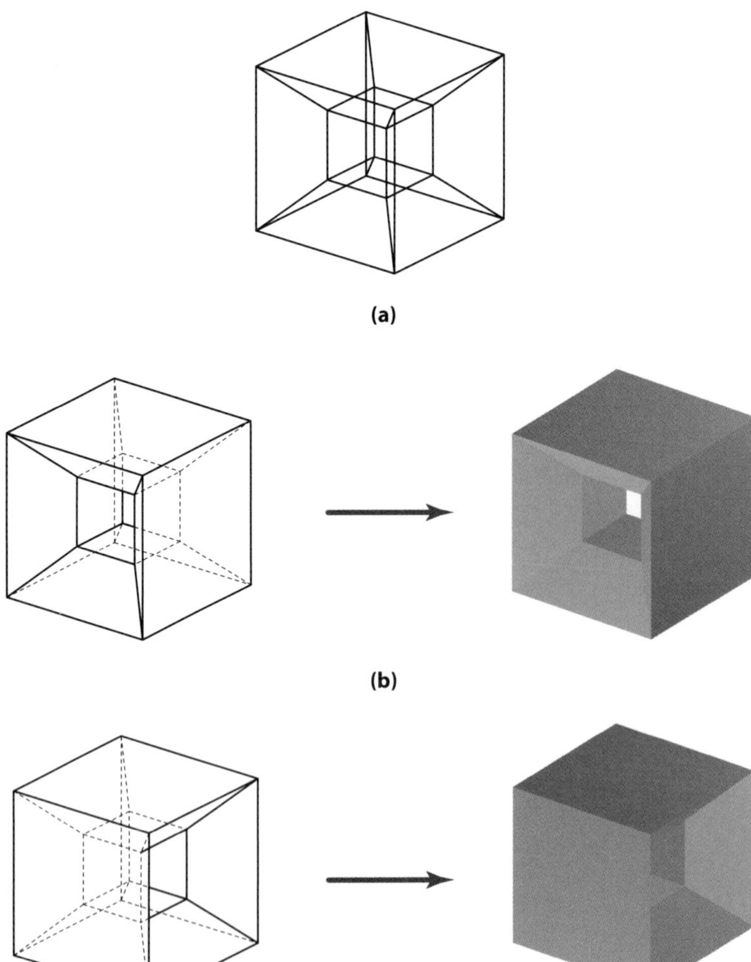

FIGURE 6.07. The wireframe model in (a) can represent the object in (b) or the object in (c).

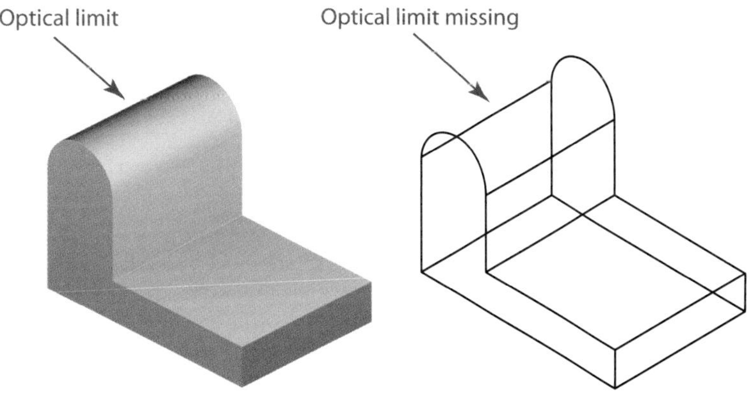

FIGURE 6.08. Wireframe models do not show the optical limit of a curved surface.

Another problem with wireframe CAD systems was that the geometry was limited to shapes with simple planar and cylindrical surfaces. Also, parts with cylindrical features, such as the one shown in Figure 6.08, generated wireframe models that did not show the optical limit or silhouette of the cylindrical surface. Even so, wireframe modeling represented a tremendous advance in technology compared to the drafting board. It is estimated that more than 75 percent of all common machined parts can be accurately represented using 3-D wireframe models.

6.02.03 Surface Modeling

As computers became more powerful and data storage capabilities increased, surface modeling techniques were developed. With a **surface model**, the designer could display the surfaces of a part, such as those shown in Figure 6.09, and use the model to perform engineering analyses such as calculating the part's mass properties. Such models also could be used to generate computer programs that controlled the fabrication of parts, for example, on a computer-controlled cutting machine called a mill.

Surface models evolved from wireframe models by mathematically describing and then displaying surfaces between the edges of the wireframe model. Thus, a surface model is a collection of the individual surfaces of the object. This modeling method is called **boundary representation**, or **b-rep**, because the surfaces "bound" the shape. The bounding entities of a simple part created using boundary representation are shown in Figure 6.10. The bounding entities can be planes, cylinders, and other surfaces in three dimensions. These surfaces are in turn bounded by simpler curve entities such as lines and arcs.

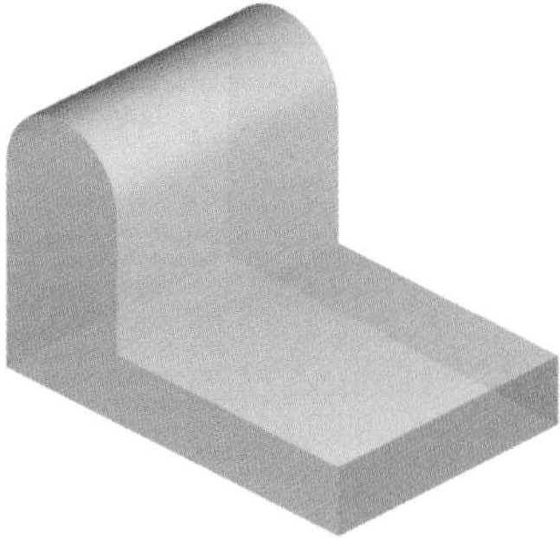

FIGURE 6.09. A surface model with semitransparent surfaces to reveal detail.

FIGURE 6.10. A surface model exploded to show individual surfaces.

The use of surface models eliminates most of the problems with visual ambiguity encountered with wireframe models.

6.02.04 Solid Modeling

Solid models are visually similar to surface models, so it is sometimes difficult to distinguish between them. With a solid model, however, the software can distinguish between the inside and outside of a part and the objects can have thickness. Thus, the information stored in the 3-D database is sufficient to distinguish between an empty shoe box and a brick. The software also easily computes information such as the object's volume, mass, center of mass, and other inertial properties. Early solid models, developed in the late 1980s, were made using a technique known as **constructive solid geometry (CSG)**. CSG models are composed of standard building blocks in the form of simple solids such as rectangular prisms (bricks), cylinders, and spheres, called **primitives**. The shapes are easy to define using a small number of dimensions. Figure 6.11 shows some of these basic solids. To create more complex solids, the primitives are assembled using Boolean operations such as addition (union), subtraction (difference), and interference. Examples of these operations are shown in Figure 6.12.

Surface and CSG models were very powerful tools for design, but their early versions were rather cumbersome to use. As computational resources improved, so did the capabilities of modeling software. Increasingly more sophisticated modeling methods, such as creating a solid model by moving or rotating a closed 2-D outline on a path

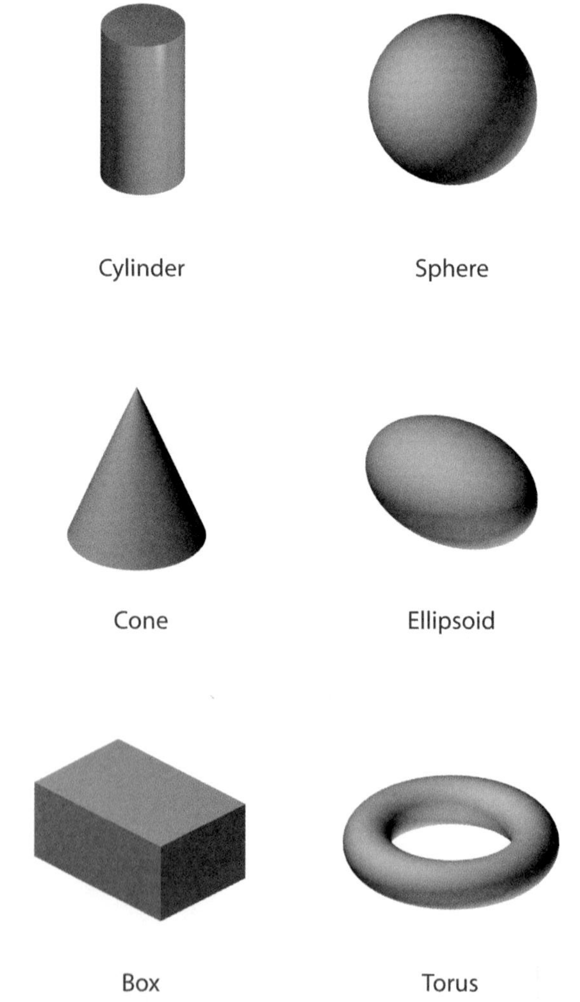

FIGURE 6.11. Some 3-D primitives used in solid modeling.

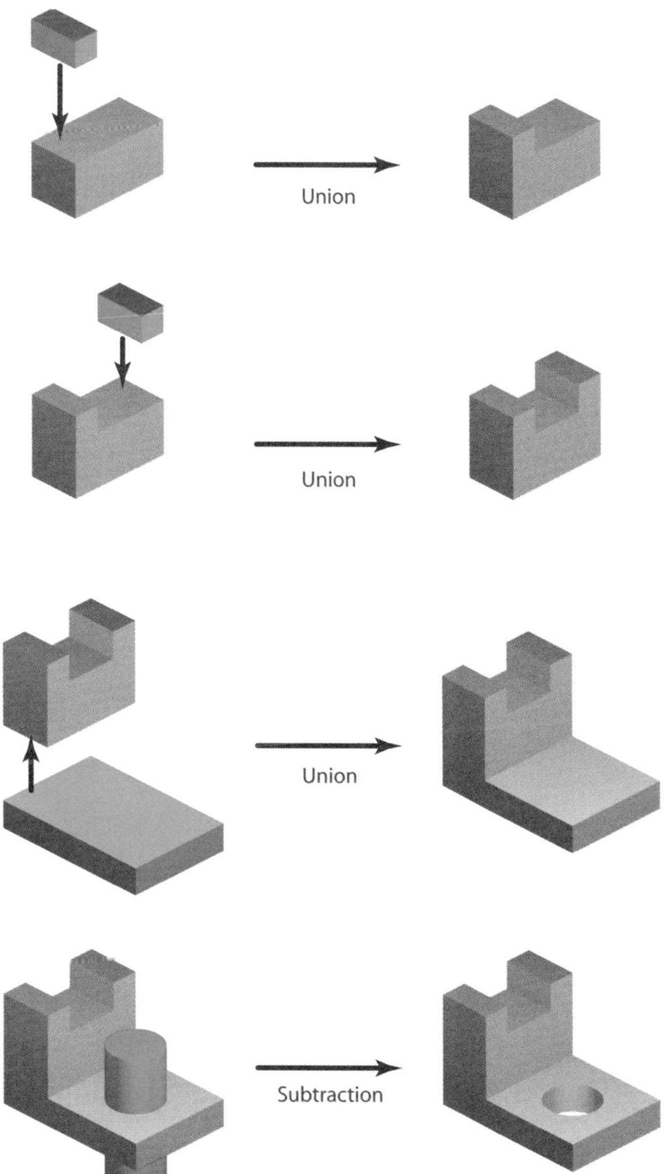

FIGURE 6.12. Steps in using solid primitives to build a more complicated solid model using Boolean operations.

through space, as shown in Figure 6.13, were developed. Further developments included software tools for taking many individual solid model parts and simulating their assembly into a larger structure, as explained in Chapter 1, and for easily creating formal engineering drawings for parts and assemblies from their solid models.

A more accurate and efficient modeling tool called **feature-based solid modeling** was developed in the mid-1990s. This modeling method permitted engineers and designers to create a more complex part model quickly by adding common features to the basic model. Features are 3-D geometric entities that exist to serve some function. One common and easily recognizable feature is a hole. Holes in a part exist to serve some function, whether it is to accommodate a shaft or to make the part lighter. Other features, such as bosses, fillets, and chamfers, will be defined later in this chapter.

Parametric solid modeling is a form of feature-based modeling that allows the designer to change the dimensions of a part or an assembly quickly and easily. Since parametric feature-based solid modeling is currently considered the most powerful 3-D CAD tool for engineers and designers, the remainder of this chapter will be devoted to this modeling method.

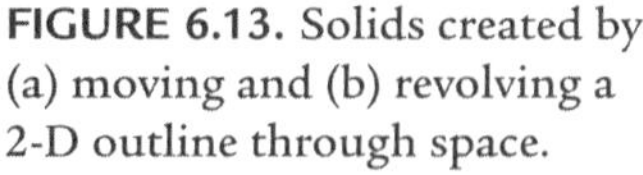

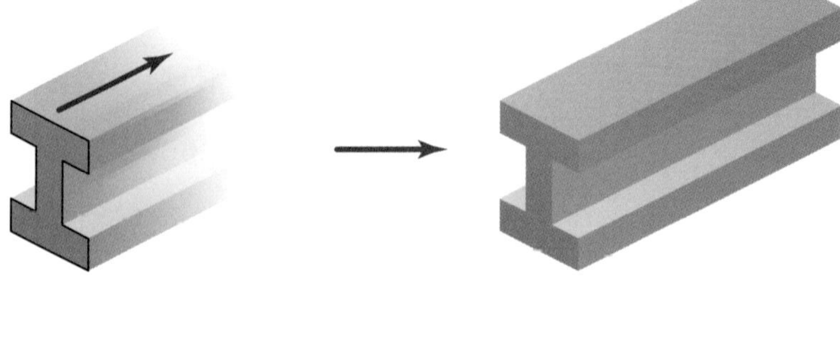

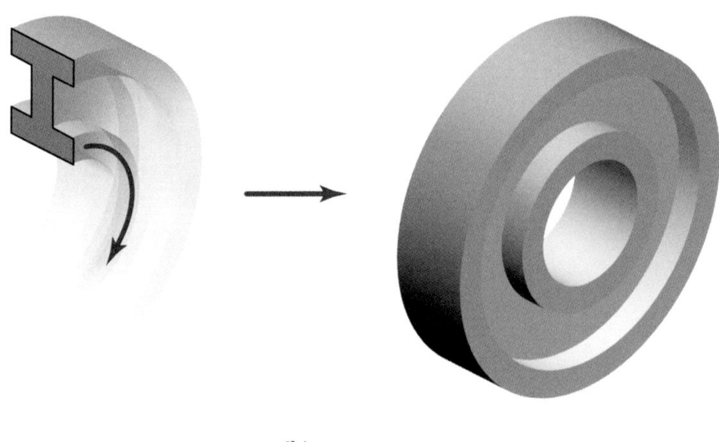

FIGURE 6.13. Solids created by (a) moving and (b) revolving a 2-D outline through space.

6.03 A Parametric Solid Model

So how does one go about creating a parametric feature-based solid model? In this section, a very simple model will be created to demonstrate basic concepts. More detail and sophistication will be presented in subsequent sections of this chapter. The tools that you need to create a parametric model are solid modeling software and a computer that is powerful enough to run the software. As you create the model, the software will display an image of the object which can be turned and viewed from any direction as if it actually existed in three dimensions.

Using the mouse and keyboard, you will interact with the software through a **graphical user interface (GUI)** on the computer's display device (i.e., the computer monitor). The GUI gives you access to various tools for creating and editing your models. GUIs differ slightly in different solid modeling software. However, most of the packages share some common approaches. When creating a new model (i.e., with nothing yet existing), you will probably be presented with a display of 3-D Cartesian coordinate x-, y-, and z-axes and the three **primary modeling planes**, which are sometimes called the **principal viewing planes** or **datum planes**. These planes help you visualize the xy, yz, and xz planes and are usually displayed from a viewing direction from which all three planes can be seen, as shown in Figure 6.14.

Nearly all solid modelers use 2-D **sketches** as a basis for creating solid features. Sketches are made on one of the planes of the model with a 2-D sketching editor similar to a drawing editor found on most 2-D CAD drafting software. When you begin a new model, you often make a sketch on the one of the basic modeling planes. When the sketching plane is chosen, some modelers will reorient the view so you are looking straight at the 2-D sketching plane. You can then begin sketching.

Line segments are usually inserted using mouse clicks, as shown in Figure 6.15(a). A sketch is initially created without much attention being paid to precise dimensions and exact orientations of the different segments. For convenience, the **sketching editor** in most solid models automatically corrects sloppy sketches by making assumptions about the intended geometry. For example, if a line segment is sketched almost vertically or almost horizontally, the sketching editor will force the line into a vertical or horizontal orientation. Figure 6.15(a) shows a sketch of a rectangle created by clicking the four corners, or vertices; Figure 6.15(b) shows the cleaned-up sketch after the sketching editor corrects the user input and reorients the line segments.

6.03.01 Valid Profiles

Before a solid feature can be created by extrusion or rotation, the final profile of the shape must be a closed loop. Extra line segments, gaps between the line segments, or overlapping lines create problems because the software cannot determine the boundaries of the solid in the model. Samples of proper and improper profiles are shown in Figure 6.16.

6.03.02 Creation of the Solid

A completed sketch that is used to create a solid is called a **profile**. A simple solid model can be created from the profile by a process known as **extrusion**, as shown in

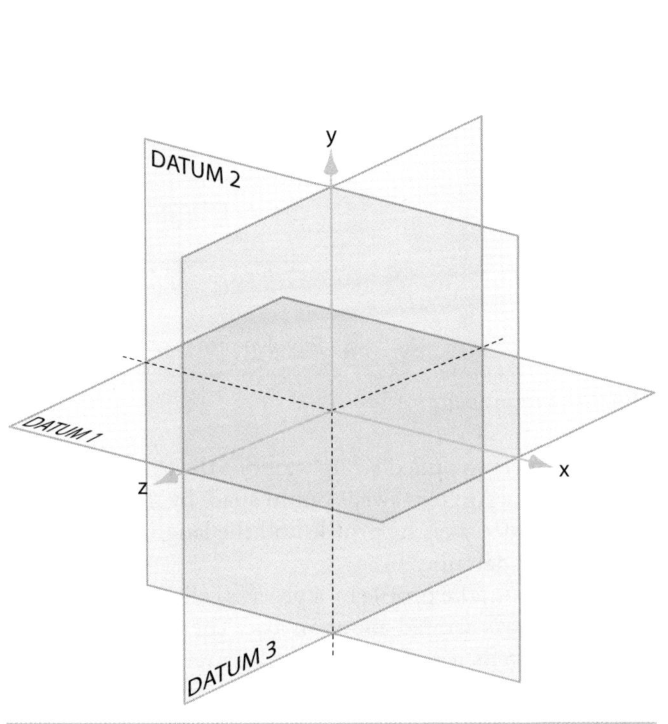

FIGURE 6.14. The primary modeling planes for solid modeling.

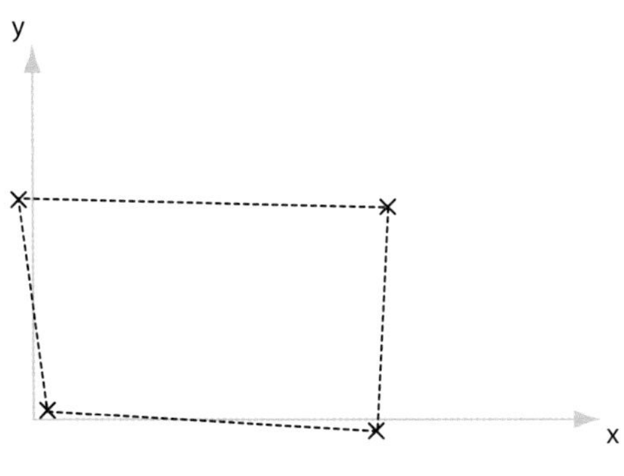

(a) corners of rectangle specified by user

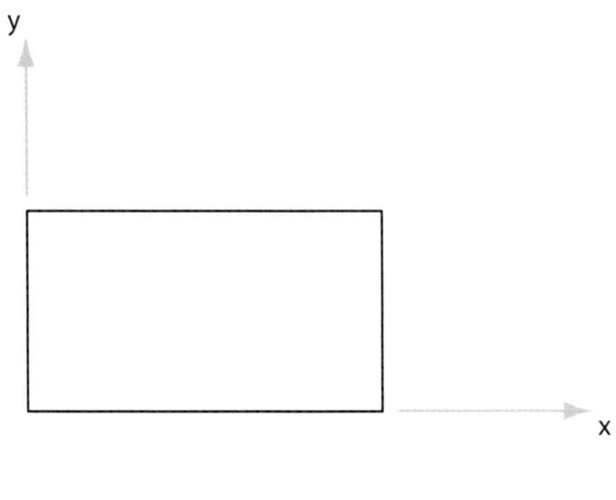

(b) rectangle corrected by software

FIGURE 6.15. 2-D sketching.

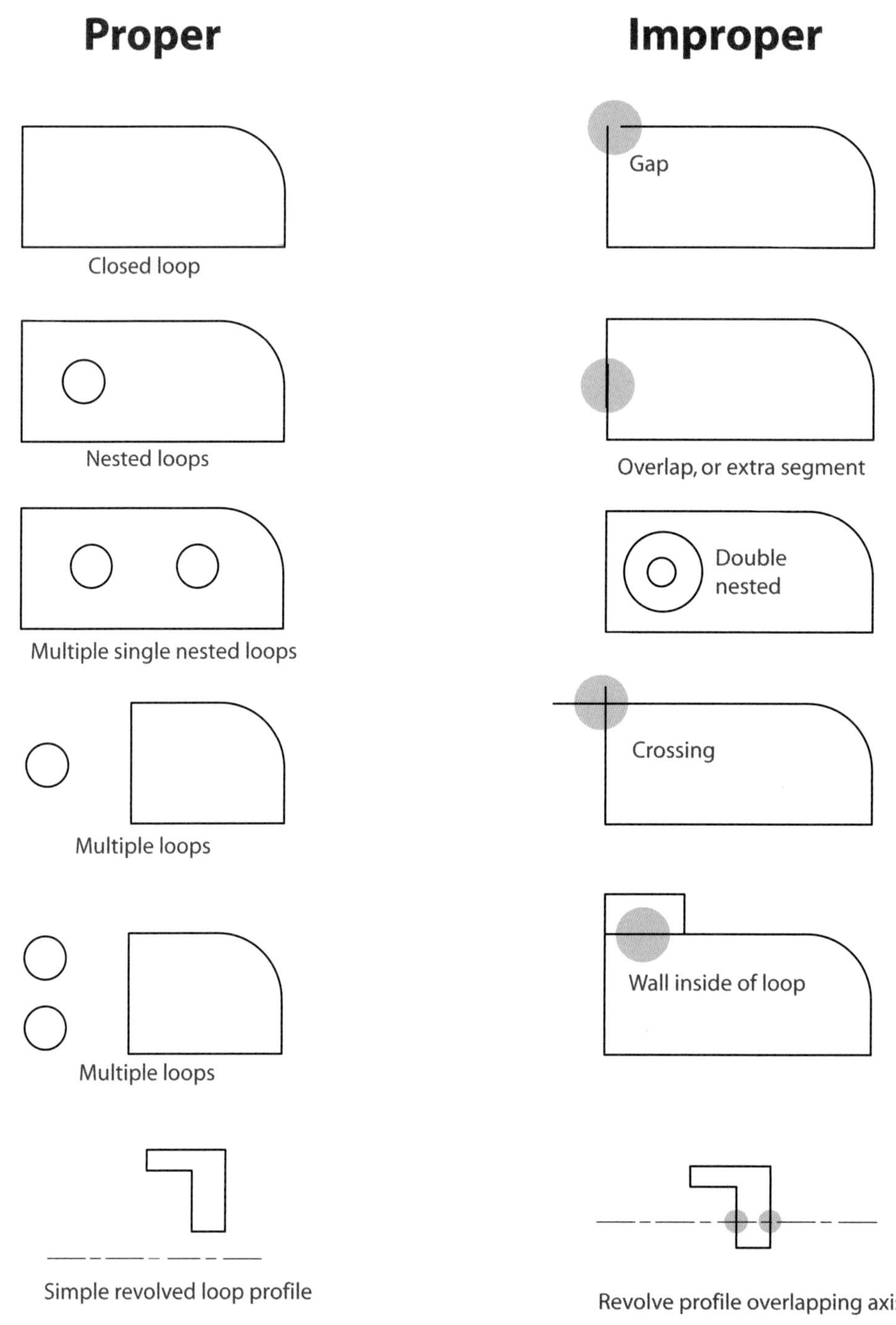

FIGURE 6.16. Examples of proper and improper profiles.

Figure 6.17. Imagine the profile curve being pulled straight out of the sketching plane. The solid that is formed is bound by the surfaces swept out in space by the profile as it is pulled along the path. Both the geometry of the profile and the length of the extrusion must be specified to define the model fully.

A different model can be created from the profile by a process called revolution. To create a **revolved solid**, a profile curve is rotated about an axis. The process is similar to creating a clay vase or bowl on a potter's wheel. The profile of a revolved part is also planar, and the axis of revolution lies in the profile plane (sketching plane). One edge of

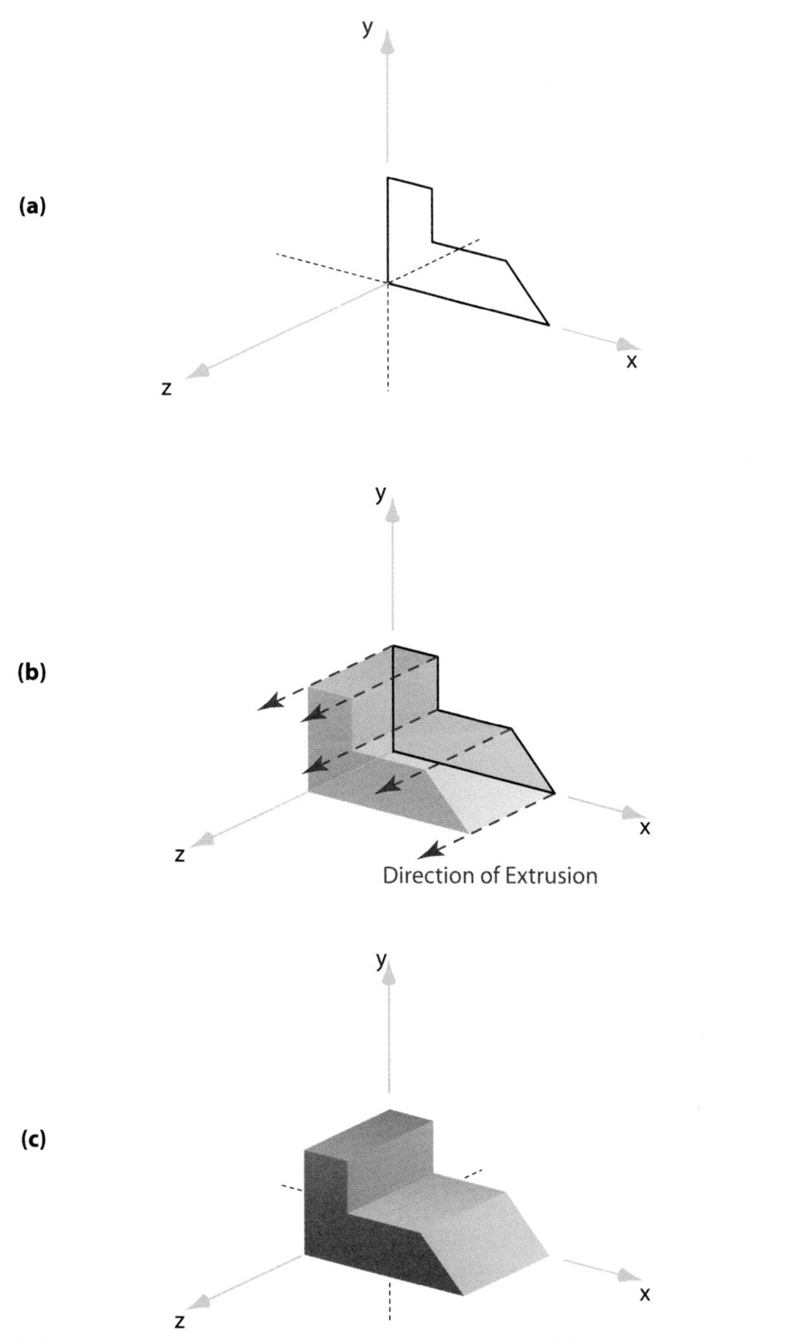

FIGURE 6.17. A solid created by extrusion of a 2-D profile.

the sketch may lie along the axis of revolution, as shown in Figure 6.18(a); or the sketch may be offset from the axis, as shown in Figure 6.18(b). It is important to make sure that the profile does not cross over the axis of revolution. This would create a self-intersecting model (i.e., a solid created inside another solid), which most solid modeling software interpret to be a geometric error. The geometry of the profile and the angle of rotation must be specified to define the model fully. The models shown in Figure 6.18 are revolved through a full 360 degrees.

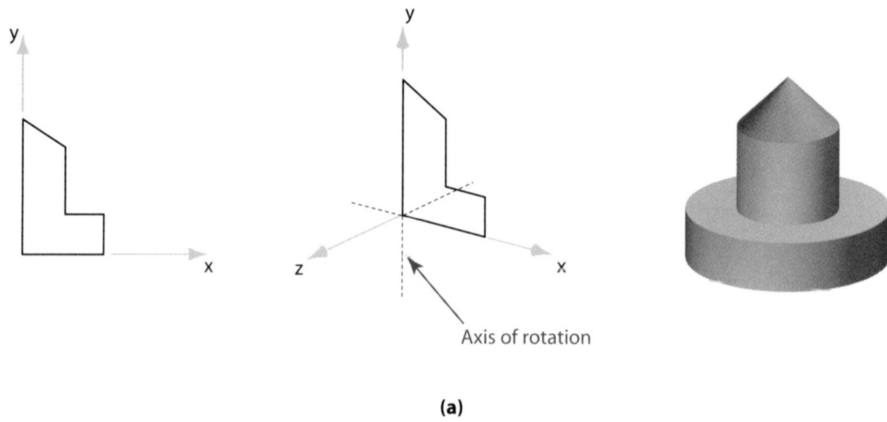

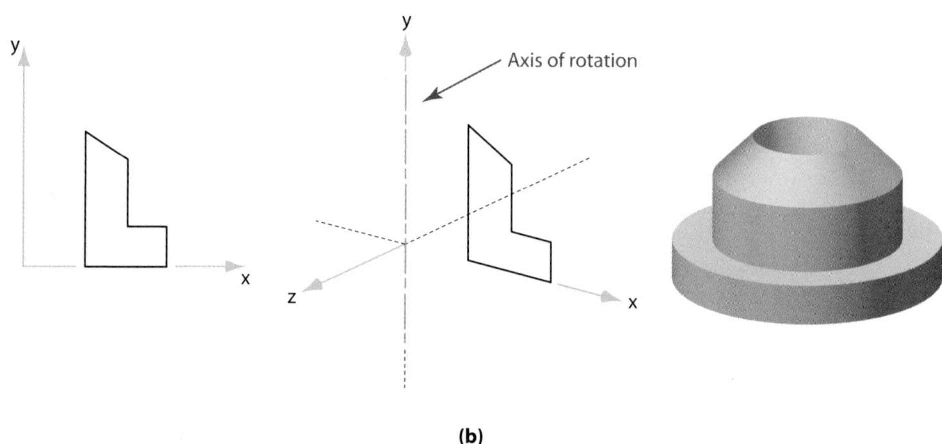

FIGURE 6.18. A solid created by rotation of a 2-D profile, with the axis on the profile in (a) and with the axis off the profile in (b).

6.04 Making It Precise

Before a part can be submitted for analysis or fabrication, the sizes and locations of all of its features must be completely specified. To see how this is done, let's back up a few steps in our discussion of the creation of the model.

6.04.01 Orientation of the Sketch

Before you begin to create the first extrusion or revolution, you must decide where to place the part in the space relative to the xyz coordinate system. With the model shown in Figure 6.17, the initial sketch was placed on one of the basic modeling planes. If the sketch was placed on one of the other basic modeling planes instead, the model would have the same geometry but with a different orientation in space, as shown in Figure 6.19.

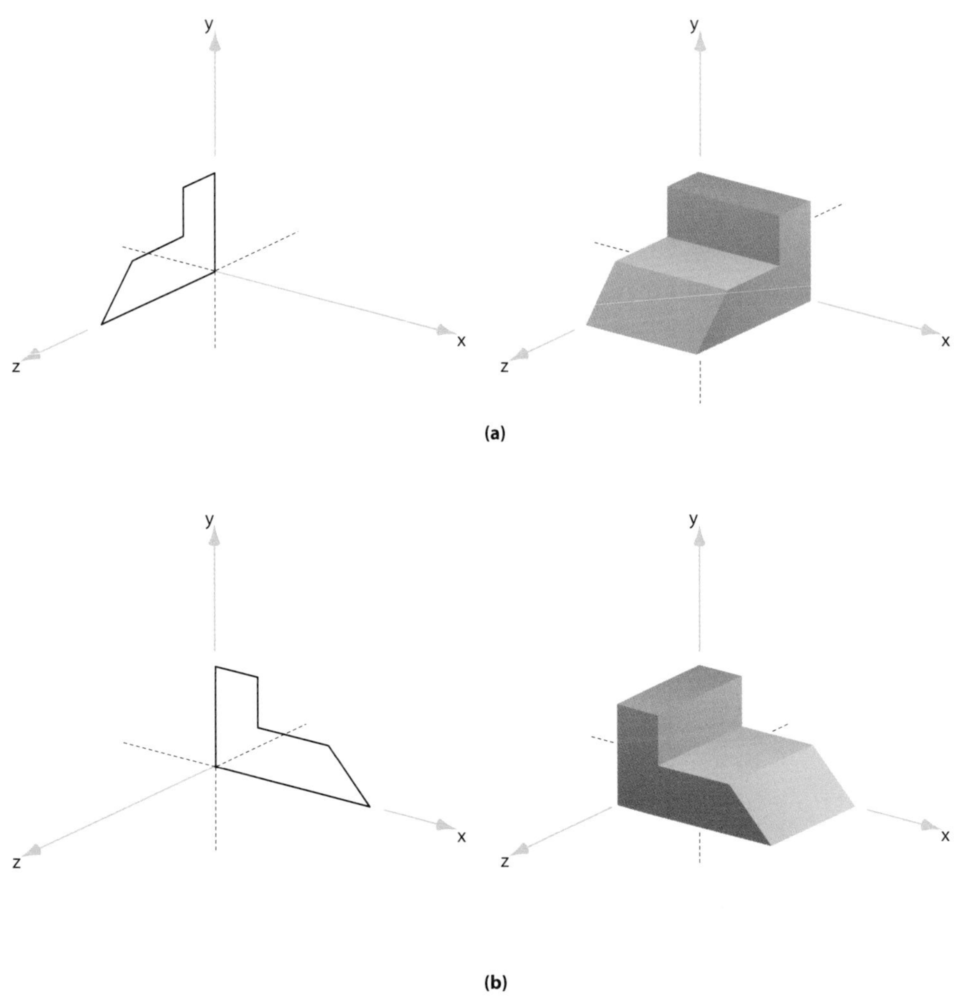

FIGURE 6.19. The same profile made in different sketching planes produces the same object but in different orientations. In (a), the profile is made in the yz plane; and in (b), the profile is made in the xy plane.

6.04.02 Geometric Constraints

Formally, **constraints** are the geometric relationships, dimensions, or equations that control the size, shape, and/or orientation of entities in the profile sketch and include the assumptions that the CAD sketcher makes about your sloppy sketching. Constraints that define the size of features will be discussed in the following section. The previous section provided a few examples of **geometric constraints** that were applied to a simple sketch: lines that were drawn as nearly horizontal were assumed to be horizontal, and lines that were drawn as nearly vertical were assumed to be vertical. Those assumptions reduce the number of coordinates needed to specify the location of the endpoints. Some solid modelers require you to constrain the profile fully and specify the sizes and locations of all of its elements before allowing the creation of a solid feature; others allow more free-form sketching. Geometric constraints may be either implicitly defined (hidden from the designer) or explicitly displayed so you can modify them. A set of geometric constraints is not unique, as demonstrated in Figure 6.20. In

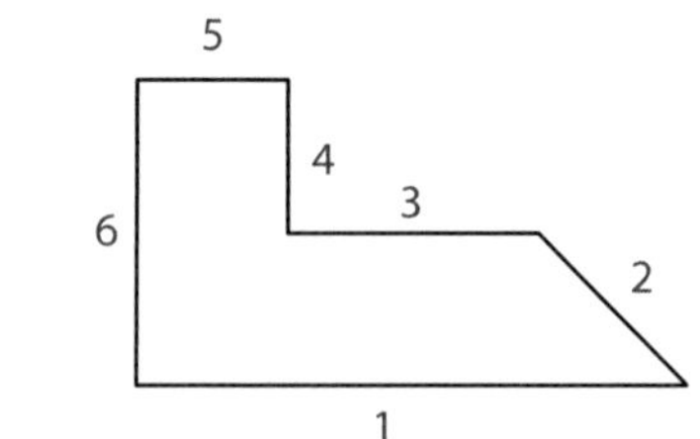

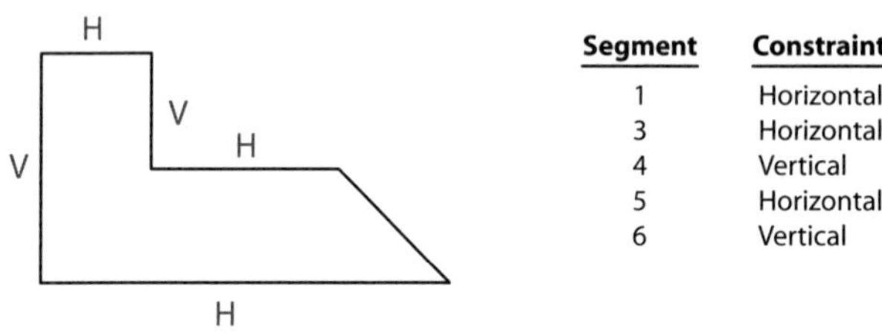

Segment	Constraint
1	Horizontal
3	Horizontal
4	Vertical
5	Horizontal
6	Vertical

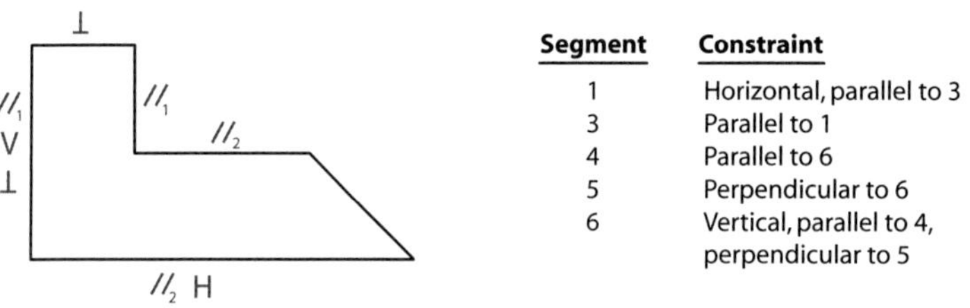

Segment	Constraint
1	Horizontal, parallel to 3
3	Parallel to 1
4	Parallel to 6
5	Perpendicular to 6
6	Vertical, parallel to 4, perpendicular to 5

FIGURE 6.20. The line segments in a profile are numbered in (a). The implied geometric constraints for each segment are shown in (b), and an equivalent set of applied constraints is shown in (c). A letter or symbol beside a segment signifies the type of geometric constraint applied to it.

this example, a set of geometric constraints that restricts some lines to being horizontal or vertical is equivalent to another set of constraints that restricts some lines to being either parallel or perpendicular to each other.

Geometric constraints specify relationships between points, lines, circles, arcs, or other planar curves. The following is a list of typical geometric constraints. The results of applying the constraints are shown graphically in Figure 6.21.

- Coincident—forces two points to coincide
- Concentric—makes the centers of arcs or circles coincident
- Point on Line—forces a point to lie on a line
- Horizontal/Vertical—forces a line to be horizontal/vertical
- Tangent—makes a line, a circle, or an arc tangent to another curve
- Colinear—forces a line to be colinear to another line
- Parallel—forces a line to be parallel to another line
- Perpendicular—forces a line to be perpendicular to another line
- Symmetric—makes two points symmetric across a centerline

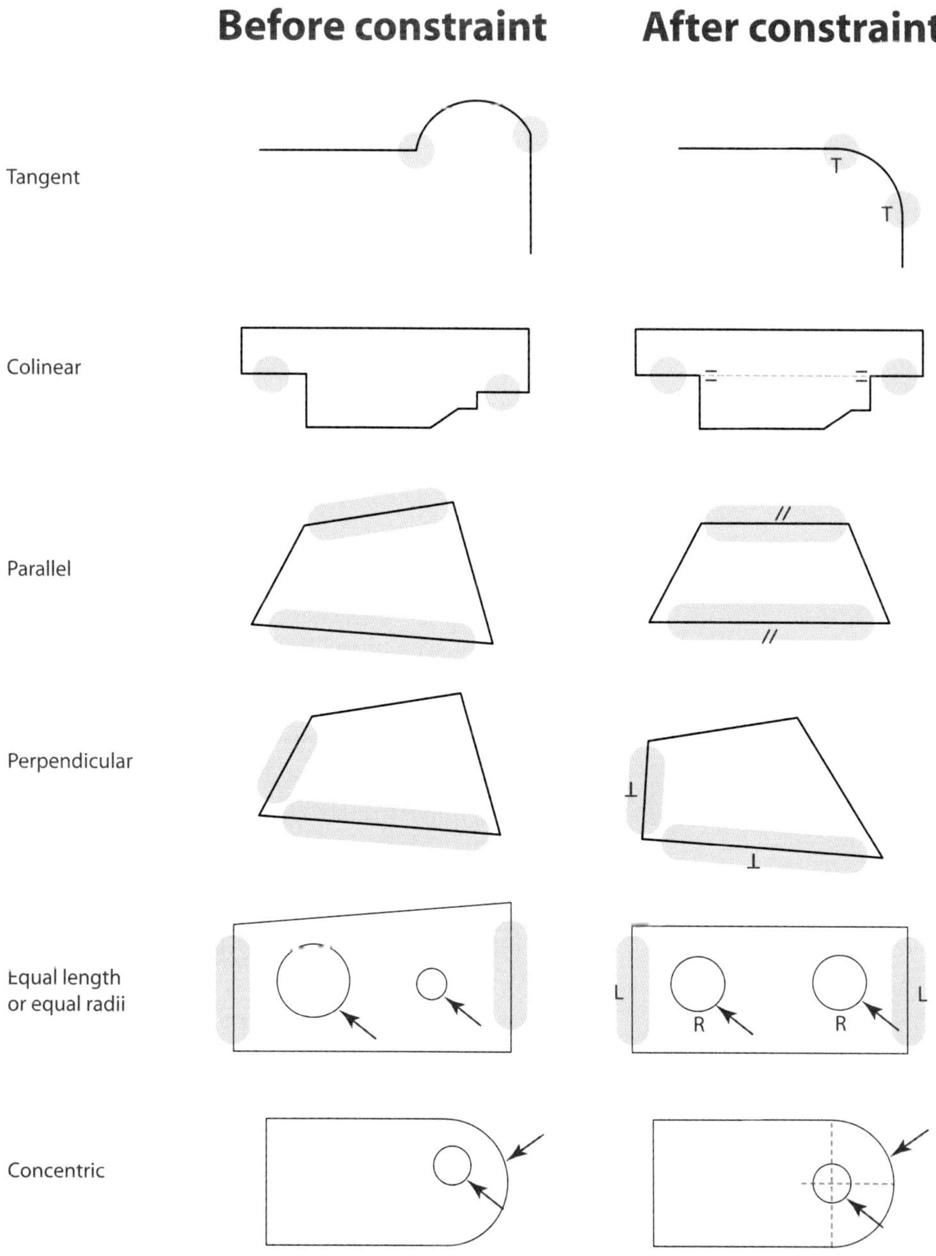

FIGURE 6.21. Geometric constraints commonly found in sketching editors.

The sketching editors in most solid modeling software are usually configured to try to interpret the user's sketching intent such that certain constraints are created automatically. In addition to adjusting nearly horizontal or vertical lines into true horizontal or vertical lines, if two lines are nearly perpendicular or parallel or an arc and a line are nearly tangent at the common endpoint, the sketching editor will impose the assumed geometric relationship. These automatically applied geometric constraints can be changed at a later time if desired.

6.04.03 Dimensional Constraints

Each of the 2-D entities in the profile must have size and position. **Dimensional constraints** are the measurements used to control the size and position of entities in your sketch. Dimensional constraints are expressed in units of length, such as

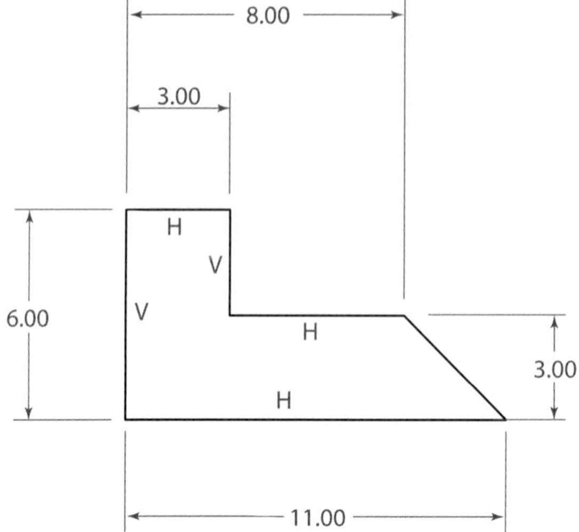

FIGURE 6.22. A profile fully constrained with geometric and dimensional constraints.

millimeters, meters, inches, or feet. For example, look at the profile in Figure 6.22, which shows dimensional constraints that define its size. If you, the designer, do not fully specify all of the necessary information, the software will default to some value that you may not want. It is better if you control the model, rather than have the software assign assumed parameters and conditions to the model.

Dimensional constraints can be created interactively while you are sketching, but also automatically as a result of a feature operation, an extrusion, or a revolution. There are three principal types of dimensional constraints:

- Linear dimensional constraints define the distance between two points, the length of a line segment, or the distance between a point and a line. Linear dimensions can be measured horizontally or vertically or aligned with the distance being measured.
- Radial and diametral dimensional constraints specify the radius or diameter of an arc or a circle.
- Angular dimensional constraints measure the angle between two lines. The lines do not need to intersect, but they cannot be parallel.

6.04.04 Uniqueness of Constraints

A set of dimensional constraints is not unique. It is possible to apply a different set of dimensional constraints on a profile to produce exactly the same geometry, as shown in Figure 6.23.

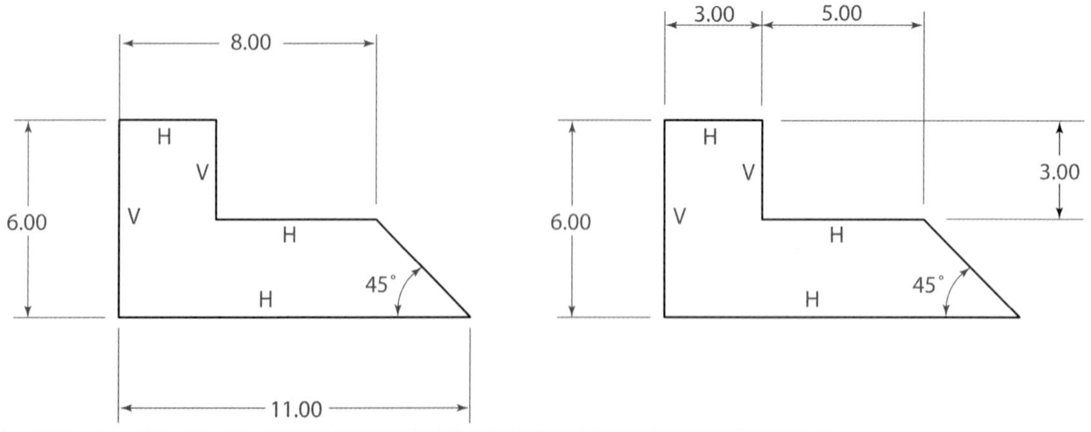

FIGURE 6.23. Two different sets of dimensional constraints that can be used to define the same geometry.

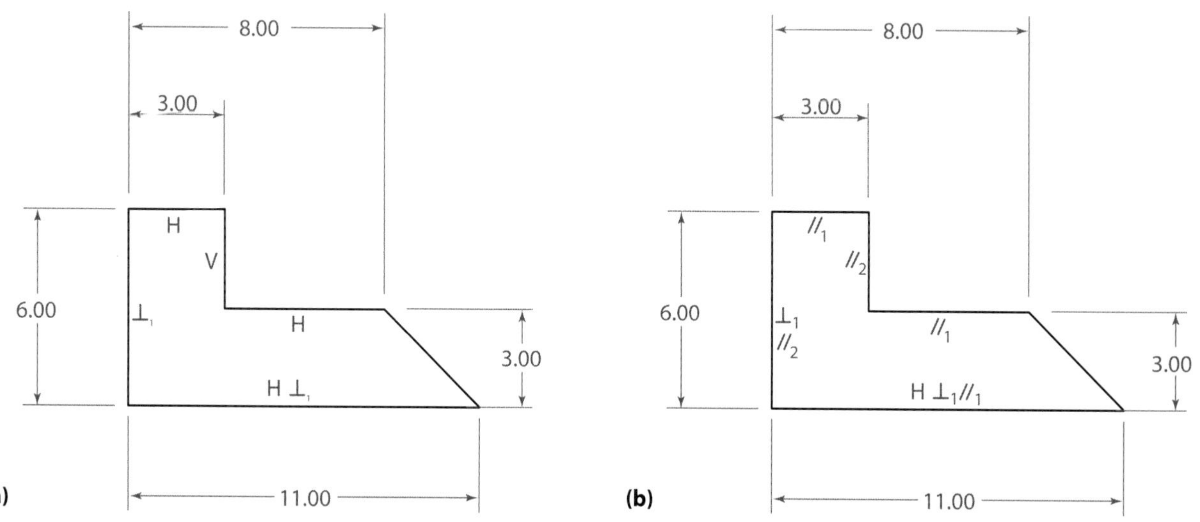

FIGURE 6.24. Two different sets of geometric constraints that define the same geometry.

Combinations of dimensional and geometric constraints also are not unique. It is possible to have different combinations of geometric constraints and dimensional constraints define exactly the same geometry, as shown in Figure 6.24.

The natural question then becomes, which set of constraints is correct or preferred? The answer depends on what the function of the part and the design intent is or how the designer wants to be able to change the model. You also should consider how the solid model will be used for analysis, manufacturing, and documentation when applying sets of constraints. One of the greatest advantages of a parametric solid model is that the model can be changed easily as the design changes. However, the constraints limit the ways in which the model can be changed.

6.04.05 Associative and Algebraic Constraints

Associative constraints, sometimes called **algebraic constraints**, can be used to relate one dimensional constraint to another. The dimensional constraints on a profile are expressed in terms of variables. Each dimensional constraint is identifiable by a unique variable name, as shown in Figure 6.25. Algebraic constraints can be used to control the values of selected variables as the result of algebraic expressions. Algebraic expressions consist of constants and variables related to each other through the use of arithmetic functions (+, −, *, absolute value, exponent, logarithm, power, square root, and sometimes minimum and maximum); trigonometric functions; and conditional expressions (if, else, or when) including inequalities comparisons (if A > B then . . .).

There are two different methods for solving sets of algebraic constraint equations. Software that uses **variational techniques** solves the equations simultaneously. A compatible solution for all of the variables can be calculated when there are a sufficient number of equations. In a system using **parametric techniques**, the equations are usually solved in sequential order. The equations will have only one unknown variable. All other variables in the algebraic expression must be known for the value of the unknown variable to be calculated, which is called the dependent or **driven dimension**. The known variables are called the **driving** dimensions. As shown in Figure 6.25, when the value of a driving dimensional constraint is changed, the value of its driven dimensional constraints are automatically changed too.

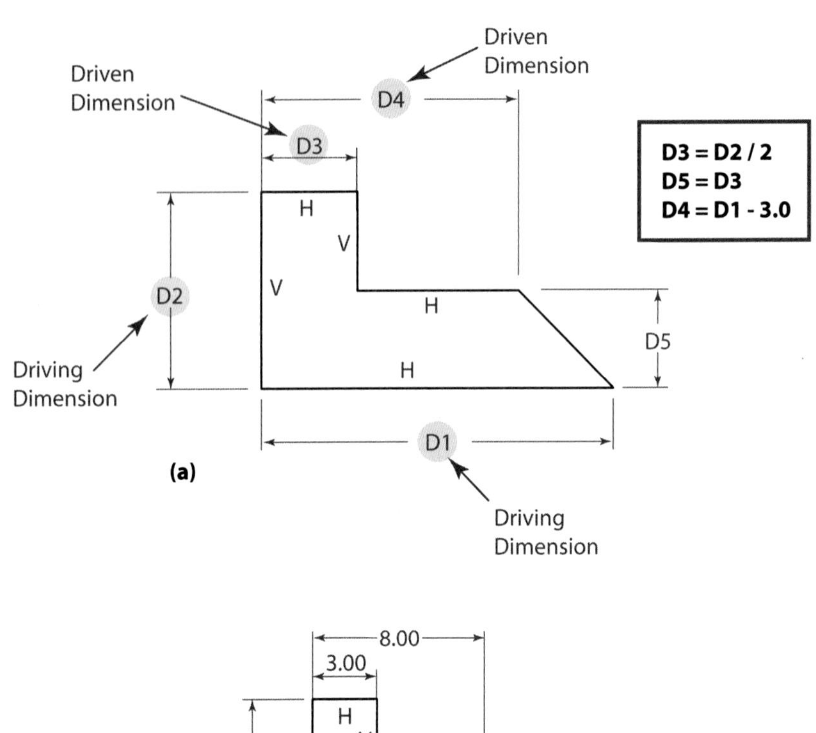

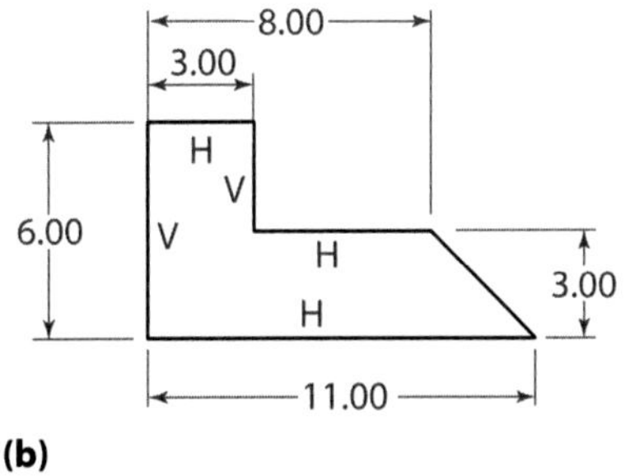

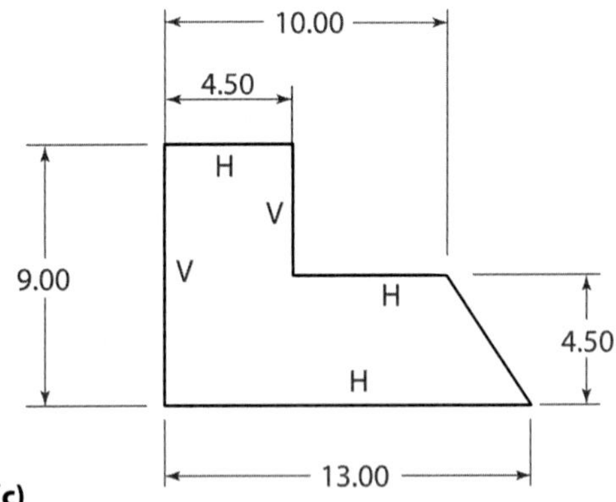

FIGURE 6.25. Dimensional constraints are shown in term of variables and a set of algebraic constraints in (a). Dimensions D3, D4, and D5 are automatically specified by specifying dimensions D1 and D2 in (b). Dimensions D3, D4, and D5 change automatically when D1 and D2 are changed in (c).

6.05 Strategies for Combining Profile Constraints

A completed profile is constrained using a combination of geometric and dimensional constraints and may include algebraic constraints as well. The constraint set must be complete for the geometry to be fully defined. If a profile is overconstrained or underconstrained, it may not be possible to create a solid feature from the profile. Some solid modeling software automatically applies constraints, but these constraints usually need to be changed to reflect the design intent. Furthermore, most software systems expect the user to apply constraints in addition to the automatically generated constraints; in particular, variational modelers allow underconstrained sketches and do not require user-applied constraints, but these systems can yield unpredictable results when the dimension values are changed. By gaining a thorough understanding of constraints (and how and when to apply them), you will be able to control the behavior of your models and capture your design intent. A strategy for applying geometric and dimensional constraints to a profile is demonstrated next.

The first constraint usually applied to a new sketch is a **ground constraint**. Ground constraints serve as anchors to fix the geometry in space. Ground constraints may have various forms. The most common type of ground constraint is a geometric entity such as a line or point on the profile having been made coincident with one of the basic modeling planes or with the origin of the coordinate system. For example, if the first feature of a model is created by extrusion, it may be convenient to place a corner of the profile on the origin of the coordinate system. This is usually done by placing one of the vertices of the sketch exactly at the origin. If the first feature is created by rotation, it may be convenient to place one of the endpoints of the center axis at the origin.

When the profile is closed and the automatically generated constraints have been applied the interactive constraint definition phase begins. Some software creates a fully constrained sketch, including both geometric and dimensional constraints; but the constraint set chosen by the software is usually not exactly what you want. Other software does not fully constrain the sketch, but leaves this task to the designer. Ground constraints should be specified if this was not already done when the profile was sketched. Next, geometric and dimensional constraints should be added and/or changed until the profile is fully constrained. Typically, your solid modeling software will alert you when the profile is fully constrained or when you try to overconstrain the profile. In particular, you should take care to delete any unwanted geometric and dimensional constraints that may have been automatically added. Finally, the profile should be changed to reflect the design intent and the dimensional constraints adjusted to the desired values. Some sketching editors automatically readjust the profile after each constraint is added; others wait until all of the constraints have been specified before readjusting the profile. Updating the profile to show its new shape after constraints are added or changed is called **regeneration**.

The way dimensional constraints are added depends largely on what the intended function of the part is, how it is to be fabricated, and how the geometry of the part may change in the future. What would a simple L-bracket look like if some of the dimensions were changed, as shown in Figure 6.26? In this case, d1 was changed from 30 to 40 and d2 was changed from 3 to 8. The result is shown in the figure. But if you want to make the bracket by bending a piece of sheet metal, the part should have a uniform thickness throughout. One way to do this is to force the length of line segments that define the thickness of both legs of the *L* to be equal. The geometric constraint shown in Figure 6.27 has this effect. The equal length geometric constraint replaces the dimensional constraint for the thickness of the vertical leg of the bracket. If you tried to apply the equal length constraint and the dimensional constraints on both line segments, the sketch would be overconstrained, a situation the software would not accept. In addition, an associative constraint needs to be added between the radius of the inside corner of the bracket and the radius of the outside corner to ensure that the thickness of the part is constant around the corner.

This constraint strategy demonstrates how to make your parts more robust. Through this simple example, you can see the importance of fully understanding the behavior of your model and the effects of your selection of dimensions and constraints. Your choices for geometric, dimensional, and algebraic constraints are not unique; but the decisions you make in selecting a set of constraints will have a big impact on the behavior of your model if you make changes to it. You must choose a modeling strategy that will reflect your design intent.

FIGURE 6.26. Changing the values of the dimensions changes the geometry of the model, without the need to reconstruct the model.

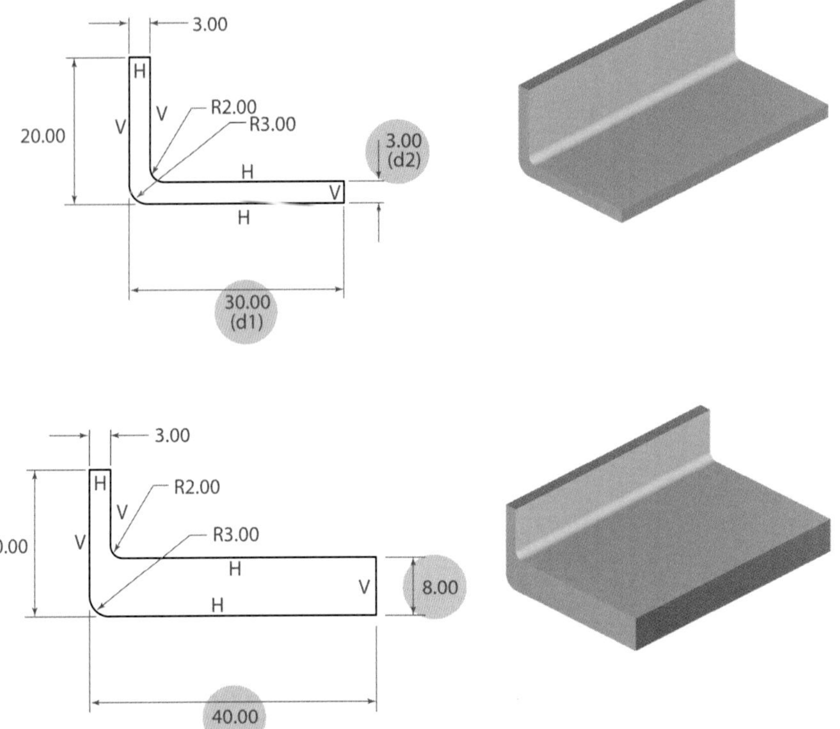

FIGURE 6.27. When compared to the original model in (a), the addition of the equality and associative constraints in (b) ensures a constant material thickness even if the dimensions are changed, thus adding functionality to the model if that is the intent.

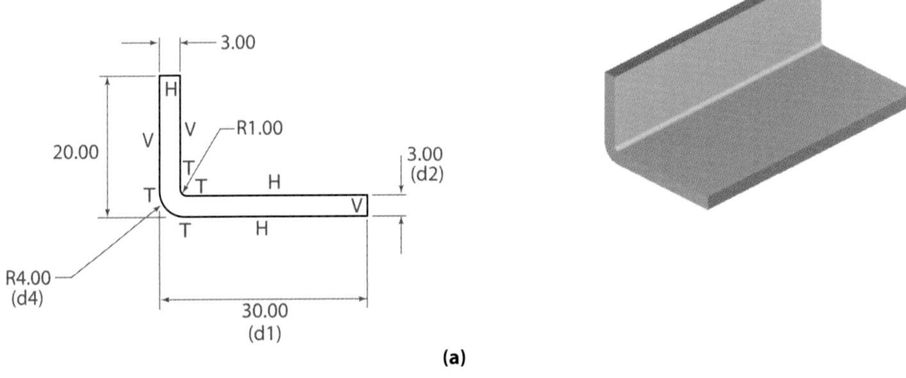

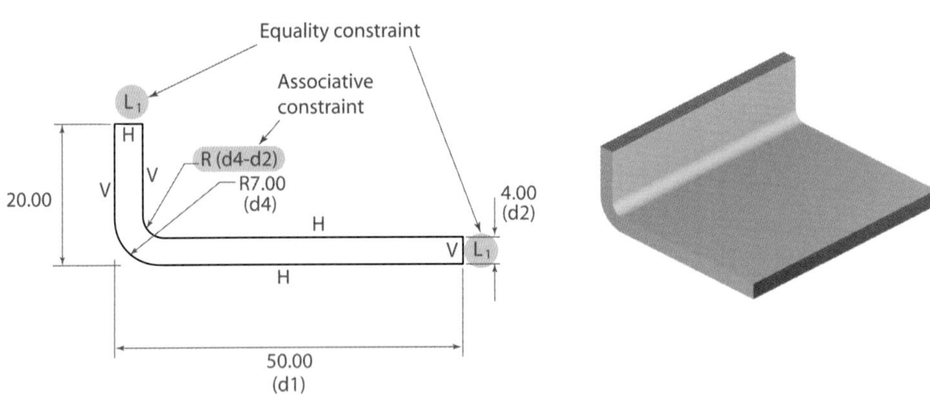

As an exercise for developing skill in the application of constraints, consider the rough sketch and the finished profile shown in Figure 6.28. For the profile to be fully constrained using only the dimensional constraints shown, certain geometric constraints are needed. Segment 1, for example, needs to be horizontal and tangent to Segment 2. Segment 2 needs to be tangent to Segment 1 as well as to Segment 3. Segment 3 needs to be tangent to Segment 2, perpendicular to Segment 4, and parallel to Segment 5. Segment 4 needs to be perpendicular to Segment 3 and equal in length to Segment 14. These constraints and the required geometric constraints on the remaining segments are shown in Figure 6.29. Keep in mind that a set of geometric constraints may not be unique. Can you specify another set of geometric constraints for this example that would create the same profile with the same dimensional constraints?

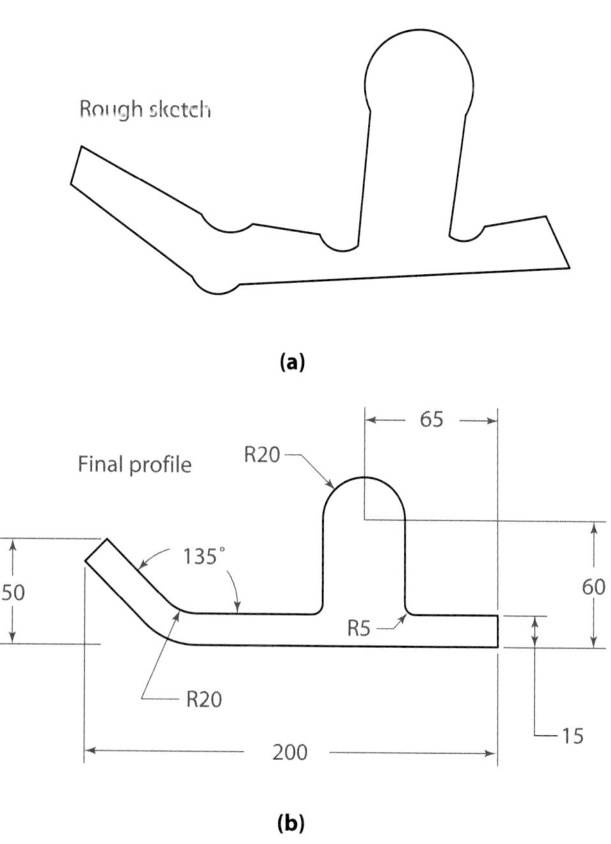

FIGURE 6.28. Geometric constraints need to be applied to the rough sketch (a) to produce the desired, fully constrained profile (b).

FIGURE 6.29. Applying geometric constraints to the first four segments of the sketch in Figure 6.28(a) to produce the finished profile in Figure 6.28(b).

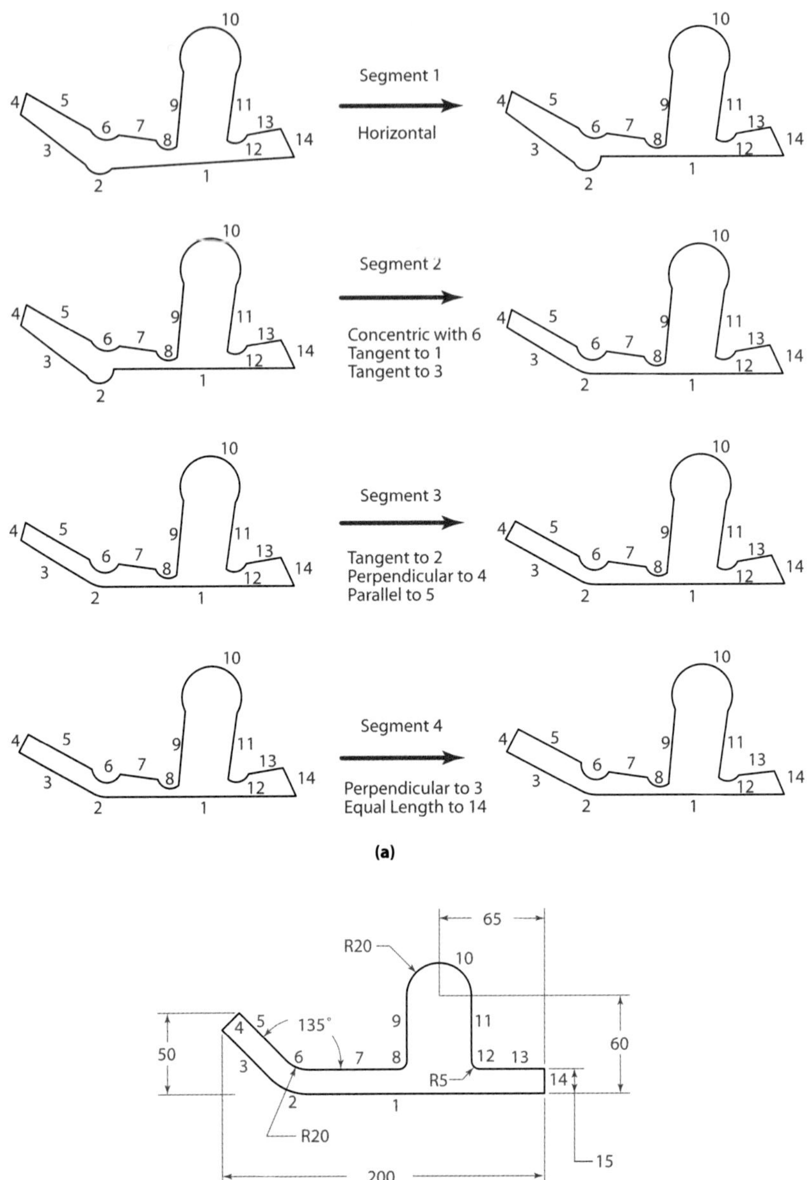

Segment	Constraint
1	Horizontal, Tangent to 2
2	Concentric with 6, Tangent to 1, Tangent to 3
3	Perpendicular to 4, Parallel to 5, Tangent to 2
4	Equal Length to 14, Perpendicular to 3
5	Parallel to 3, Tangent to 6
6	Concentric with 2, Tangent to 5, Tangent to 7
7	Horizontal, Tangent to 6, Tangent to 8
8	Tangent to 7, Tangent to 9
9	Vertical, Tangent to 8, Tangent to 10
10	Tangent to 9, Tangent to 11
11	Vertical, Tangent to 10, Tangent to 12
12	Tangent to 11, Tangent to 13
13	Horizontal, Tangent to 12
14	Vertical, Equal Length to 4

(b)

6.06 More Complexity Using Constructive Solids

You have seen how to create solid models by sketching a 2-D profile on one of the basic modeling planes and then using a single extrusion or a single rotation to create a 3-D model. Adding material to or removing material from the original model can create a more complex model. When material is added, a **protrusion** feature is created. When material is removed, a **cut** feature is created. Both protrusions and cuts begin with sketched profiles that are then extruded or revolved to form solid shapes that are added to or removed from the existing body of the model. For an extruded feature, the profile lies in the sketch plane and is extruded in a direction perpendicular to the sketching plane. For a revolved feature, the profile and the axis of revolution must be coplanar so both will lie on the sketch plane.

When protrusions or cuts are made on an existing model, sketches and profiles are no longer restricted to be located on one of the basic modeling planes. Instead, any planar surface on the model can be selected and used as a **sketching plane** on which sketches and profiles can be created. Once a sketching plane has been selected, any 2-D element that is created will be forced to lie on that plane. After a sketching plane is selected, the model can be reoriented to look directly into the sketching plane. Although you can sketch when not looking directly into the sketching plane, you need to be very careful when viewing from a different orientation. Edges of your sketch may not be shown in their true shape, and angles may appear distorted. Most people find it easier to create 2-D profiles when they are looking directly into the sketching plane, just as it is easier for someone to draw straight lines and angles with correct measurements when the paper is oriented straight in front of her.

Examples of profiles on various sketching planes on a model and resulting extruded protrusions are shown in Figure 6.30; examples of extruded cuts are shown in Figure 6.31. Examples of revolved protrusions are shown in Figure 6.32, and examples of revolved cuts are shown in Figure 6.33.

As with the first extrusion or revolution that created the main body of the model, the profiles for the added protrusions or cuts must be fully defined by geometric, dimensional, and algebraic constraints before they can be extruded or revolved. A common geometric constraint for protrusions or cut features is to make one or more edges or vertices of the new profile coincident with edges of the surface used as the sketching plane. In Figure 6.30(a), notice that one surface of the original object has been selected as a sketching plane and a rectangular profile has been sketched on the selected plane. The top and bottom edges of the sketched profile are coincident with edges of the sketching surface. The direction of extrusion is, by default, perpendicular to theselected sketching plane.

The length of the extrusion or angle of rotation also must be specified. There are several options for defining the length of the extrusion, as shown in Figure 6.30 and Figure 6.31. The simplest is to specify a **blind extrusion**. A blind extrusion is one that is made to a specified length in the selected direction, analogous to specifying a dimensional constraint, as shown in Figure 6.30(b). If your extrusion is the first feature used to create your initial model, it will be a blind extrusion. For a cut such as a hole, a blind extrusion creates a hole of a specified depth, as shown in Figure 6.31(b).

Another way to determine the length of the extrusion is to use existing geometry. One option for specifying an extrusion length is to **extrude to the next surface**. With this option, the extrusion begins at the profile and the protrusion or cut stops when it intersects the next surface encountered, as shown in Figure 6.30(c) and Figure 6.31(c). Another option is to **extrude to a selected surface**, where the protrusion or cut begins at the profile and stops when it intersects a selected surface, which may not necessarily be the first one encountered. See Figure 6.30(d) and Figure 6.31(d). For extruded cuts, there is an option to **extrude through all**. This option creates a cut or protrusion that starts at the profile and extends in the selected direction through all solid features, as shown in Figure 6.31(e). A **double-sided extrusion** permits the

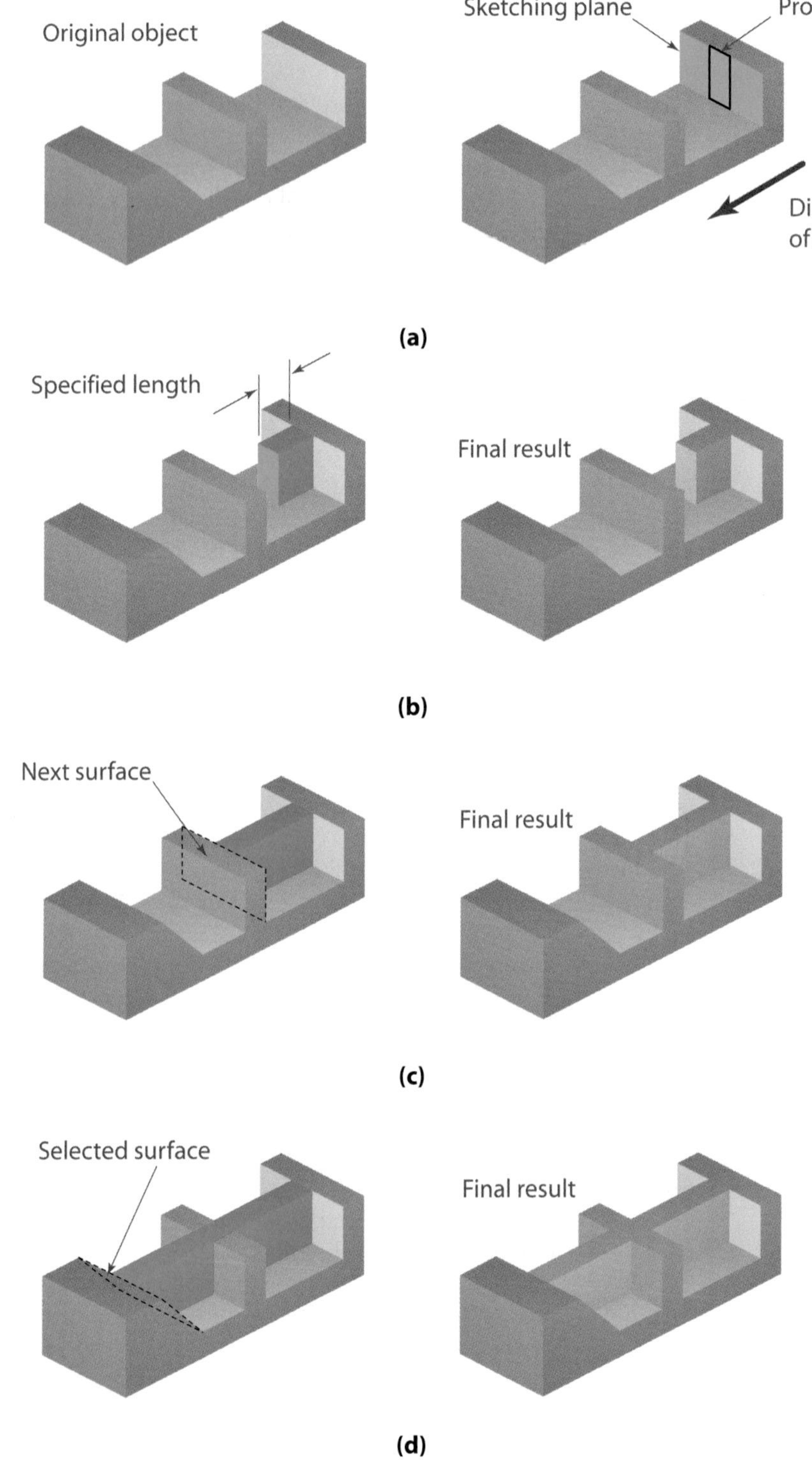

FIGURE 6.30. Different ways to terminate an extruded protrusion from the profile in the sketching plane in (a). Blind extrusion in (b), extrude to next surface in (c), extrude to a selected surface in (d).

protrusion or cut to extend in both directions from a profile. The method of termination in each direction can then be specified independently. Other methods of terminating the extrusion length may be available depending on the specific solid modeling software used. You also can specify the angle of rotation of a revolved protrusion or cut in a similar manner by using a specified angle (blind revolution) or by revolving up to next or selected surfaces.

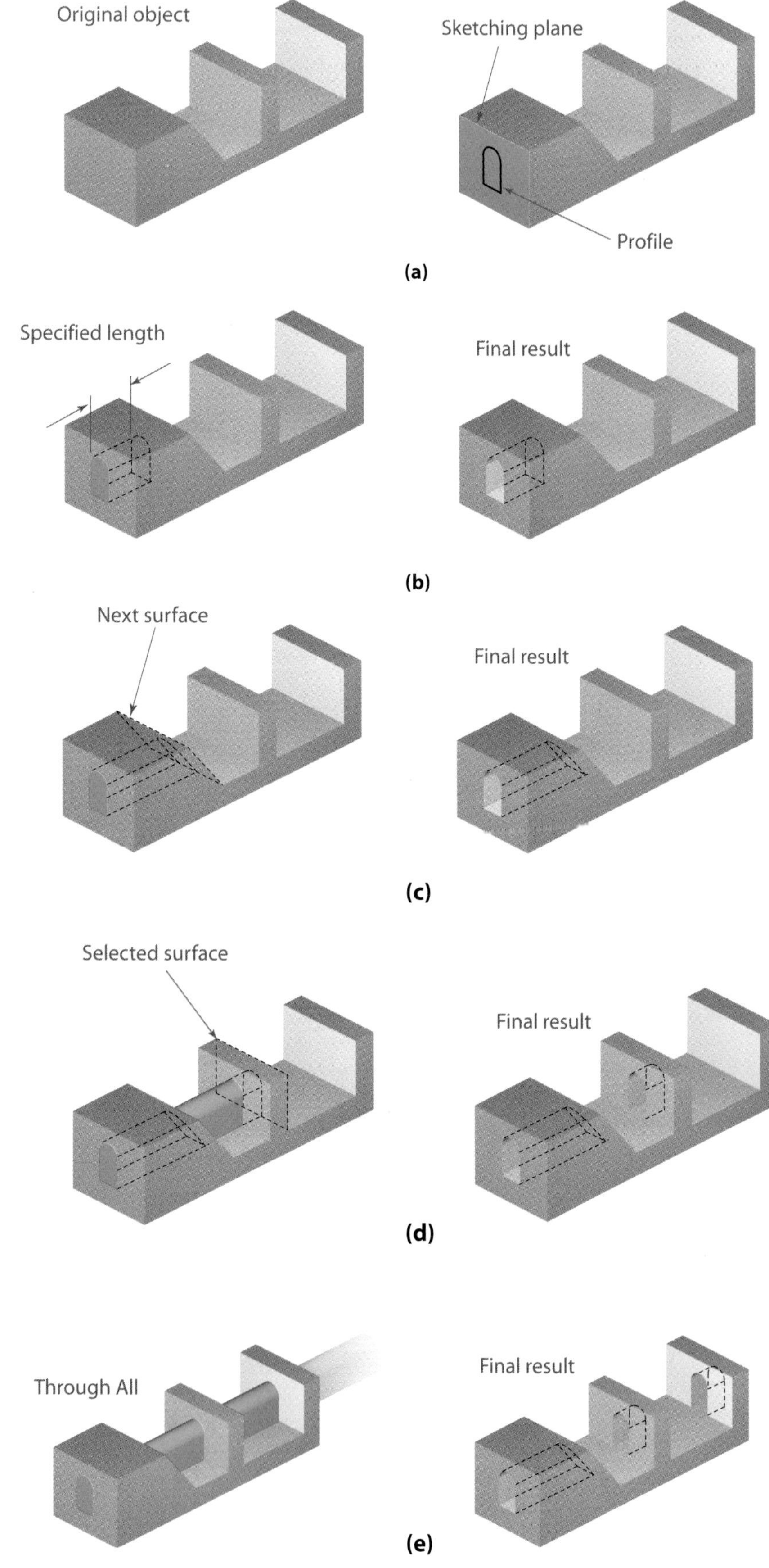

FIGURE 6.31. Different ways to terminate an extruded cut from the profile in the sketching plane in (a). Blind cut in (b); cut to next surface in (c). Cutting to a selected surface (d) and cutting through all (e).

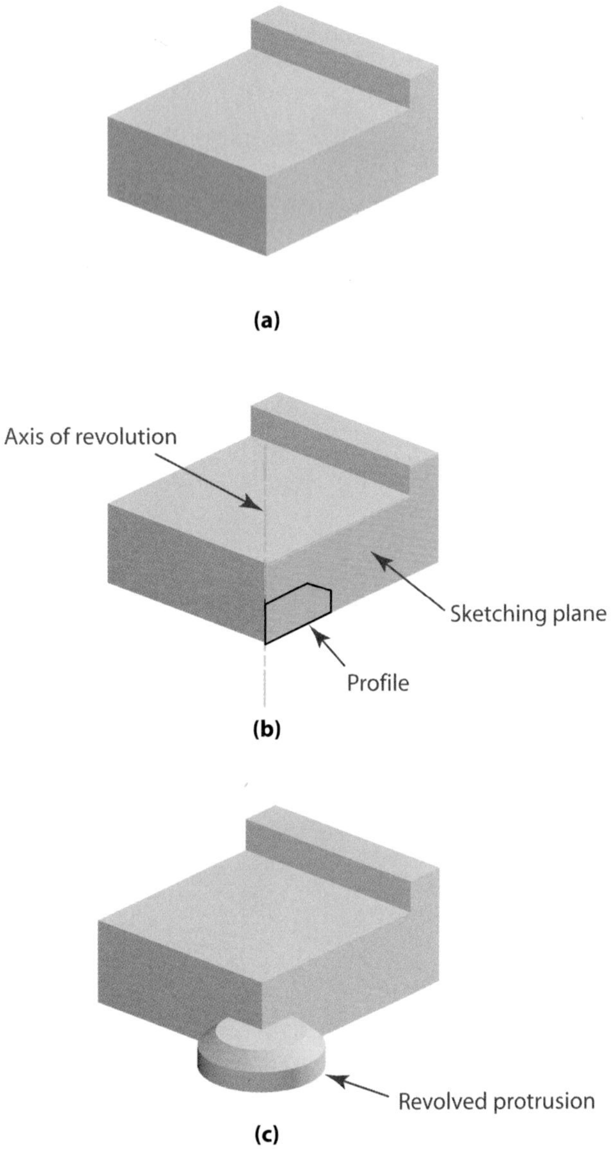

(a)

Axis of revolution

Sketching plane

Profile

(b)

Revolved protrusion

(c)

FIGURE 6.32. The addition of a revolved protrusion to an existing base in (a) by using one of its surfaces as a sketching plane to create a centerline and profile in (b) and revolving it to produce the final result in (c).

(a)

Profile

Sketching plane

Axis of revolution

(b)

Revolved cut

(c)

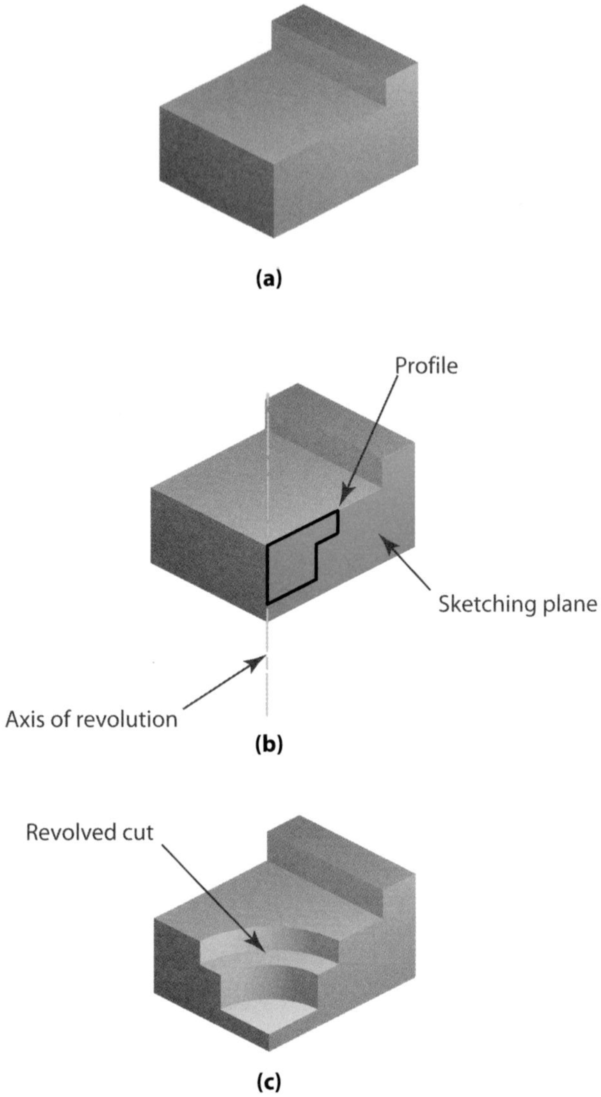

FIGURE 6.33. The addition of a revolved cut to an existing base in (a) by using one of its surfaces as a sketching plane to create a centerline and profile in (b) and revolving it to produce the final result in (c).

6.07 Breaking It Down into Features

When you build a solid model, you need to decide how to create the various shapes that compose the part. Very few parts can be modeled as a single extrusion or revolution. The various protrusions and cuts on the main body of a model are called **features**. What are features? If you consider your face, you might say that its features are your eyes, nose, lips, and cheeks. It is not much different on a manufactured part; a feature can be any combination of geometric shapes that make up the part and are distinctive in shape, size, or location. Features are characteristic elements of a particular object, things that stand out or make the object unique. Features often have characteristic geometric shapes and specific functions. A simple hole, for example, is a cylindrical cut that is often used as a receptacle for a fastener such as a bolt or screw. A manufactured part may have many different types of features. Since these features are the foundation of contemporary solid modeling systems, you must be able to recognize them.

Engineered parts also have features that are composed of repeated combinations of shapes. Most feature-based modelers have a collection of standard built-in features and may also allow you to define your own features. This can be handy when your products are designed with a particular shape that varies in size, such as gear teeth, airfoils, or turbine blades. The challenge for designers is to identify part features and build solid models that reflect the function of the part and design intent.

6.07.01 The Base Feature

All of your parts will be created from a collection of features, but you need to start your model with a basic shape that represents the general shape of the object. Your first step should be to study the part and identify the shape that you will use as the **base feature**. The base feature should be something that describes the overall shape of the part or something that gives you the greatest amount of functional detail that can be created with a single extrusion or rotation. Figure 6.34 shows several parts with the base features used to create the solid models.

After the base feature is created, you can modify the shape by adding or subtracting material to it to create form features. A **form feature** is a recognizable region or area on the part geometry that may have a specific function and/or method of manufacture. The geometric components or shapes within the feature usually have some geometric relationships or constraints. Different CAD systems use various names for these features, but you should become familiar with some of the common terms. The following section discusses common feature types.

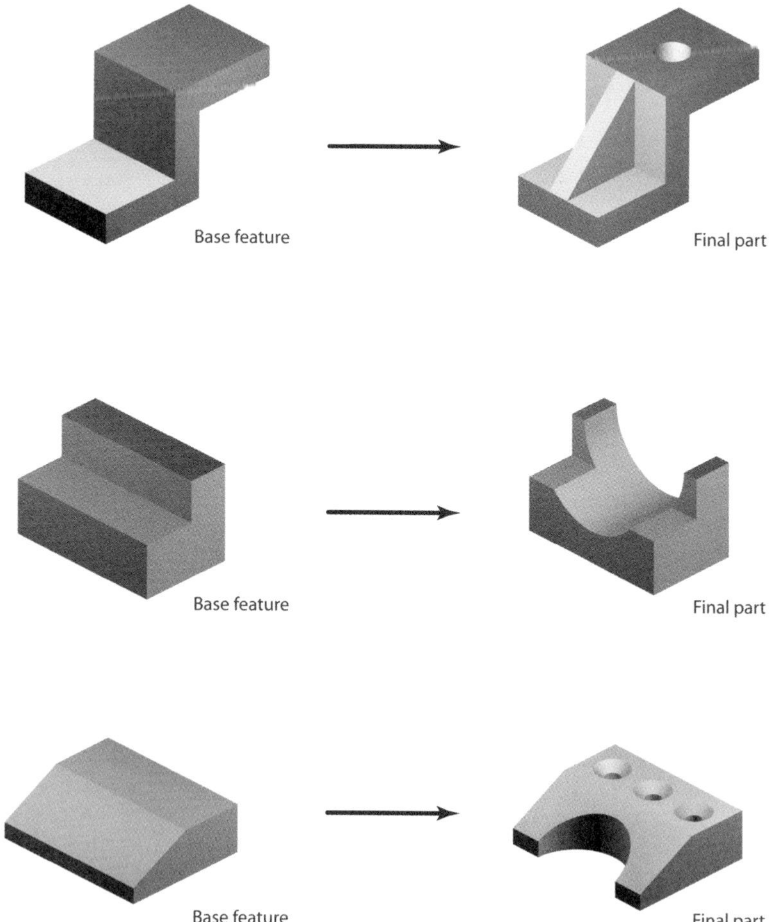

FIGURE 6.34. Parts and their base features.

FIGURE 6.34. (CONTINUED)
Parts and their base features.

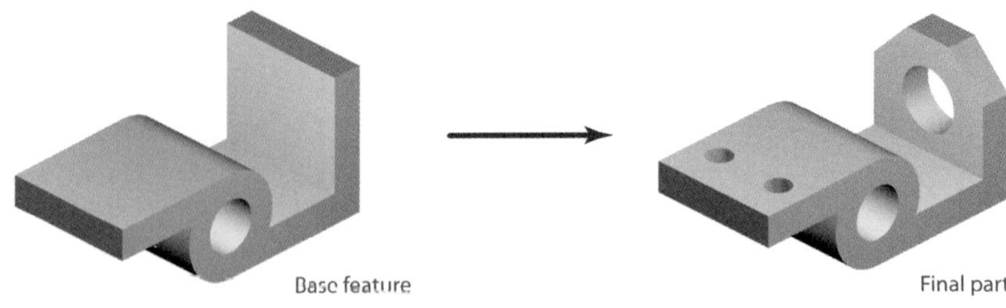

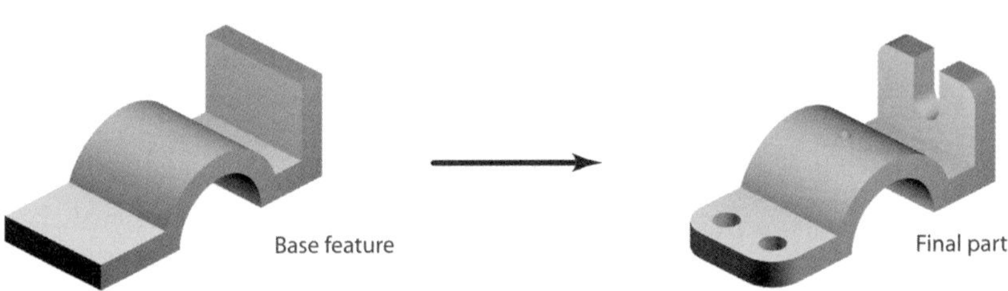

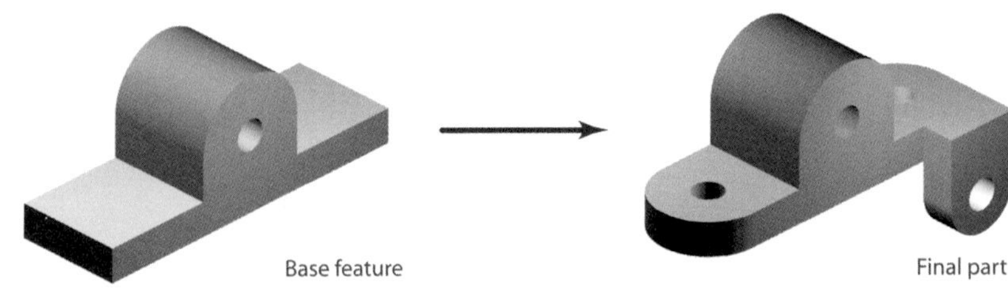

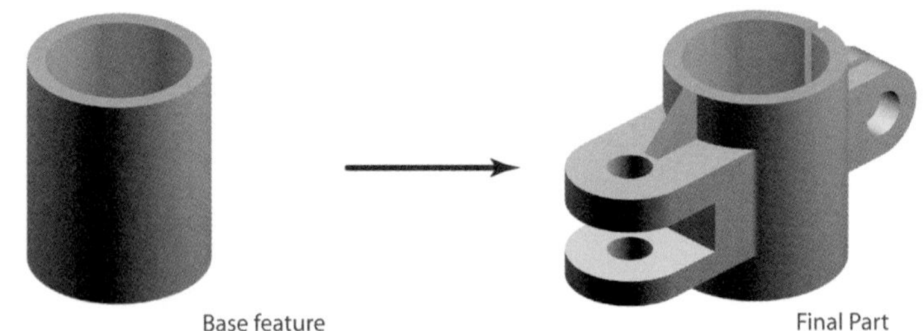

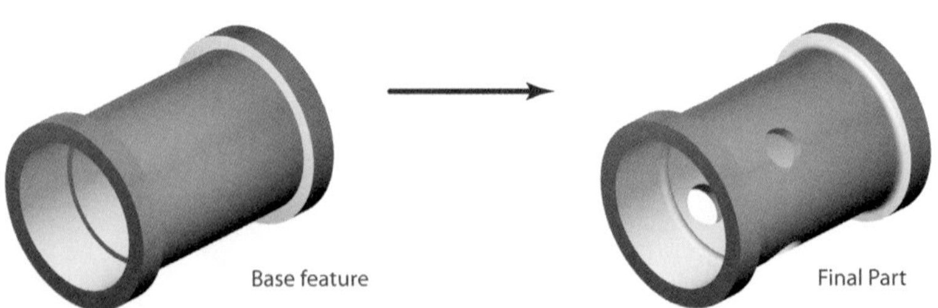

6.07.02 Chamfers, Rounds, and Fillets

Unless otherwise specified, adjoining surfaces on a virtual part can intersect to form sharp corners and edges, but real parts often have smooth transitions along the edges of these surfaces. The most common edge transitions are **rounds**, **fillets**, and **chamfers**. A round is a smooth radius transition of the external edge created by two intersecting surfaces. A fillet is a smooth transition of the internal edge created by two intersecting surfaces. Geometrically, the rounds and fillets are tangent to both intersecting surfaces. Examples of rounds and fillets are shown in Figure 6.35. Fillets and rounds are specified by the size of their radii and the edge(s) that are rounded.

Chamfers also provide a transition between two intersecting surfaces, but the transition is an angled cut instead of a radius. Examples of chamfers are shown in Figure 6.36. Chamfers can be specified by the distance along each intersecting surface to the original edge or by the distance along one of the original surfaces and the angle made with that surface.

Functionally, on an inside edge, a fillet may be necessary to facilitate fabrication or to reduce stresses at the corner so the part does not break as easily. On an outside edge, rounds and chamfers are usually used to eliminate sharp edges that can be easily damaged or that can cause injury or damage when the part is handled. Rounds, fillets, and chamfers are generally small when compared to the overall size of the associated base or parent feature.

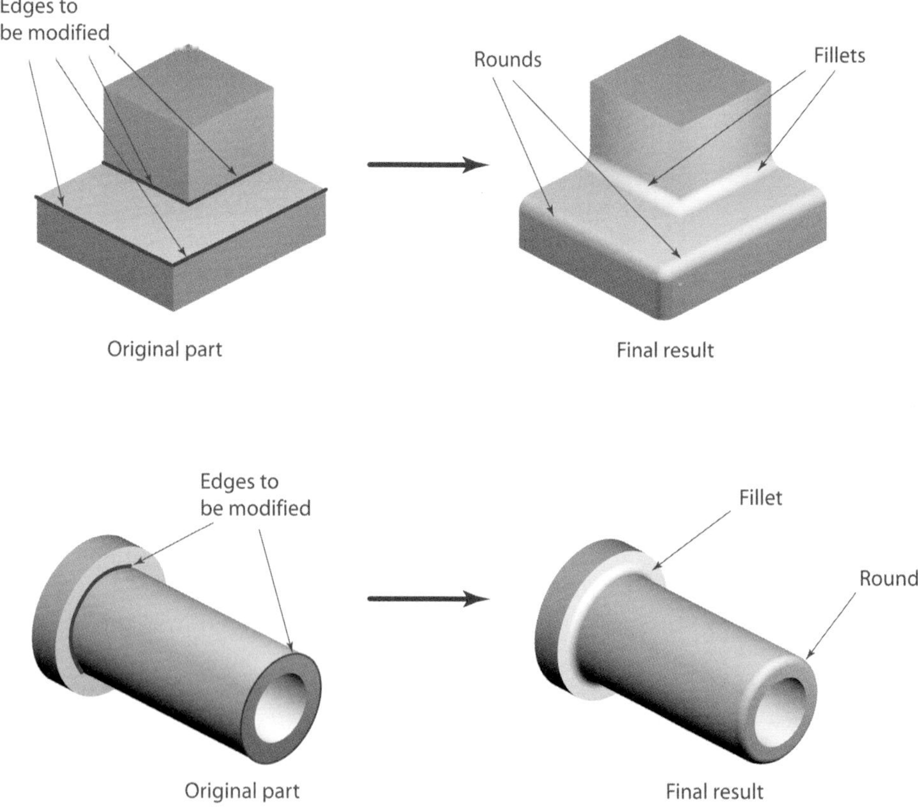

FIGURE 6.35. Examples of rounds and fillets applied to the edges of a part.

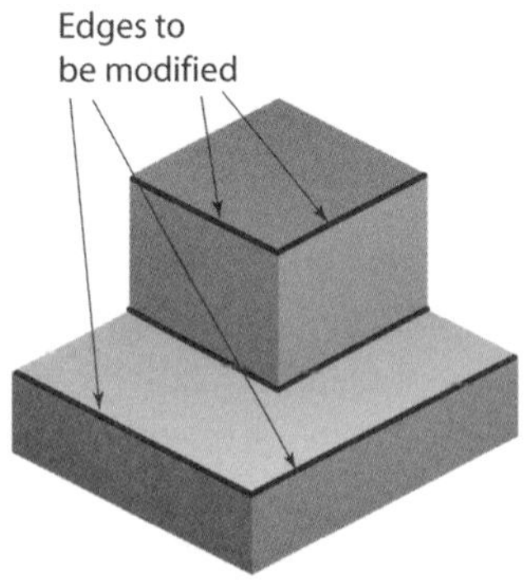

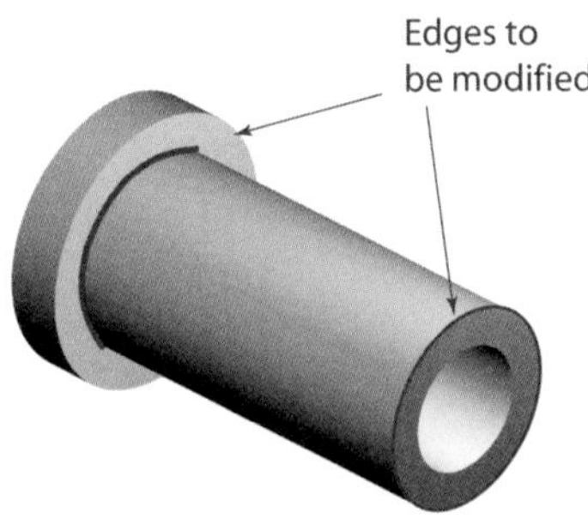

FIGURE 6.36. Examples of chamfering applied to the edges of a part.

6.07.03 Holes

Holes are ubiquitous in nearly all manufactured parts and, therefore, can be inserted into a model as features by most solid modeling software. Holes are often used with bolts or screws to fasten parts together. Many different types of holes can be used with specific fasteners or can be created using different manufacturing processes. Some special types include holes that are blind, through, tapped, counterbored, or countersunk, as shown in Figure 6.37. Each type of hole has a particular geometry to suit a specific function. You should study the hole types so you recognize them when you model your parts.

Many solid modeling software packages include standard or built-in features to help you with your modeling task. When you use a standard hole feature, the solid modeling software makes certain assumptions about the geometry of a hole so you do not need to specify all of the dimensions and constraints that make up the feature. A countersunk hole, for example, can be made as a revolved cut. What do you need to do to create this feature? You begin by selecting a sketching plane, then create and constrain the sketch and revolve the sketch about a specified axis. Many things can go wrong if you are not careful. There might not be a plane on which to sketch. Your sketch might not have the proper shape, or the axis might not be perpendicular to the desired surface. However, a countersunk hole feature can often be created from a standard feature by selecting the location of the axis of the hole on the desired surface, the diameter of the hole, the diameter and angle of the countersink, and the depth of the hole. The shape of the profile, axis of revolution, and angle of revolution are included automatically in the feature definition. No sketching plane is needed.

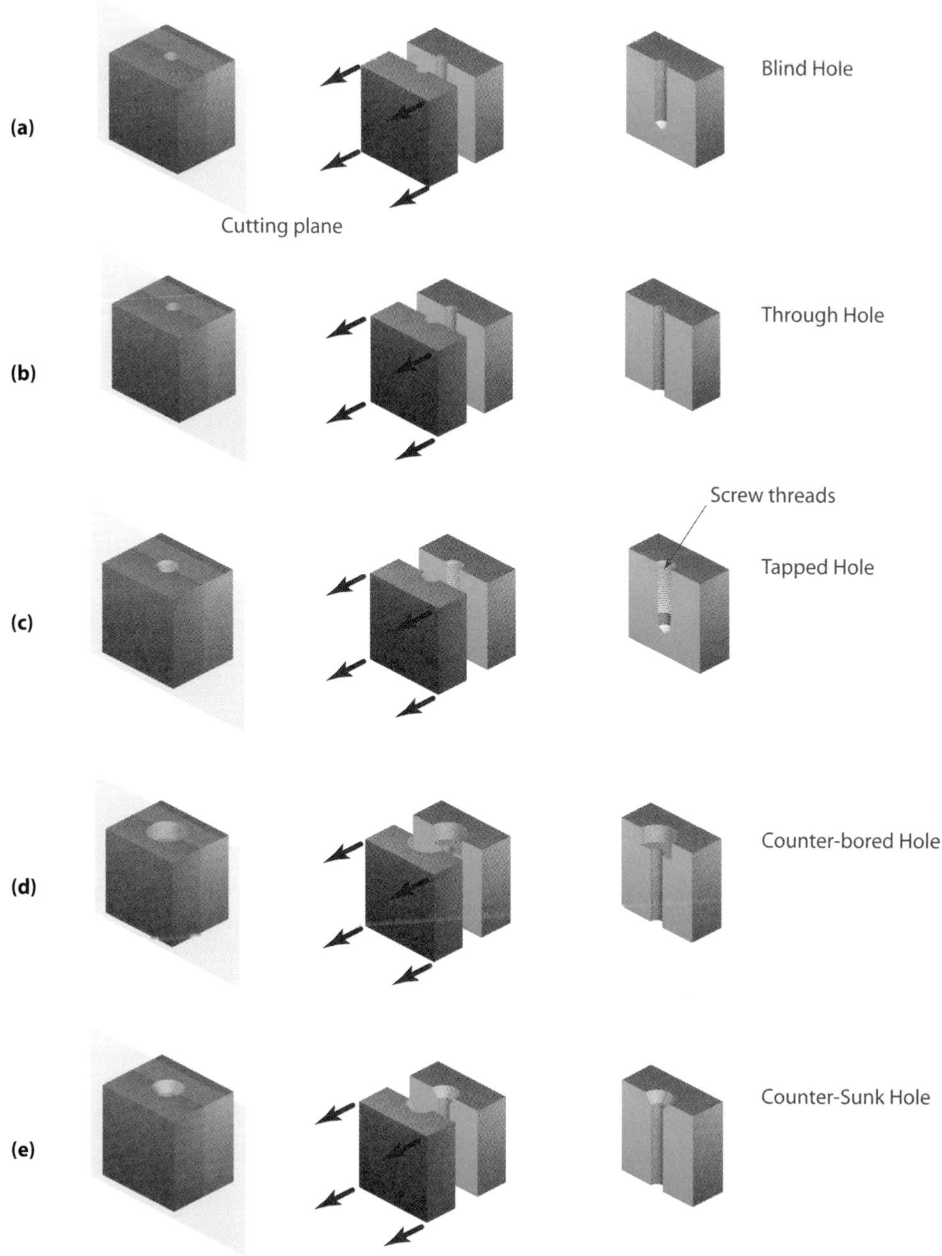

FIGURE 6.37. Cross sections of various types of holes to reveal their geometry.

Figure 6.38 illustrates the use of a cut feature compared to the use of a built-in hole feature to create a countersunk hole. In most cases, it is more desirable to create a hole using the built-in hole feature instead of a general purpose cut feature. Besides being a more natural way to place a hole in a model, you avoid potential errors in creating the desired geometry. Furthermore, using a general cut feature does not incorporate the specific geometry and function of a "hole" in the knowledge base of the model, which may be useful in downstream applications such as process planning for manufacturing the hole.

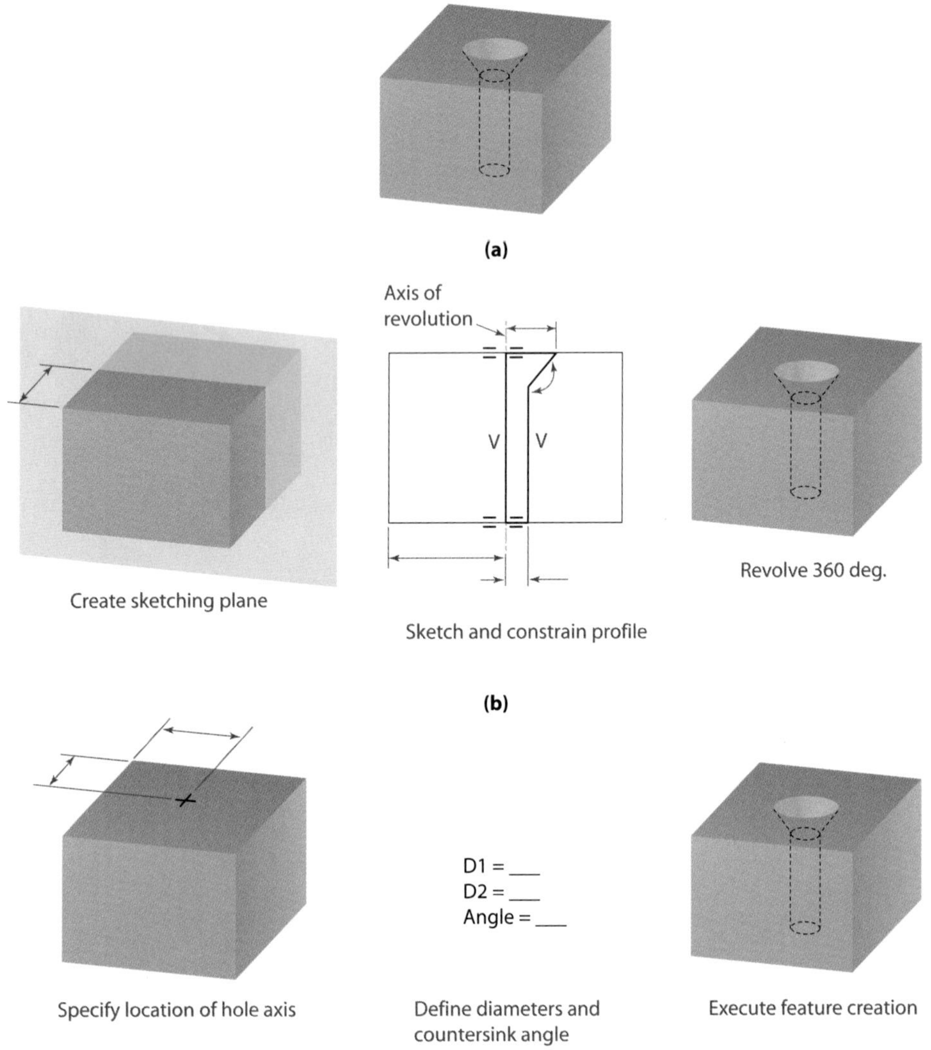

(a)

Axis of revolution

V V

Create sketching plane

Sketch and constrain profile

Revolve 360 deg.

(b)

D1 = ___
D2 = ___
Angle = ___

Specify location of hole axis

Define diameters and countersink angle

Execute feature creation

FIGURE 6.38. The countersunk hole shown in (a) can be created by using a general revolved cut, as shown in (b), or by specifying the hole as a built-in feature, as shown in (c).

6.07.04 Shells

The process of creating a shell, or **shelling**, removes most of the interior volume of a solid model, leaving a relatively thin wall of material that closely conforms to the outer surfaces of the original model. Shelled objects are often used to make cases and containers. For example, a soda bottle is a shell, as are cases for electronic products such as cell phones and video displays. The walls of a shell are generally of constant thickness, and at least one of the surfaces of the original object is removed so the interior of the shell is accessible. Figure 6.39 shows examples of a model that has been shelled.

Shelling is sometimes considered an operation rather than a feature. It is usually performed on the entire model, including all of its features, by selecting the surfaces to be removed and the thickness of the shell wall. Any feature not to be shelled should be added to the model after the shelling operation is complete. The order of feature creation and shelling operations may have a dramatic effect on the shape of the part, as will be shown later in this chapter.

FIGURE 6.39. Examples of shelling.

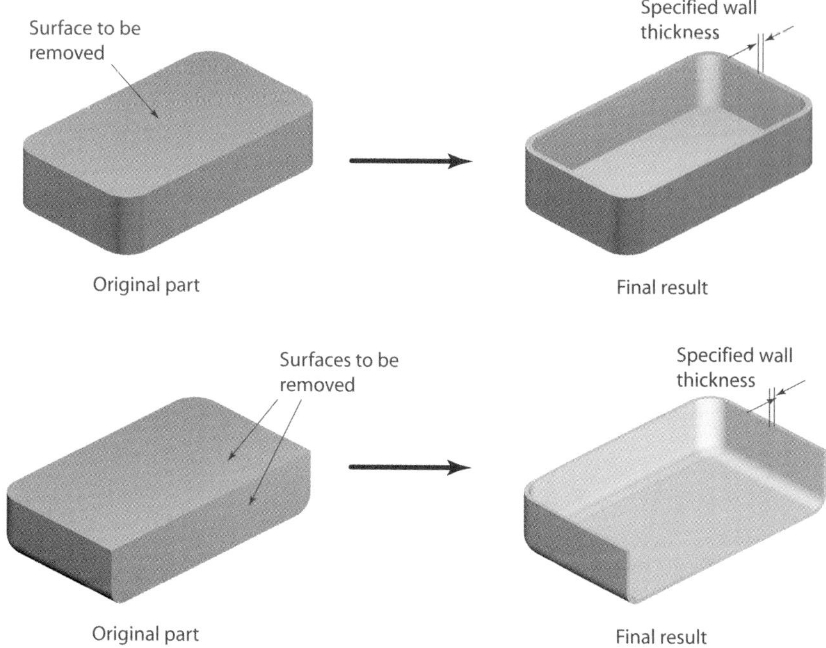

6.07.05 Ribs and Webs

Ribs are small, thin, protrusions of constant thickness that extend predominantly from the surface of a part. Ribs are typically added to provide support or to stiffen a part. Sometimes they are added to improve a part's heat transfer ability. **Webs** are areas of thin material that connect two or more heavier areas on the part. Examples of ribs and webs are shown in Figure 6.40. These features are usually specified by their flat geometry, thickness, and location.

FIGURE 6.40. Ribs (a) added to parts to reinforce them, and webs (b) connect thicker sections on parts.

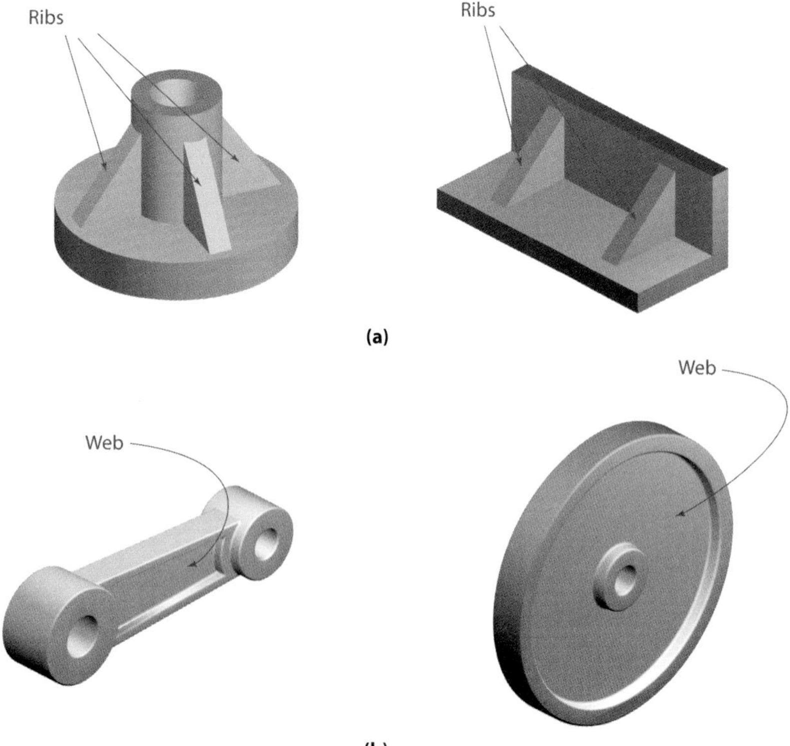

6.07.06 Other Feature Types

The features that follow (and that are shown in Figure 6.41) are less commonly found in solid modelers. When available, they should be used as needed. When such features are not available in the solid modeler, the geometric shapes can still be created from sketched profiles as protrusions or cuts. Note that special feature types usually imply a particular shape, function, manufacturing process, or other feature attribute.

- Boss—a slightly raised circular area, usually used to provide a small, flat, clean surface
- Draft—a slight angle in the otherwise straight walls of a part, usually used to facilitate its removal from a mold
- Groove—a long, shallow cut or annulus
- Island—an elongated or irregularly shaped raised area, usually used to provide a flat, clean surface
- Keyseat—an axially oriented slot of finite length on the outside of a shaft
- Keyway—an axially oriented slot that extends the entire length of a hole
- Slot—a straight, long cut with deep vertical walls
- Spot face—a shallow circular depression that has been cut, usually used to provide a small, flat, clean surface
- Taper—a slight angle in the otherwise cylindrical walls of a part, usually used to facilitate its insertion or removal into another part

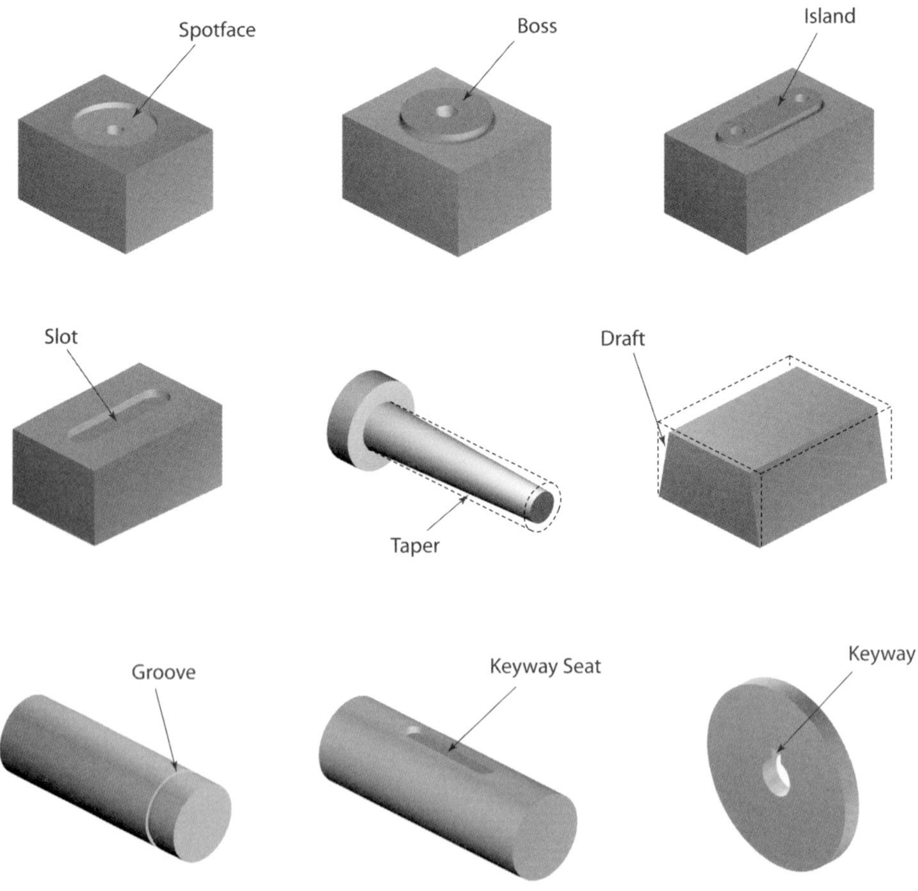

FIGURE 6.41. Various features with specific functions that may be added to a solid model.

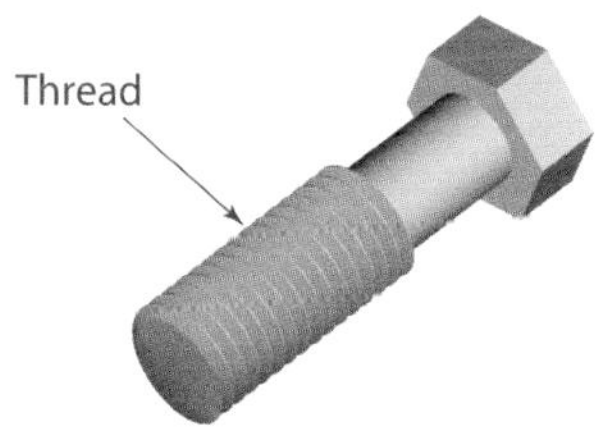

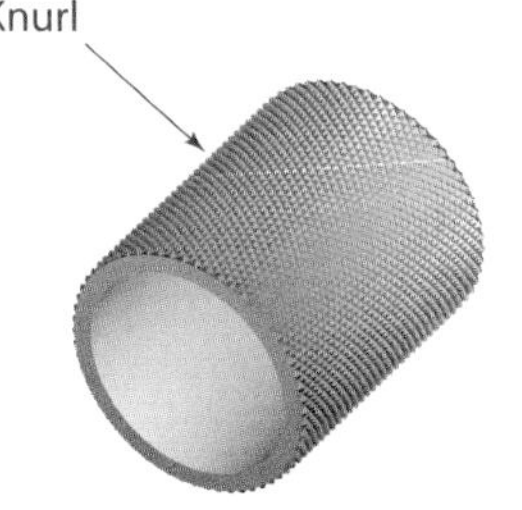

FIGURE 6.42. Some cosmetic features.

6.07.07 Cosmetic Features

Parts can be modified by altering their surfaces characteristics. These characteristics are called **cosmetic features** because they generally modify the appearance of the surface but do not alter the size or shape of the object, just like lipstick or hair coloring. Cosmetic features are necessary to the function of the part and may be included in the model so they can be used in later applications, such as fabrication. Some common cosmetic features include threads and knurls. Since the geometric changes are small and detailed, the cosmetic features usually are not modeled in their exact geometric form in the database of the object, but are included as notes or with a simplified geometric representation. You will learn more about simplified representations on drawings in later chapters. Some cosmetic features are shown in Figure 6.42.

6.07.08 An Understanding of Features and Functions

As a design engineer, you need to become familiar with the different types of features on various parts. Doing so will help you communicate with other engineers as well as imbed more of a part's engineering function into your models. For example, if you look at Figure 6.43, you will notice a rectangular cut on the edge of the hole. This cut is a geometric feature called a keyway. Why is it there? What purpose does it serve? In Figure 6.44, the gear is mounted to a shaft, which also has a rectangular cut. A small part called a key is used to line up the shaft and the gear and transmits torque from the shaft to the gear. If you were to create a feature-based solid model of the gear, you could identify the rectangular cut as a keyway feature. If the model parts were to be assembled with assembly modeling software (which is explained in detail in a subsequent chapter), the computer and software would recognize the models as mating parts and orient the gear, key, and shaft automatically.

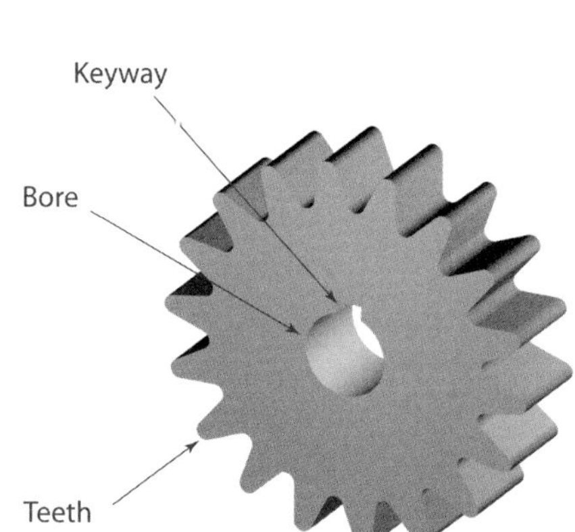

FIGURE 6.43. A gear with teeth, a bore, and a keyway as functional features.

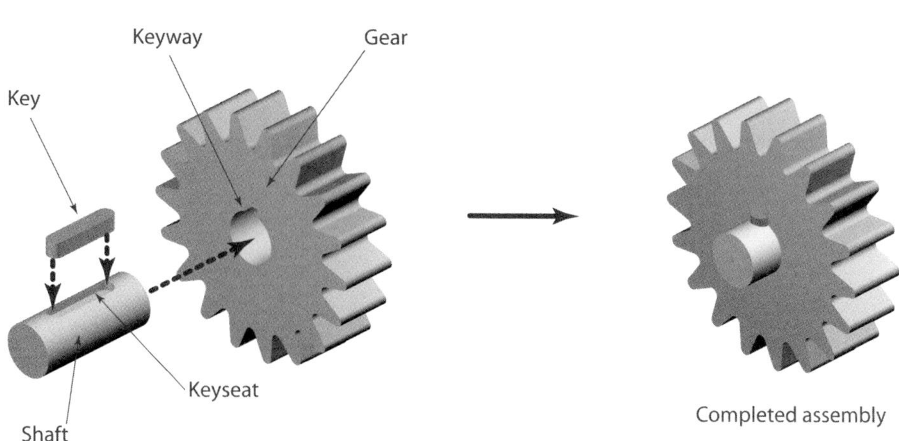

FIGURE 6.44. A gear and shaft assembly. The key functions to transmit torque. The keyseat receives the key in the shaft, and the keyway receives the key in the gear.

6.08 More Ways to Create Sophisticated Geometry

Creating protrusions and cuts by extending the sketch profiles made on either the basic modeling plane or one of the existing surfaces of the model results in a wide variety of possible models. Even more sophisticated models, however, can be created by using reference geometries called datums, which can be added to the model, displayed, and used to create features. Generally, solid modelers offer at least three types of **datum geometries** that can be placed into a model: datum points, datum axes, and datum planes. These datum geometries do not actually exist on the real part (i.e., they cannot be seen or felt) but are used to help locate and define features. Consider, for example, the part shown in Figure 6.45(a). The angled protrusion with the hole would be easy to create if an angled sketching plane could be defined as shown. The extrusion could be made to extend from the sketching plane to the surface of the base feature. This feature would be more difficult, although not impossible, to define using extruded protrusions and cuts that extended only from the basic modeling planes or one of the surfaces on the existing model. In Figure 6.45(b), the uniquely shaped web would be easy to create if a sketching plane could be placed between the connected features as shown. An extruded protrusion could extend from both sides of the sketching plane to the surfaces of the connected features.

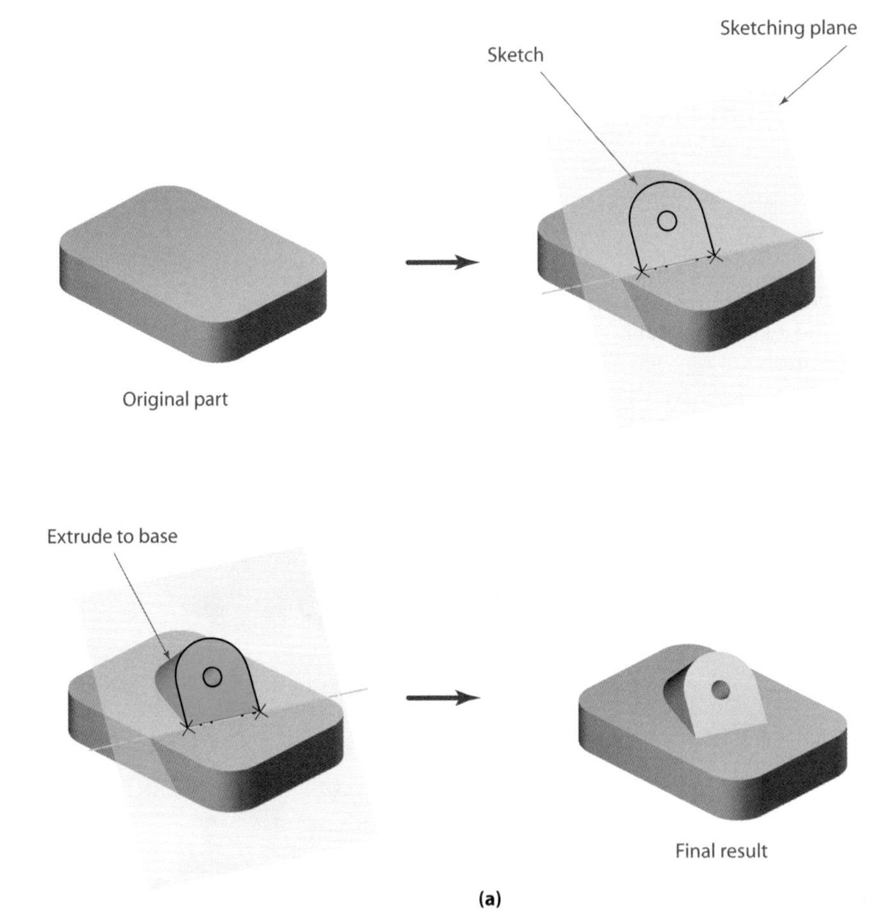

FIGURE 6.45. Using a sketching plane and profile, which are not on an existing surface of the object, to create a protrusion feature (a).

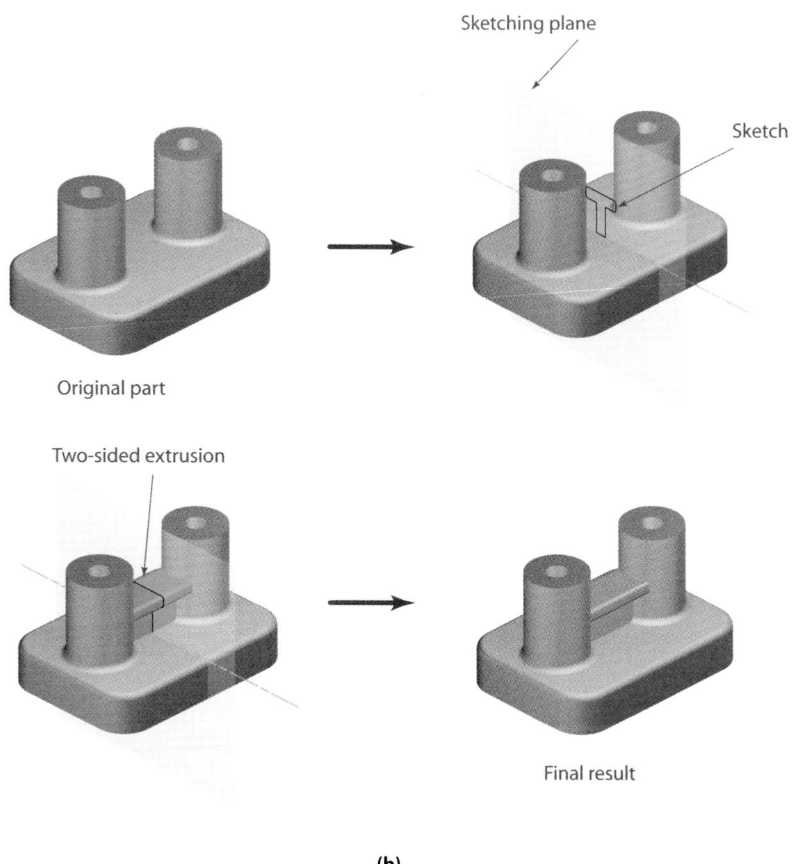

FIGURE 6.45. (CONTINUED) Using a sketching plane and profile, which are not on an existing surface of the object, to create a web feature (b).

The next few sections will describe methods in which the three different types of datums can be defined geometrically and how the datums can be used to create a variety of new types of features. Depending on the specific solid modeling software being used, some of the methods described here for datum definition may or may not be available.

6.08.01 Defining Datum Points

Following are some of the different ways a datum point can be defined and created. The definitions are shown graphically in Figure 6.46.

- At a vertex
- On a planar surface at specified perpendicular distances from two edges
- At the intersection of a line or an axis and a surface that does not contain the line

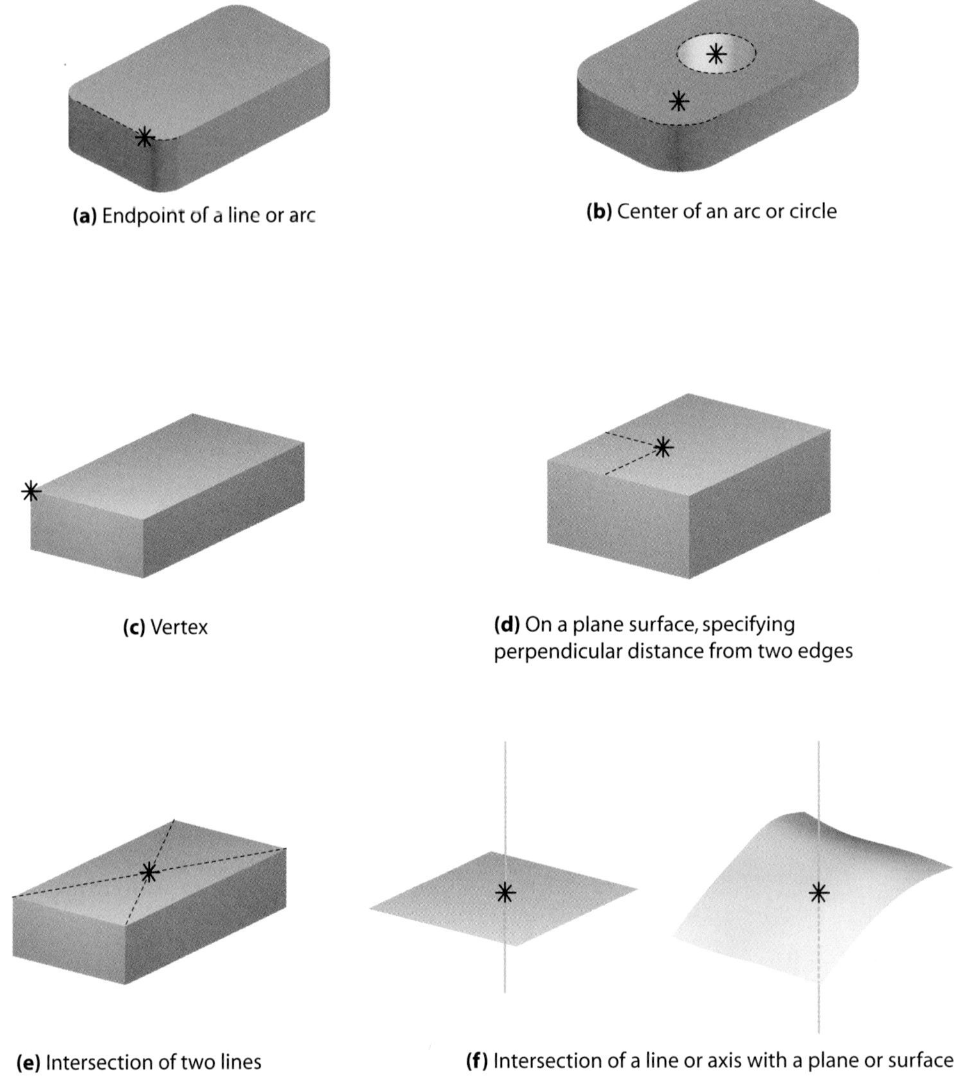

(a) Endpoint of a line or arc
(b) Center of an arc or circle
(c) Vertex
(d) On a plane surface, specifying perpendicular distance from two edges
(e) Intersection of two lines
(f) Intersection of a line or axis with a plane or surface

FIGURE 6.46. Various ways to define a datum point.

6.08.02 Defining Datum Axes

Following are some of the different ways a datum axis can be defined and created. The definitions are shown graphically in Figure 6.47.

- Between two points (or vertices)
- Along a linear edge
- At the intersection of two planar surfaces
- At the intersection of a cylinder and a plane through its axis
- Along the centerline of a cylinder or cylindrical surface

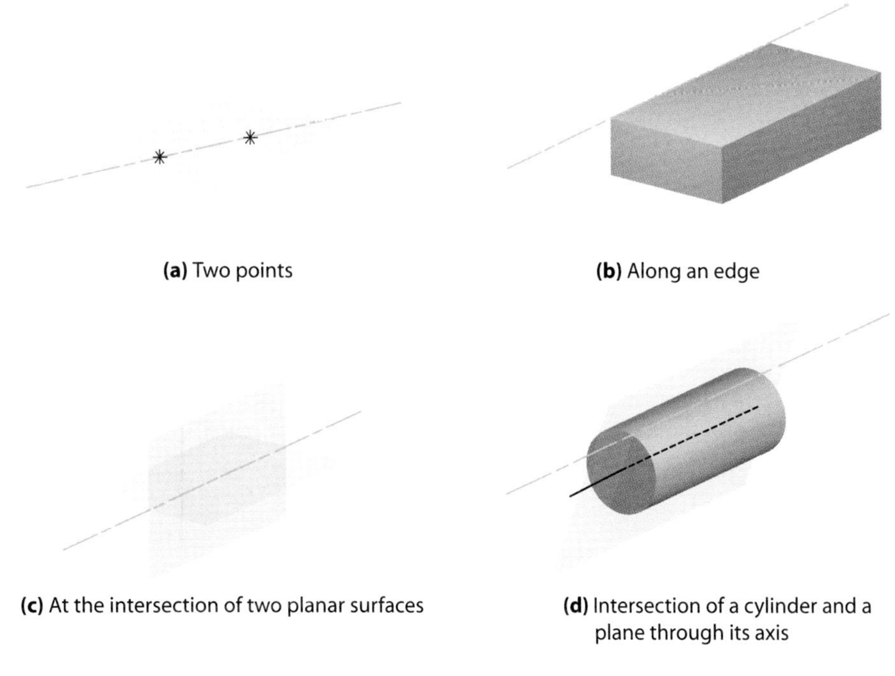

(a) Two points

(b) Along an edge

(c) At the intersection of two planar surfaces

(d) Intersection of a cylinder and a plane through its axis

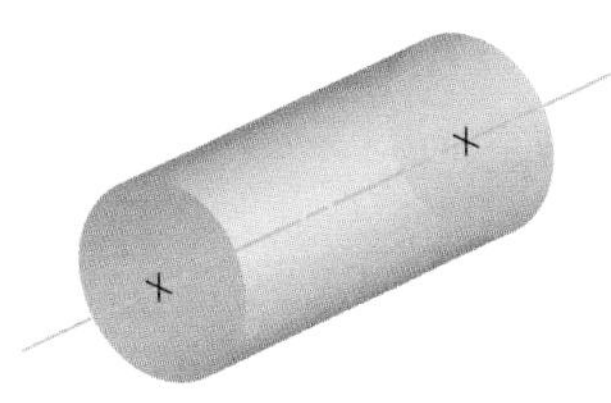

(e) At the centerline of a cylinder or cylindrical surface

FIGURE 6.47. Various ways to define a datum axis.

6.08.03 Defining Datum Planes

Following are some of the different ways a datum plane can be defined and created. The definitions are shown graphically in Figure 6.48.

- Through three noncolinear points
- Through two intersecting lines
- Through a line and a noncolinear point
- Offset from an existing flat surface at a specified distance
- Through an edge or axis on a flat surface at an angle from that surface
- Tangent to a surface at a point on that surface
- Perpendicular to a flat surface and through a line parallel to that surface
- Perpendicular to a flat or cylindrical surface through a line on that surface
- Tangent to a cylindrical surface at a line on that surface

FIGURE 6.48. Various ways to define a datum plane. More ways to define a datum plane.

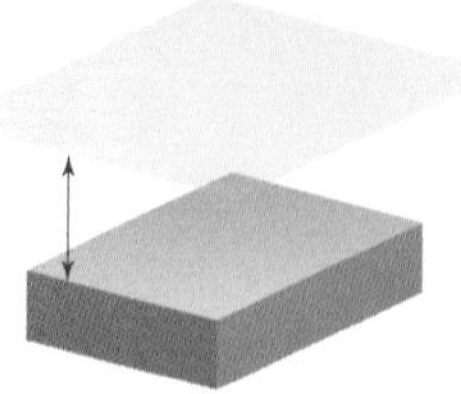

(a) Offset from Existing flat surface, and a distance

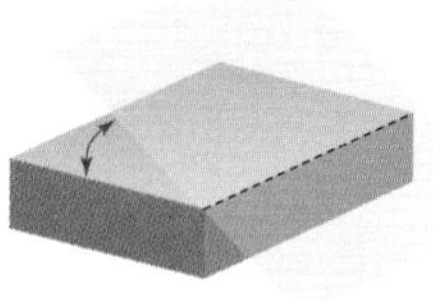

(b) A flat surface, through an edge or axis on that surface, and an angle from that surface

(c) Tangent to a surface at a point on that surface

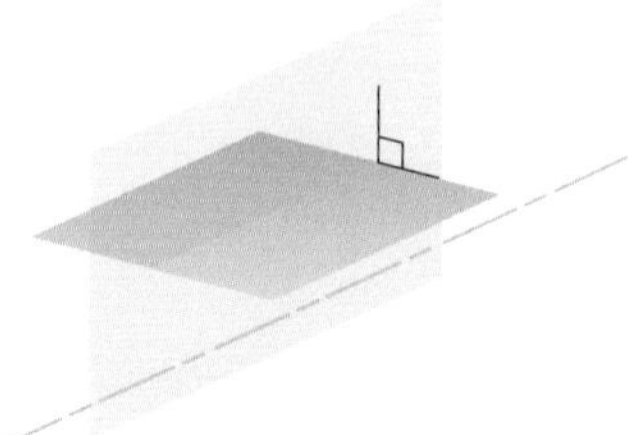

(d) Perpendicular to a flat surface, and through a line parallel to that surface

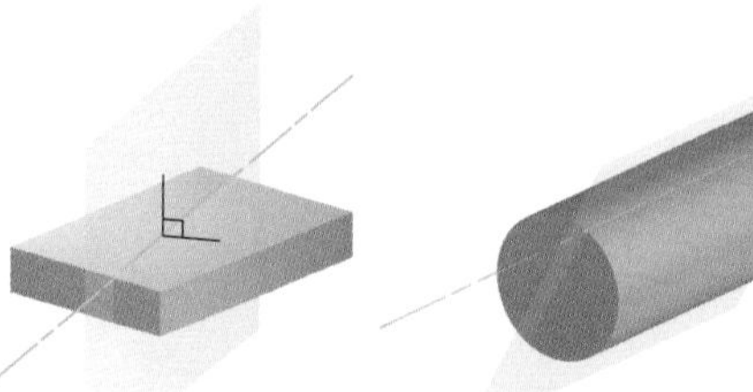

(e) Perpendicular to a flat or cylindrical surface and through a line on that surface

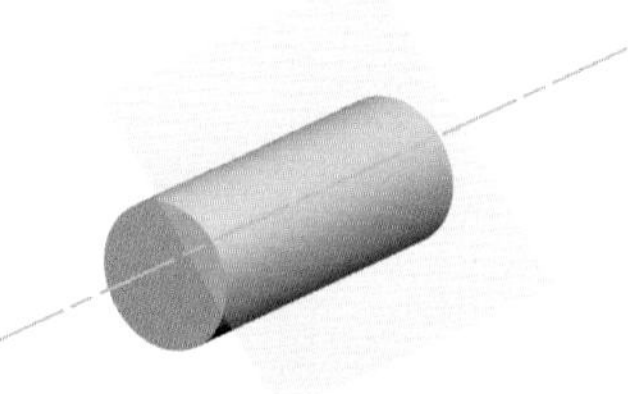

(f) Tangent to a cylindrical surface at a line on that surface

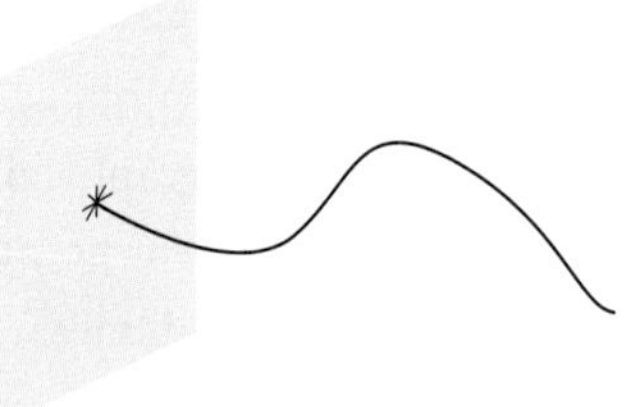

(g) Perpendicular to a curve at a point on that curve

(h) Three points

(i) Two intersecting lines

(j) A Line and a point

6.08.04 Chaining Datums

Series of simply defined datums are often used for creating more complex datums. In the example shown in Figure 6.45(a), the angled protrusion was created in this manner: On the top surface of the base extrusion, two datum points were created by defining each of their locations from the edges of the base extrusion. A datum axis was then created using the two datum points as the endpoints of the axis. Finally, the desired datum plane was defined using the top surface of the base extrusion, using the datum axis created in that plane, and specifying the angle that the new datum plane makes with the top surface.

Another example is shown in Figure 6.49, where a datum plane is created to be tangent to the surface of a cylindrical extrusion. An intermediate datum plane is defined by one of the basic planes; the axis of the cylinder, which lies on that basic plane; and the angle the intermediate datum plane makes with the basic plane. The final datum plane is then created to be tangent to the surface of the cylindrical extrusion at its intersection with the intermediate datum plane. A datum plane tangent to the surface of a cylinder is commonly used to create cuts that extend radially into a cylindrical surface, such as holes or slots, and protrusions that extend radially from the cylindrical surface, such as spokes or vanes. Note that with protrusion from a tangent datum plane, the extrusion must be specified to extend in both directions from that datum; otherwise, there will be a gap between the extrusion and the curved surface.

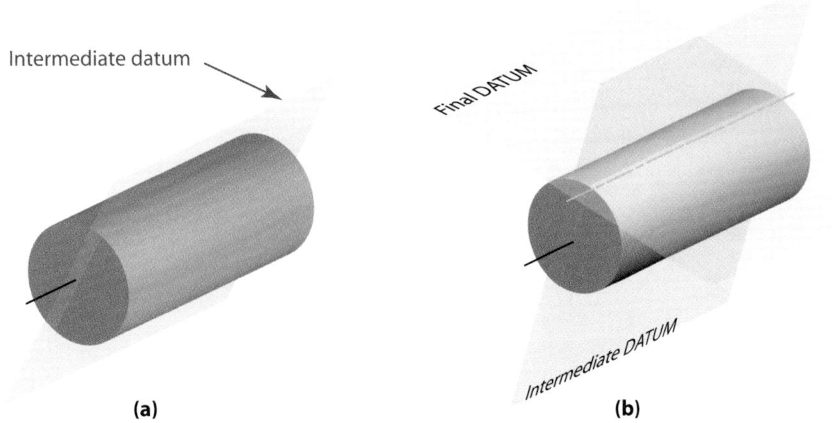

FIGURE 6.49. To create a datum plane that is tangent to a cylindrical surface at a specific location, an intermediate datum plane, shown in (a), can be created through the centerline of the cylinder. The intersection of the intermediate datum with the cylinder creates a datums axis that is used to locate the final datum plane.

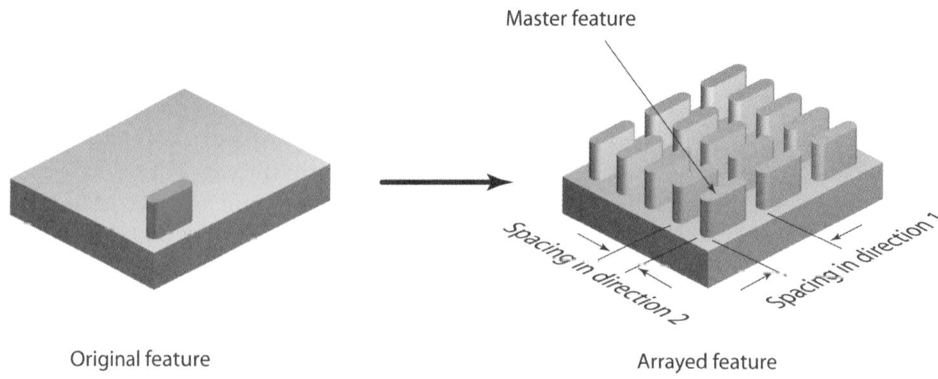

FIGURE 6.50. A rectangular array of protrusions created from a master feature.

6.08.05 Using Arrays (Rectangular and Circular)

One method of creating multiple identical copies of a feature in a model is to create a **feature array**, which is sometimes called a **feature pattern**. A feature array takes one feature, called the **master feature**, and places copies of it on the model at a specified spacing. The copied features are identical to the master feature, and changing the geometry of the master at a later time also changes the geometry of the copies at that time. Including features in this manner can save time and effort in creating the entire model, especially when the features are rather complex. An example of a model with a rectangular array of features is shown in Figure 6.50. An array of rectangular cuts is shown in Figure 6.51. As shown, rectangular arrays can generate copied features in two directions. These directions must be specified, as well as the spacing of the copied features in each direction. Finally, the number of copies in each direction must be specified. Care must be taken to ensure that there is enough room on the model to accommodate all of the copied features.

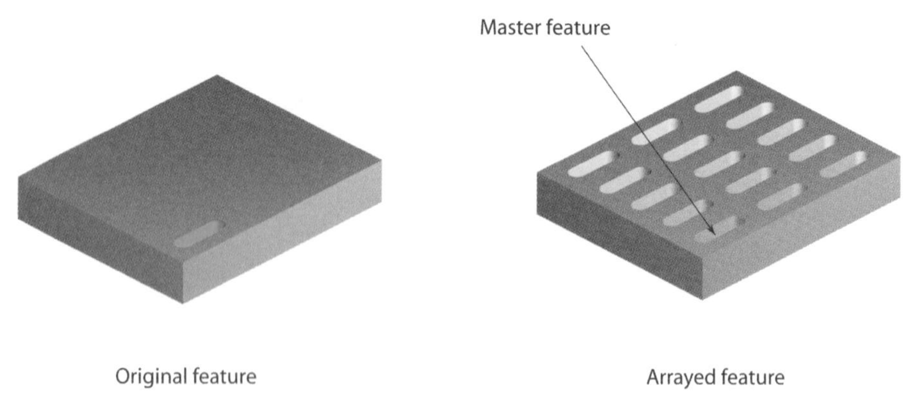

FIGURE 6.51. A rectangular array of cuts created from a master feature.

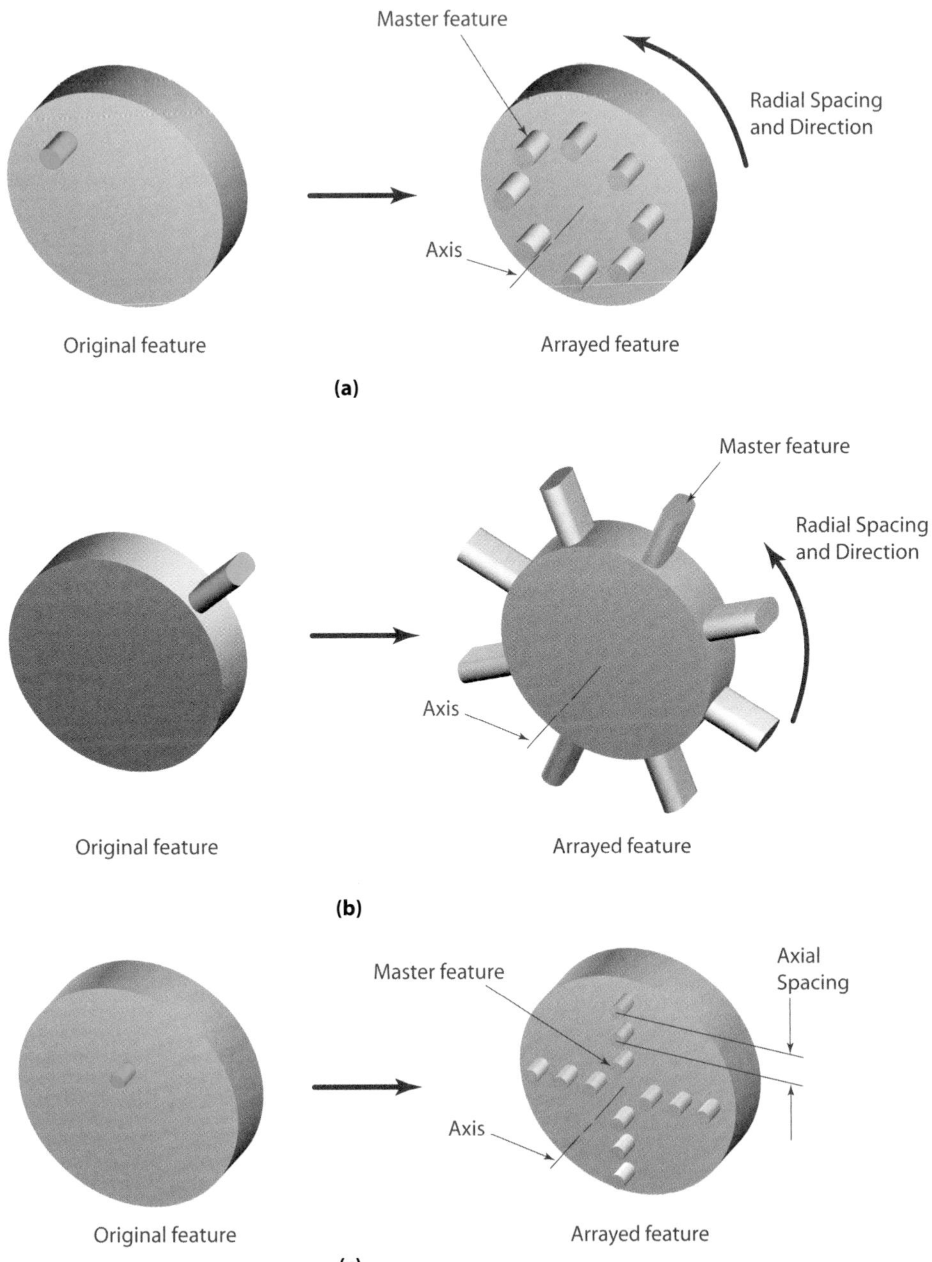

FIGURE 6.52. A circular array of protrusions created from master features in the axial direction (a) and (b), both in the axial and radial direction (c).

Examples of models with radial arrays of protrusions are shown in Figure 6.52. Radial arrays can extend radially or axially. For radial arrays, in addition to the master feature being selected, the axis of revolution for the array must be selected. If such an axis does not already exist on the model, one must be created from an added datum axis. The number of copies, the direction of the array, and the radial and axial spacing of the copies must be specified.

6.08.06 Using Mirrored Features

Another method of creating a feature, when applicable, is to create its mirrored image. To create a **mirrored feature**, you must first identify a mirror plane. You can use an existing plane or define a new datum plane to use as the mirror plane, as shown in Figure 6.53. A mirrored duplicate of the master feature can then be created on the model on the opposite side of the mirror plane. Mirrored features can be cuts or protrusions; however, keep in mind that the copied feature will be a mirror image of the master, not an identical copy. Changing the master feature at a later time also will change the mirrored feature correspondingly. As with arrayed features, using mirrored features can save a great deal of time in model creation, especially when the mirrored feature is complex.

6.08.07 Using Blends

Not all models can be created using just extruded or revolved features. One complex feature is a **blend**. Figure 6.54 shows models with blended surfaces. A blend requires at least two profile sketches, and the model is formed by a smooth transition between

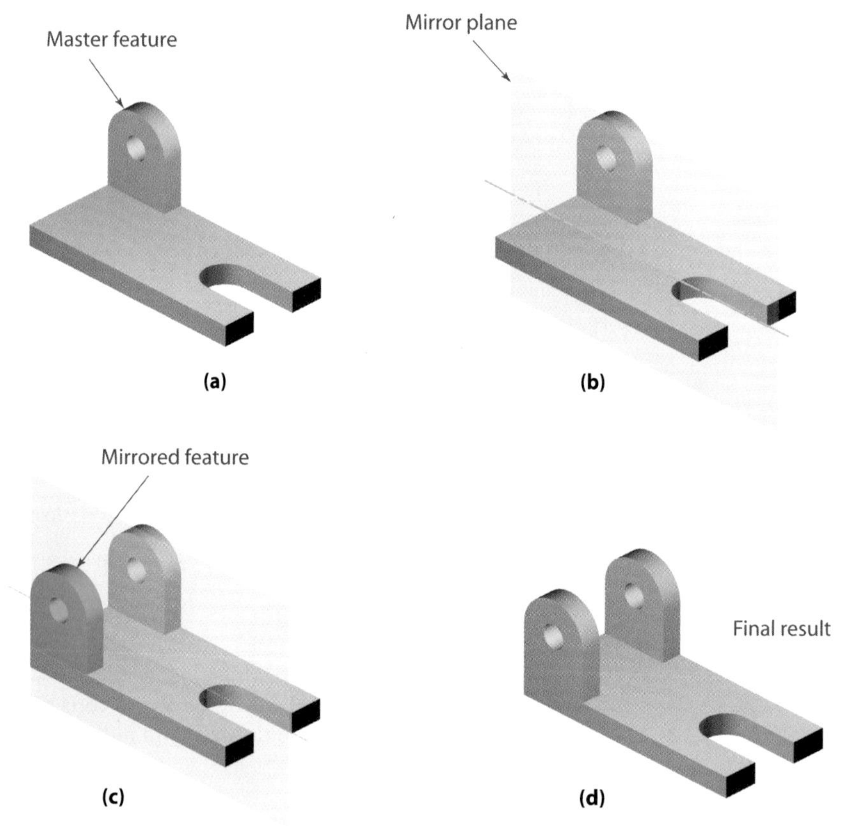

FIGURE 6.53. Creation of a mirrored feature. The master feature in (a) is mirrored by creating a datum plane as a mirror plane (b). A mirror image of the feature is produced on the opposite side of the datum plane in (c), and the final result is shown in (d).

these profiles. The profiles can be sketched on the basic modeling planes, on surfaces of an existing model, or on datum planes. In the simplest blends, the profiles are sketched on parallel planes. Many software packages require the number of vertices on each of the sketched profiles to be equal. If your profiles do not have the same number of vertices, you will have to divide one or more of the entities to create additional vertices. In some sketching editors, circles include four vertices by default. The vertices in all profiles are usually numbered sequentially, and the software usually tries to match the vertices to create an edge between vertices with the same number, as shown in (a) and (b) of Figure 6.54. Rotating the profiles or redefining the vertex numbering can control twisting of the blended transition, as shown in Figure 6.54(c). Further control on the model transition usually can be performed by specifying the slope of the transition at each vertex for each shape.

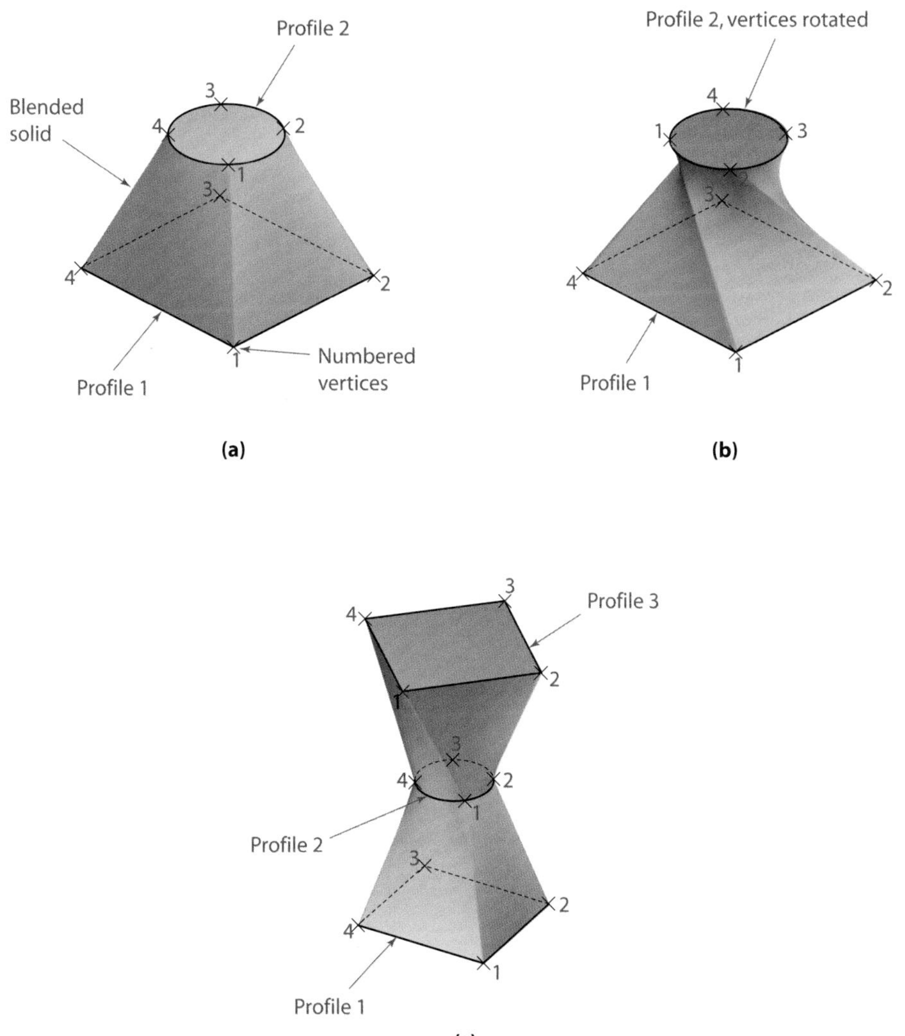

FIGURE 6.54. Blended solids created with two profiles in (a), with the same profiles but rotated vertices in (b), and with three profiles in (c).

6.08.08 Sweeps

Swept features, as with simply extruded or revolved features, are created with a single profile. The difference is that a swept feature does not need to follow a linear or circular path, but can follow a specified curve called a **path** or **trajectory**. The profile is created at an endpoint of the path on a sketching plane that is perpendicular to the path at that endpoint. In sweeping out a solid volume, the profile is imagined to travel along the path. Usually the profile is constrained to remain perpendicular to the path. A good example of a swept solid is a garden hose. The cross section or profile is a simple circle, but the path can be curved. Figure 6.55(a) shows the path and profile of a swept feature where the path is open. Figure 6.55(b) shows a swept feature where the path is closed. Care must be taken in defining the profile and path of a swept solid. Just as you cannot bend a garden hose around a sharp corner without creating a kink, if the path of your sweep contains a sharp corner or a small radius, the feature may fail by trying to create a self-intersecting solid. A special case of a swept solid is a coil spring. In this case, the path is a helix, as shown in Figure 6.56. Many solid modelers include a hel-

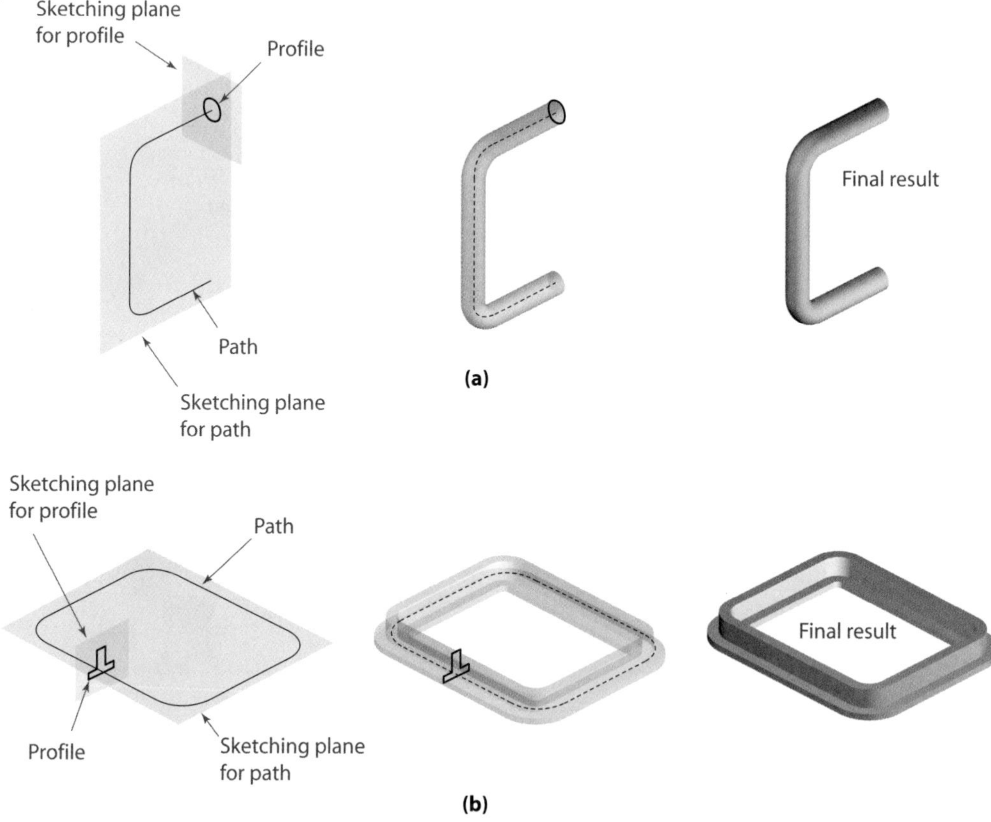

FIGURE 6.55. Features created by sweeps. The sketching plane is perpendicular to the path. The path in (a) is open, and the path in (b) is closed.

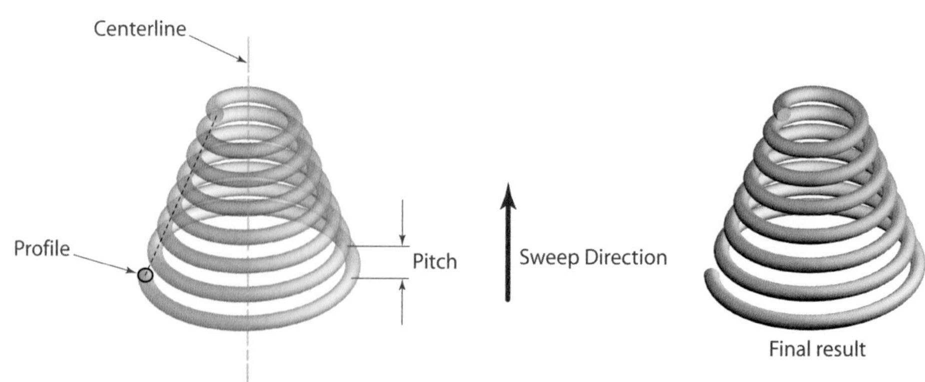

FIGURE 6.56. A tapered spring created by sweeping a circular profile on a helical path.

ical sweep as a special feature so you do not have to sketch the helix. In this case, you sketch the profile and specify an axis on the sketching plane. The helix is specified by a pitch dimension, which is the distance between coils, and the direction of the sweep. To avoid self-intersection, the pitch must be larger than the maximum size of the profile in the sweep direction.

6.09 The Model Tree

An extremely useful editing tool included in most solid modeling software is the **model tree**, sometimes called the **feature tree**, **design tree**, or **history tree**. The model tree lists all of the features of a solid model in the order in which they were created, providing a "history" of the sequence of feature creation. Further, any feature in the model tree can be selected individually to allow the designer to edit the feature. An example of a model tree and its associated solid model are shown in Figure 6.57. Usually new features are added at the bottom of the model tree. Some software allows the designer to "roll back" the model and insert new features in the middle of the tree. In this case, the model reverts to its appearance just before the insertion point, so any inserted feature cannot have its geometry or location based on features that will be created after it.

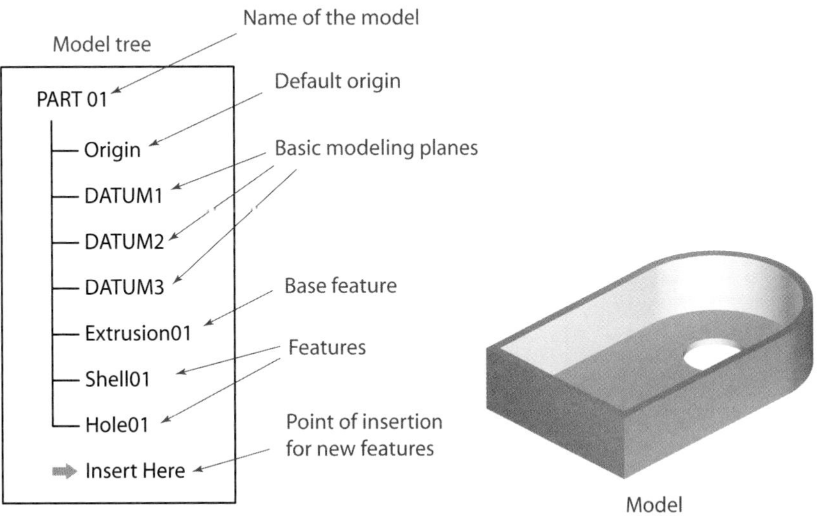

FIGURE 6.57. A typical model tree showing the features of a model in the order in which they were created.

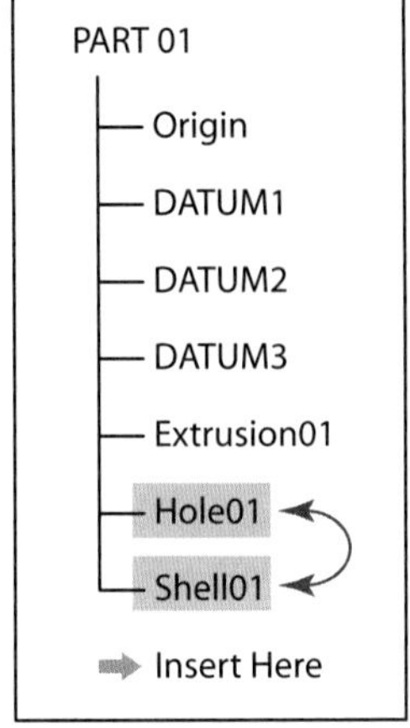

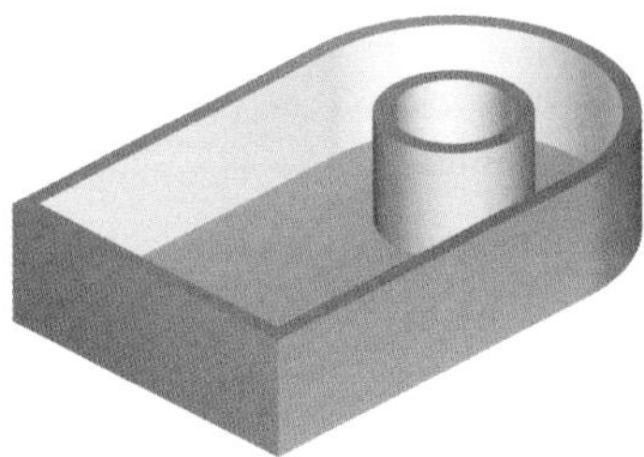

FIGURE 6.58. The result of reversing the order of creating the hole and shell features.

The order in which features are created may have a profound effect on the results. In the previous example, a shell feature, which has the effect of hollowing out a part, was performed with the top surface of the part removed from the feature. A hole was then added to the model after the shelling operation. If the hole was added to the block before the shelling operation, the result would be different, as shown in Figure 6.58, because the surface around the hole through the block would have been considered a part of the shell. In most solid modeling software, removing the feature from one location in the model tree and inserting it a new location changes the order of creation of the feature.

The model tree also provides access to the editing of features. Each feature item on the model tree can be expanded. The base extrusion in the previous example is composed of a fully constrained rectangular sketch profile that has been extruded to a specified length. The feature can be expanded in the model tree, as shown in Figure 6.59, to give access to the profile so it can be selected for editing. The sketch can then be edited by restarting the sketching editor. The dimensional constraints can be changed by selecting and editing their numerical values. Access to the sketching editor and feature parameters may vary with different software, and changes made through the model tree may be one of several different ways to modify your model.

In many models, certain features are dependent upon the existence of other features. For example, consider the features shown in the model in Figure 6.60. The location of the counterbored hole is measured from the edges of the rectangular base.

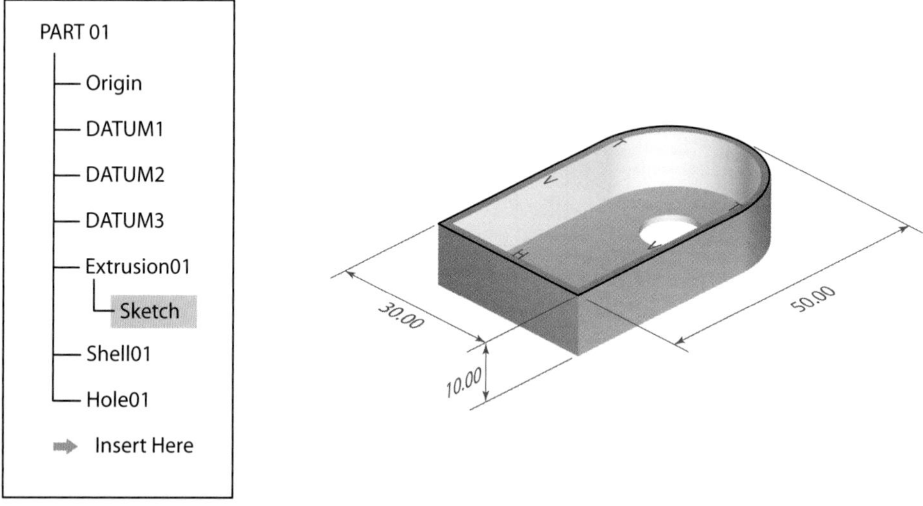

FIGURE 6.59. Use of the model tree to access and edit the sketch used to create the base feature (Extrusion01).

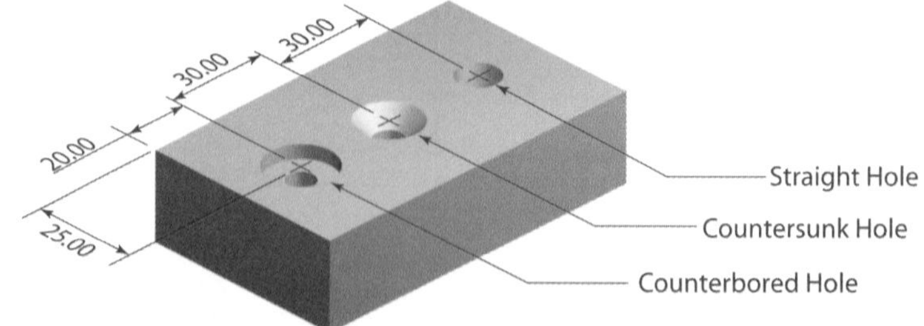

FIGURE 6.60. The holes in the model show parent-child dependencies. The existence of the straight hole depends on the existence of the countersunk hole, which depends on the existence of the counterbored hole. Elimination of a parent also eliminates its child.

However, the location of the countersunk hole is measured from the location of the counterbored hole and the location of the straight hole is measured from the location of the countersunk hole. Imagine what would happen to the straight hole if the countersunk hole were deleted. There would be no reference for placing the straight hole; therefore, it could not be created. Similarly, if the counterbored hole was deleted, neither the countersunk hole nor the straight hole could be created. This relationship is often referred to as a parent-child relationship. The straight hole is considered the **child feature** of the countersunk hole, and the countersunk hole is considered the child of the counterbored hole. The counterbored hole is considered the **parent feature** of the countersunk hole, and the countersunk hole is considered the parent of the straight hole. Just as you would not be reading this text if your parents did not exist, neither can features in a solid model exist without their parent (or grandparent) features. On the model tree, if you try to delete a particular feature, its progeny also will be deleted. However, different software behaves differently; and while some software provides specific warnings about the deletion of features, other software does not.

Understanding parent-child relationships in solid models is important if your model needs to be flexible and robust. As a designer, you undoubtedly will want to change the model at some time. You might need to add or delete features to accommodate a new function for the part or reuse the model as the basis of a new design. If you minimize the number of dependencies in the feature tree (like a family tree), it will be easier to make changes to your model. When it is likely that some features will be deleted or suppressed in a future modification of the part, those features should not be used as parents for other features that must remain present. The most extreme example of this strategy is called **horizontal modeling**, where the feature tree is completely flat; that is, there are no parent features except the base feature. This type of modeling strategy was patented by Delphi and has been used successfully by many companies. In Figure 6.61, the locations of three holes have been redefined so they are measured from the edge of the rectangular base instead of relative to one another. The base then becomes the parent to all three holes, and deleting any one of the holes does not affect the others.

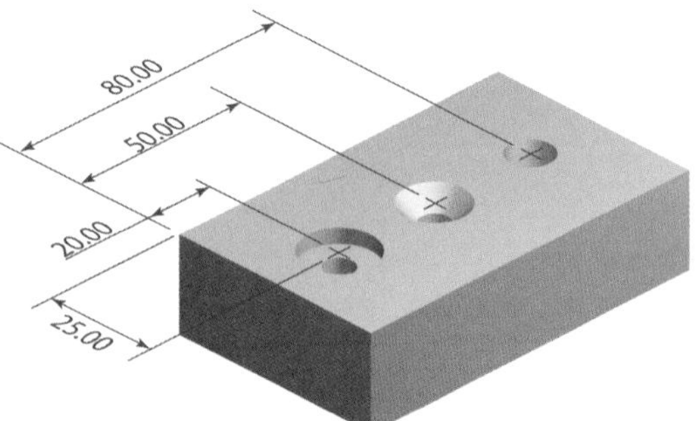

FIGURE 6.61. This model demonstrates horizontal modeling. Each hole has no parent-child dependencies except to the base feature.

6.10 Families of Parts

Groups of engineered parts often have very similar geometry. An everyday example is bolts and screws. A group of bolts may have the same head and thread geometries, but differ in their available length. Another example is the family of support brackets shown in Figure 6.62. Each bracket has a rough L-shaped base feature, holes, and a support rib (except for Version 3). Only the size and number of holes are different for each version.

When a group of parts is similar, it is possible to represent the entire group with a **family model**, with different versions of that model selected to specify particular parts. Such a model includes a **master model**, which has all of the features that are in any of the members of the group, and a **design table**, which lists all of the versions of that model and the dimensional constraints or features that may change in any of its versions. The attributes that may change are sometimes called **parameters**. The first

Version 1

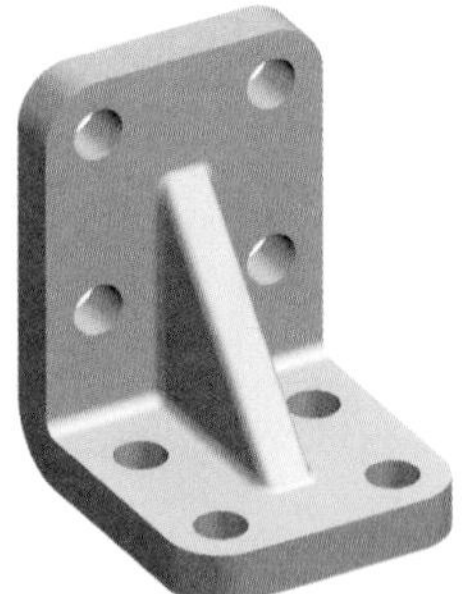

Version 2

Version 3

FIGURE 6.62. A family of three parts with similar features and geometry.

step in building a family model is to identify all of the features and parameters that can be varied in the members of the family. In addition to a numerical value, every dimensional constraint in a model has a unique **dimension name**, which can be shown by selecting the appropriate display option. In Figure 6.63, all of the dimensional constraints have been changed to show their dimension names and the features have been identified by the feature names that appear on the design tree.

The next step is to select the option for the construction of a design table, which is usually an internal or external spreadsheet, in the solid modeling software. The spreadsheet table should look similar to that shown in Figure 6.64. The first column usually contains the names of the different versions of the model. In Figure 6.64, these versions are called Version 1, Version 2, and Version 3 for convenience. The first row usually contains the names of the parameters that can change with each version. The individual cells of the spreadsheet show what the corresponding numerical values are of the

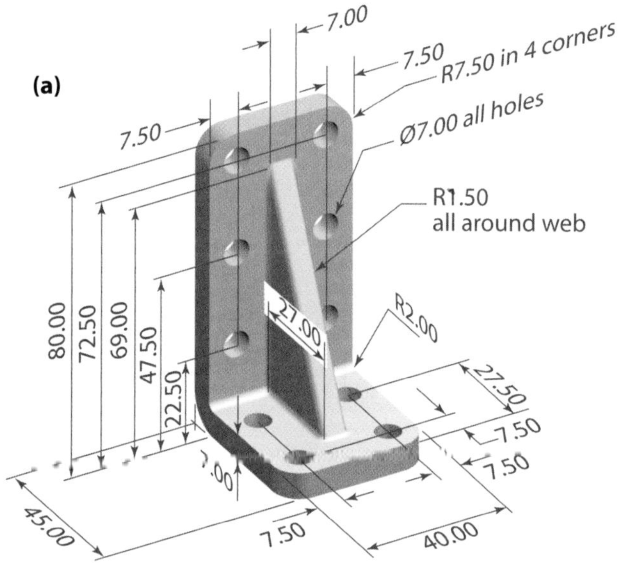

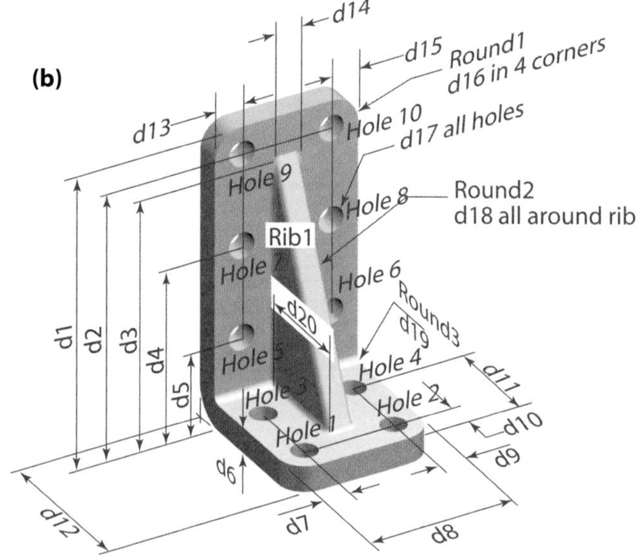

FIGURE 6.63. The master model showing the numerical values of its dimensions in (a) and the names of the features and dimensions in (b).

	Hole 3	Hole 4	Hole 7	Hole 8	Hole 9	Hole 10	Rib1	d1	d3	d12
Version 1	U	U	U	U	U	U	U	80.00	69.00	45.00
Version 2	U	U	U	U	S	S	U	60.00	50.00	45.00
Version 3	S	S	S	S	S	S	S	30.00	N/A	30.00

S = Suppressed
U = Unsuppressed

FIGURE 6.64. The design table for the parameters that change within the three versions of the family of parts in Figure 6.62 and Figure 6.63.

dimensional constraints for each version and whether a particular feature is present in that version. When a particular feature is present, it is specified as being **unsuppressed**. When that feature is not present, it is specified as being **suppressed**. When the version of the part to be displayed has been selected, the corresponding model with its specified parameters is shown.

With the existence of a design table, editing the values in the table can change the numerical values of those dimensional constraints for any model version. In Figure 6.65, selecting and editing the appropriate cell in the design table changed the height of the L-bracket. In Figure 6.66, the support rib is no longer present because it was suppressed in the design table. When suppressing a feature, remember to be cautious, because suppressing a feature will also suppress its entire progeny.

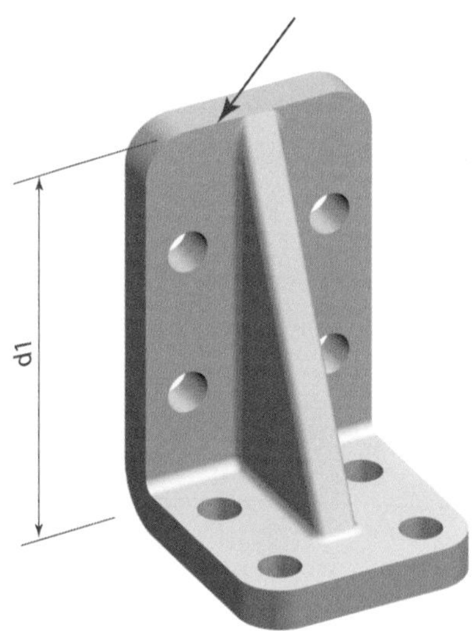

	Hole 3	Hole 4	Hole 7	Hole 8	Hole 9	Hole 10	Rib1	d1	d3	d12
Version 1	U	U	U	U	U	U	U	69.00	69.00	45.00
Version 2	U	U	U	U	S	S	U	60.00	50.00	45.00
Version 3	S	S	S	S	S	S	S	30.00	N/A	30.00

S = Suppressed
U = Unsuppressed

FIGURE 6.65. The height of the L-bracket in the model has been changed by changing the value of the cell in the design table associated with this parameter.

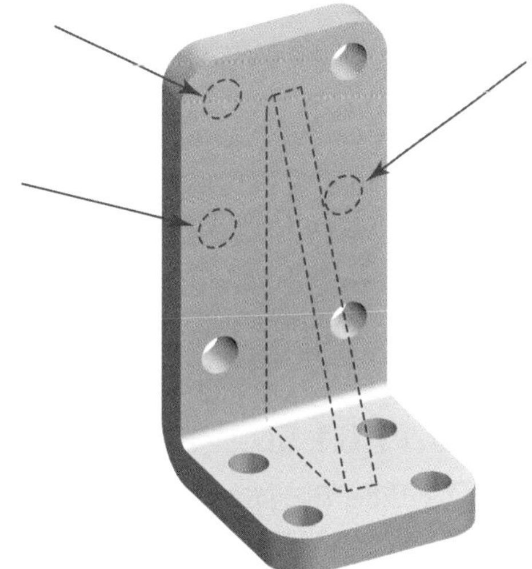

	Hole 3	Hole 4	Hole 7	Hole 8	Hole 9	Hole 10	Rib1	d1	d3	d12
Version 1	U	U	S	S	S	U	S	80.00	69.00	45.00
Version 2	U	U	U	U	S	S	U	60.00	50.00	45.00
Version 3	S	S	S	S	S	S	S	30.00	N/A	30.00

S = Suppressed
U = Unsuppressed

FIGURE 6.66. Features in the model can appear or not appear by changing their suppression states in the design table.

6.11 Strategies for Making a Model

You have a blank computer display in front of you, and your solid modeling software is running. So where do you start? The first step in modeling a solid part is to decompose it into features. Study the part and try to identify the base feature. Expanding on a previous statement, the base feature should be something that describes the overall shape of the part or something that gives you the greatest amount of functional detail that can be created with a single extrusion, rotation, sweep, or blend. Next, break the rest of the part into subsections that can be created using extruded, revolved, swept, or blended shapes. Look for standard features such as holes and slots that are manufactured using a particular process. Identify the edge features such as chamfers and rounds. Once you have studied the part, you can create the model using the following eight-step procedure:

1. Create any datum geometries or paths required to create the base geometry.
2. Sketch and constrain profiles needed for the base.
3. Extrude, rotate, sweep, or blend to create the base.
4. Create any necessary datum geometries or paths to create the next feature.

5. For sketched features, sketch and constrain or otherwise specify the feature profiles; then extrude, rotate, sweep, or blend to create the feature.
6. For standard features such as holes and edge features, specify the desired parameters and placement on the existing geometry.
7. Array or mirror the feature if necessary to create identical features.
8. Repeat steps 4–9 until the model is complete.

Once the model is complete, it can be modified to become more robust. For example, additional associative constraints may be added in place of dimensional constraints or design tables may be created for families of parts.

6.11.01 Step-by-Step Example 1—The Guide Block

Consider the guide block in Figure 6.67 as an example. How would you build a solid model for this part? What should be its base feature? What are its secondary features?

One reasonable base feature would be an extrusion made with the profile shown in Figure 6.68. This extrusion would capture many details of the part in a single operation and is representative of the general shape of the part. The sketch is made on one of the basic planes and is geometrically and dimensionally constrained. Note the use of horizontal and vertical geometry constraints, which would likely be applied automatically if the segments were sketched approximately in the orientations shown. Also note that a corner of the profile is grounded by constraining the vertex to be coincident with the origin of the coordinate system; therefore, dimensional constraints locating the profile on the plane are not needed. Note the use of the colinear constraint on the two short horizontal sketch segments, which eliminates the need to place separate dimensional constraints on the height of the segments. Once the profile is complete, it can be extruded to the width of the part, as shown in Figure 6.68(d).

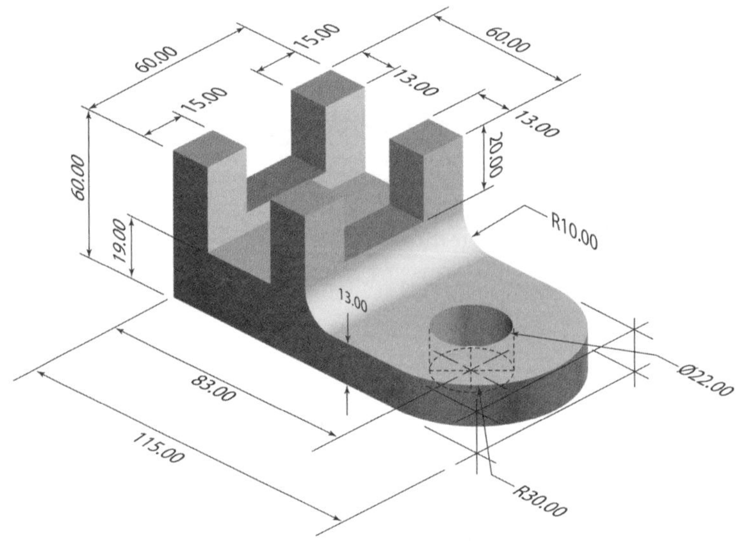

FIGURE 6.67. A solid model of this part is to be created. What operations should be performed, and in what sequence should they be made?

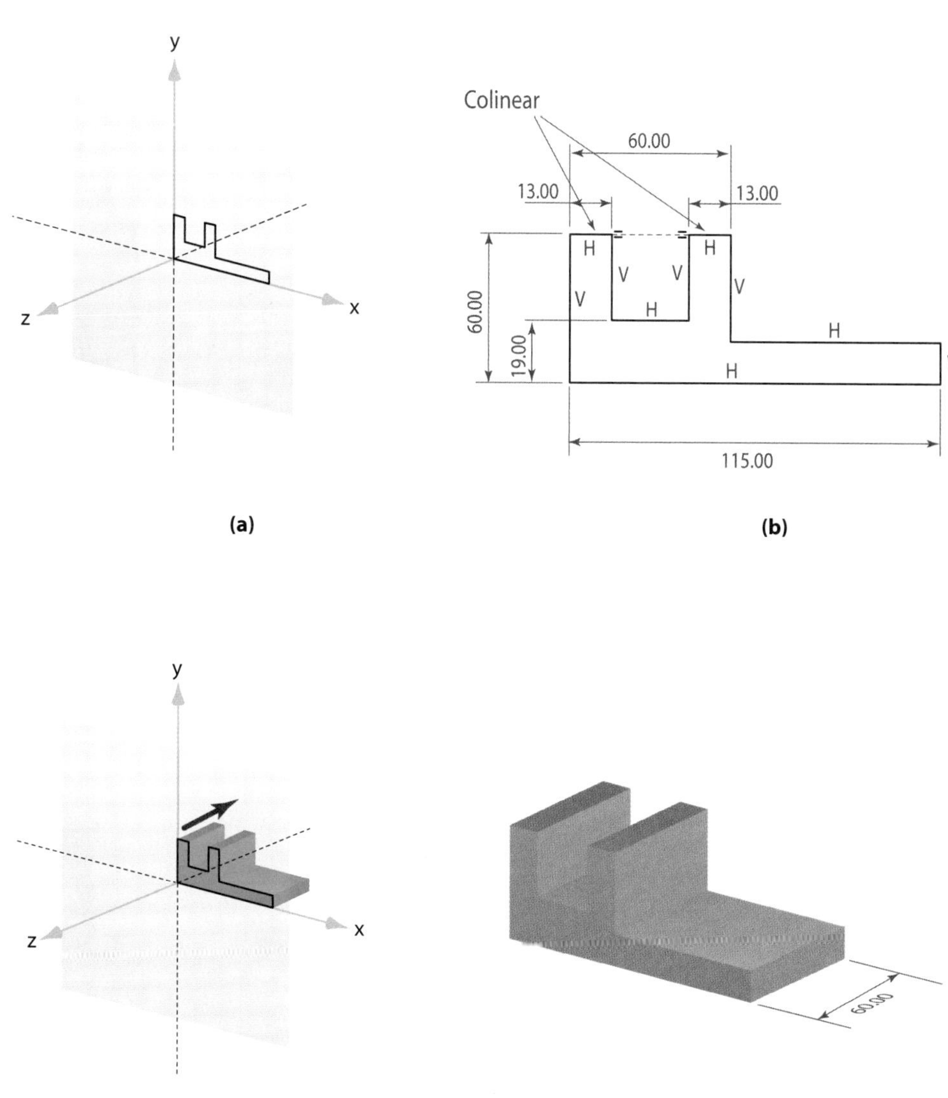

FIGURE 6.68. The base feature is created by sketching on one of the basic modeling planes (a), constraining the sketch to create a profile (b), and extruding the profile (c) to the required depth to obtain the desired result (d).

The first feature to be added is the slot across the upper portion of the part. The slot can be made on the model by an extruded cut using a rectangular profile on the sketching plane shown in Figure 6.69. Geometric and dimensional constraints are added to the sketch. One edge of the sketch is constrained to be colinear with the top edge of the base feature, guaranteeing that the slot always will be a slot (and not a square hole) if the height of the part increases. An extruded cut is then made by extruding the completed profile to the limit of the part or through the entire part. If this extruded cut was made to a specific length just beyond the limit of the part (e.g., with a blind extrusion extending past the part), the resulting model would appear identical to the desired model. However, blind extrusions like this are usually considered poor modeling practice, because if selected length constraints of the part are increased, the slot no longer extends entirely through the part.

Next, the hole is added. Simple through holes can be created as a cut feature by selecting the desired plane or face of the existing solid, sketching a circle, and extruding a cut or negative feature. However, a better way to make a hole is to use the standard

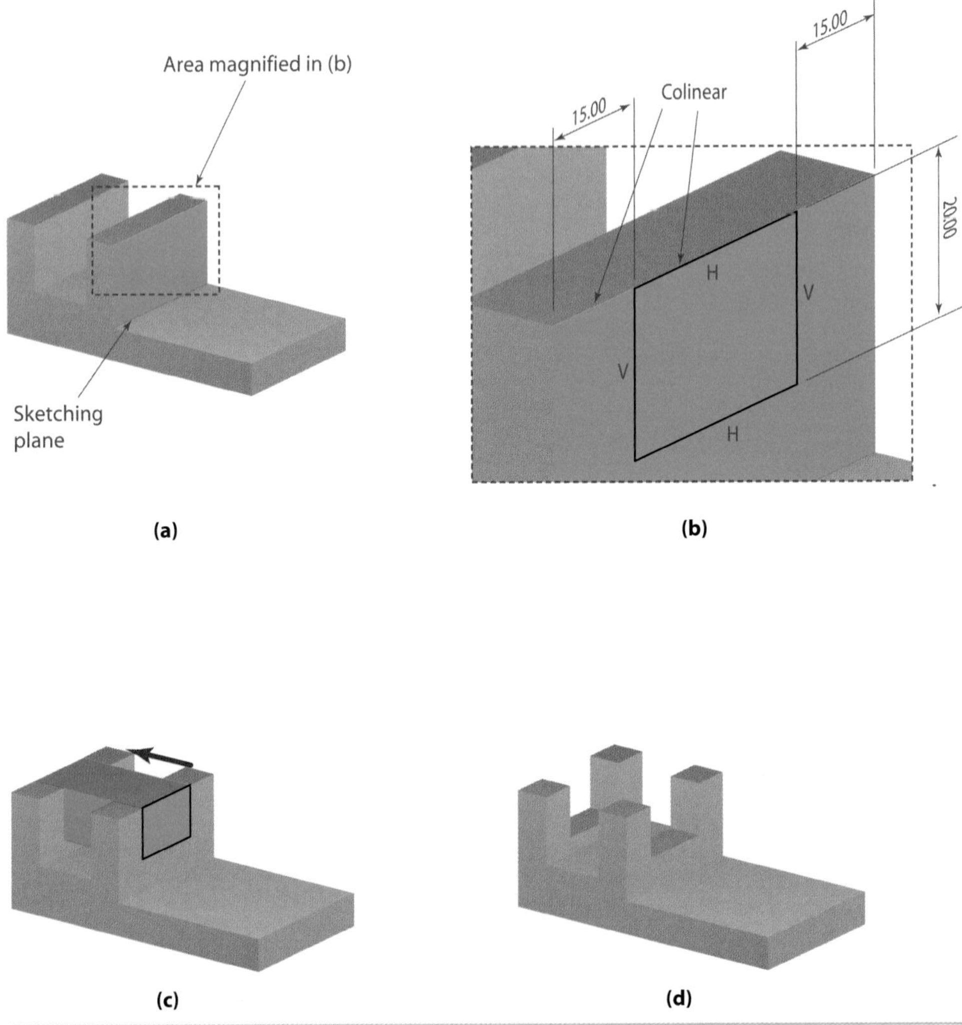

FIGURE 6.69. A slot is created by selecting a surface on the model to be the sketching plane in (a), on which a sketch is created and constrained in the magnified view in (b). The profile is extruded to the end of the part in (c), and the material is removed to create the result in (d).

hole feature, which will identify the feature as a hole in the database. Therefore, if your part is to be manufactured by an automated production system, the holes can be recognized and automatically drilled or bored. The way that the location dimensions of the hole are included in the model is also important. Why does this matter? Looking at the part, you might assume that the hole should stay centered on the width of the part. What will happen if the width of the part changes? Wouldn't you want the hole to remain centered on the width of the part? Using an associative constraint on the width location of the hole, making it always equal to one-half the part's width, ensures that the design intent will be maintained. No matter who uses your model or changes the dimensions, the intended symmetry will remain embedded in the part. Adding the hole with its associative constraint is shown in Figure 6.70.

Finally, the round and fillet features need to be added to the model geometry. Rounds and fillets are associated with particular edges, so no sketching is involved for this step. Simply pick the desired edge and apply the round or fillet feature, specifying the desired radius. The result of adding the rounds is shown in Figure 6.71, and the fillet is shown in Figure 6.72. There are no array or mirror features, so the model is now complete.

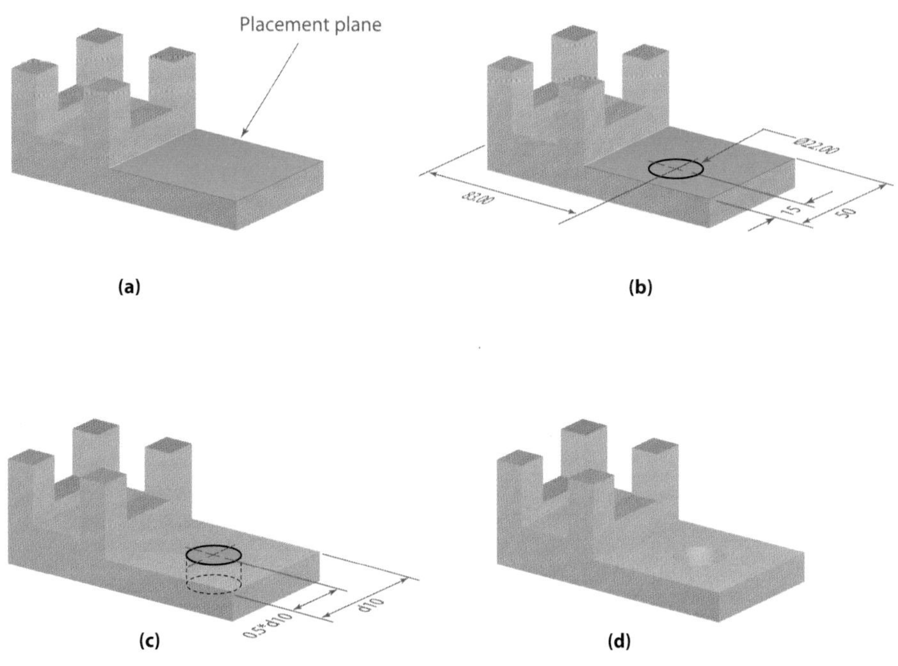

FIGURE 6.70. The surface to which the hole is to be added is selected as the placement plane in (a). The location and diameter of the hole are specified in (b). An associative constraint is used on the variable names to ensure that it remains centered in (c), and the final result is shown in (d).

FIGURE 6.71. The edges to be rounded are selected in (a), and the radius is specified in (b) as an associative constraint to ensure a full radius across the part. The result of the rounding operation is shown in (c).

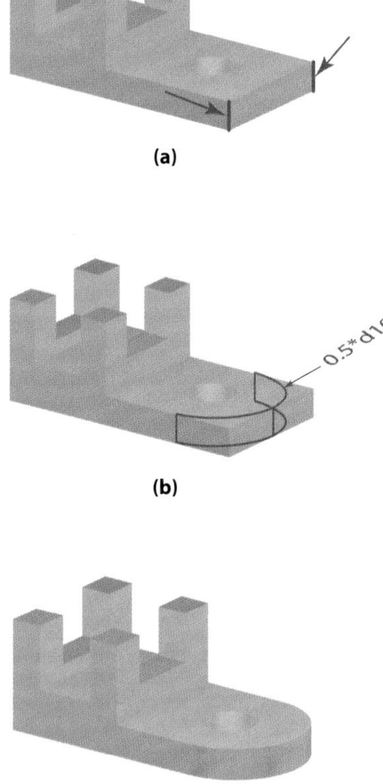

FIGURE 6.72. The edge to be filleted is selected in (a), and the radius is specified in (b). The result of the fillet operation is shown in (c).

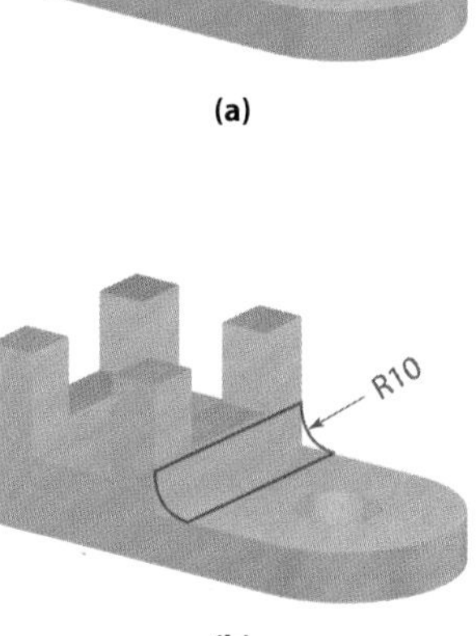

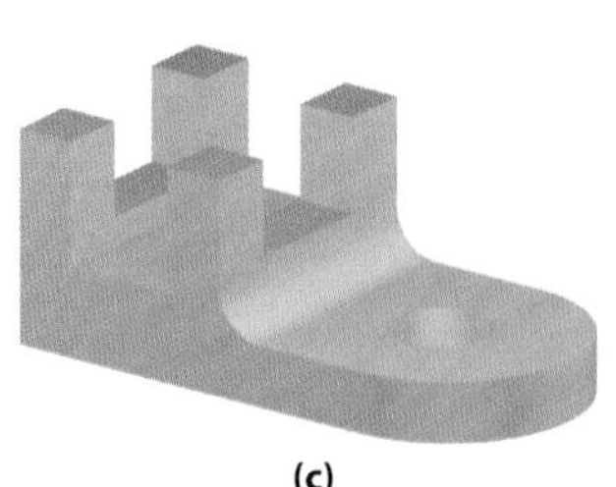

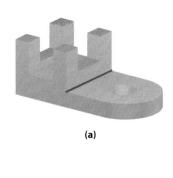

Depending on the design intent, several different modeling strategies can be used for the same part. If the designer wants the entire part to be symmetrical, another way to achieve the desired symmetry would be to create a two-sided extrusion of the base profile from the original sketching plane. By creating a two-sided extrusion, no extra datum planes are needed to ensure symmetry. Then the slot is constrained to be symmetrical across the plane, and the hole center is constrained to lie in the symmetry plane.

6.11.02 Step-by-Step Example 2—The Mounting Brace

The mounting brace shown in Figure 6.73 includes two mounting plates with holes connected by a cross-shaped web. The base feature is not as easily identifiable as in the previous example. Also, one mounting plate is set at an angle with respect to the other mounting plate, which is located on the bottom of the part. Note that no dimensions are given for the location of the web on either mounting plate because the web is assumed to be centered on both plates. The height of the web is measured at the center of the cross section. Creating the model of this part provides a good example of how datum axes and datum planes are used to help create and locate geometry.

A rectangular block that will help form the bottom mounting plate is selected as the base feature. Even though this rectangular block does not dominate the overall part shape, it is easy to locate and create; and the other features can be easily located and created from it. The sketch, profile, and extrusion of this simple base using the basic modeling planes are shown in Figure 6.74. The initial sketch is made on datum plane 1. Note that the origin is not used to constrain the location of the profile. The sketch is to be symmetrical across datum planes 2 and 3. The strategy used to achieve symmetry may vary depending on your software. Some software allows you to place symmetry constraints across the basic datum planes that appear as edges in the sketch view, such as datums 2 and 3 in this example. Otherwise, a datum axis must be used as a centerline. With some solid modeling software, datum axes are present at the intersections of the basic modeling planes and need not be created. Sometimes you may need to create datum axis 1 at the intersection of datum planes 1 and 3 and datum axis 2 at the

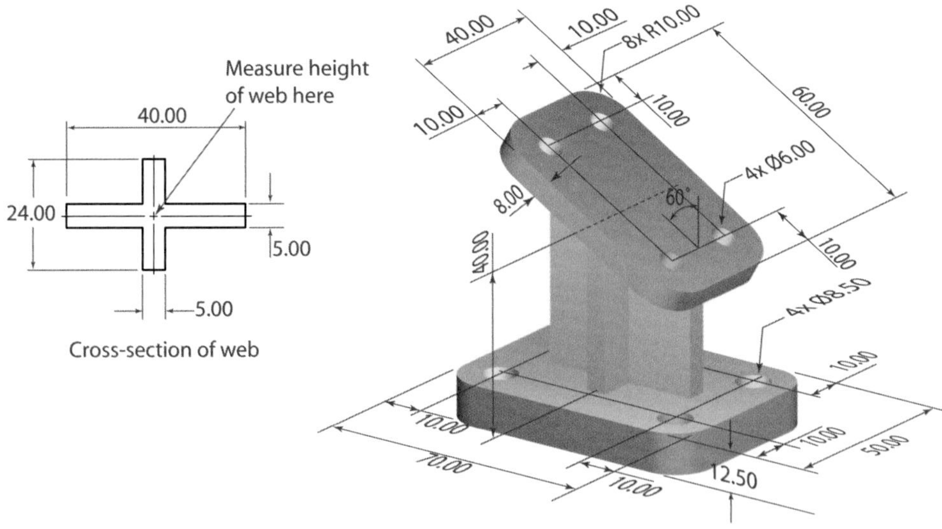

FIGURE 6.73. A solid model of this part is to be created. What operations should be performed, and in what sequence should they be made?

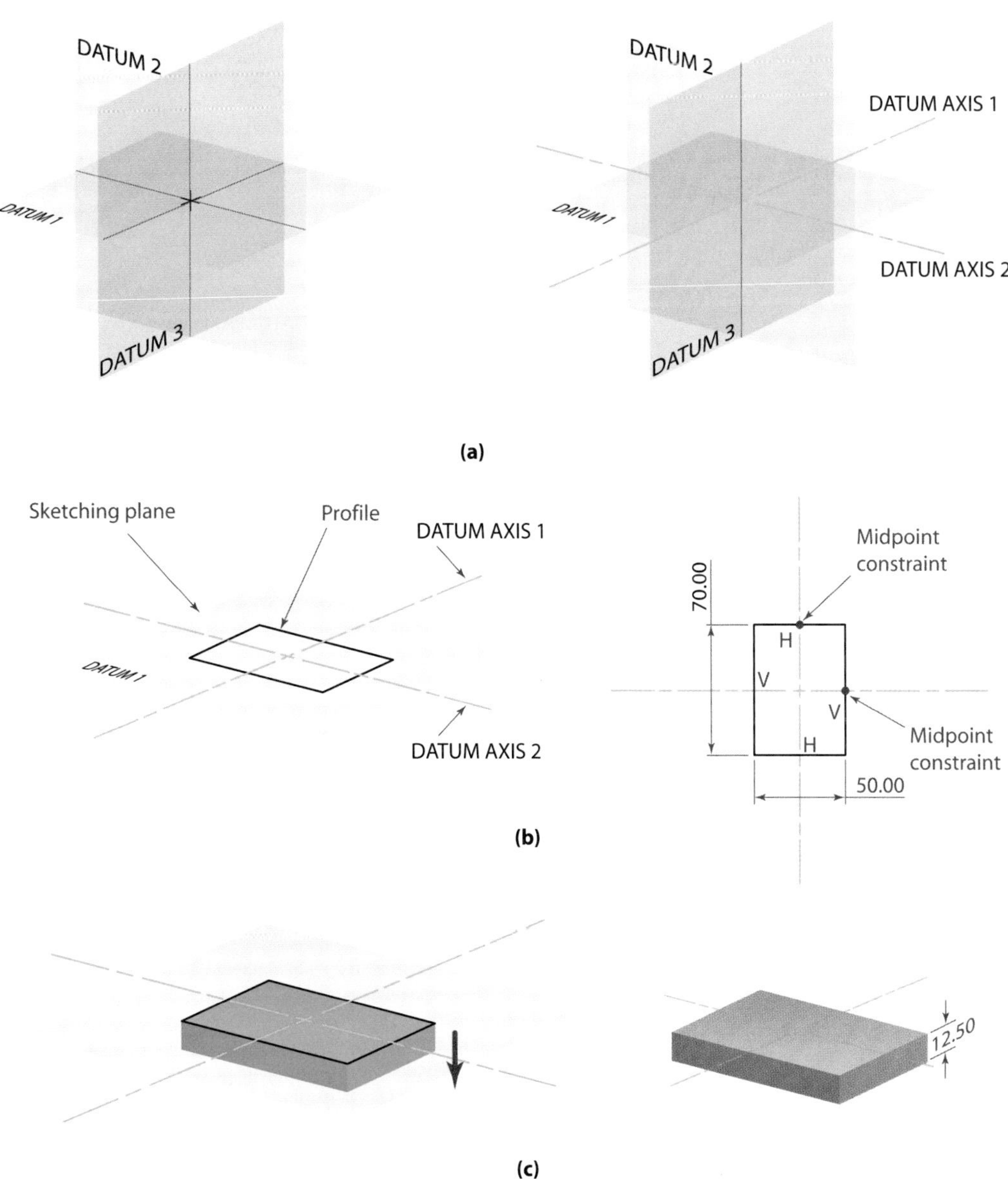

FIGURE 6.74. To create the base feature, datum axes are added to the intersections of the basic modeling planes in (a). Those axes are used to help center the base profile in (b) by constraining the midpoints of the segments to be colinear with the axes. The base feature is extruded to the required depth in (c).

intersection of datum planes 2 and 3. Now the midpoints of two adjacent legs of the rectangle can be constrained to lie on the datum axes using either coincidence or colinearity constraints. With these constraints, the rectangular profile will be centered about the axes and will remain centered about the axes if the design is modified in the future. The extrusion direction is chosen to be downward from datum plane 3, thus putting that datum plane on top of the base.

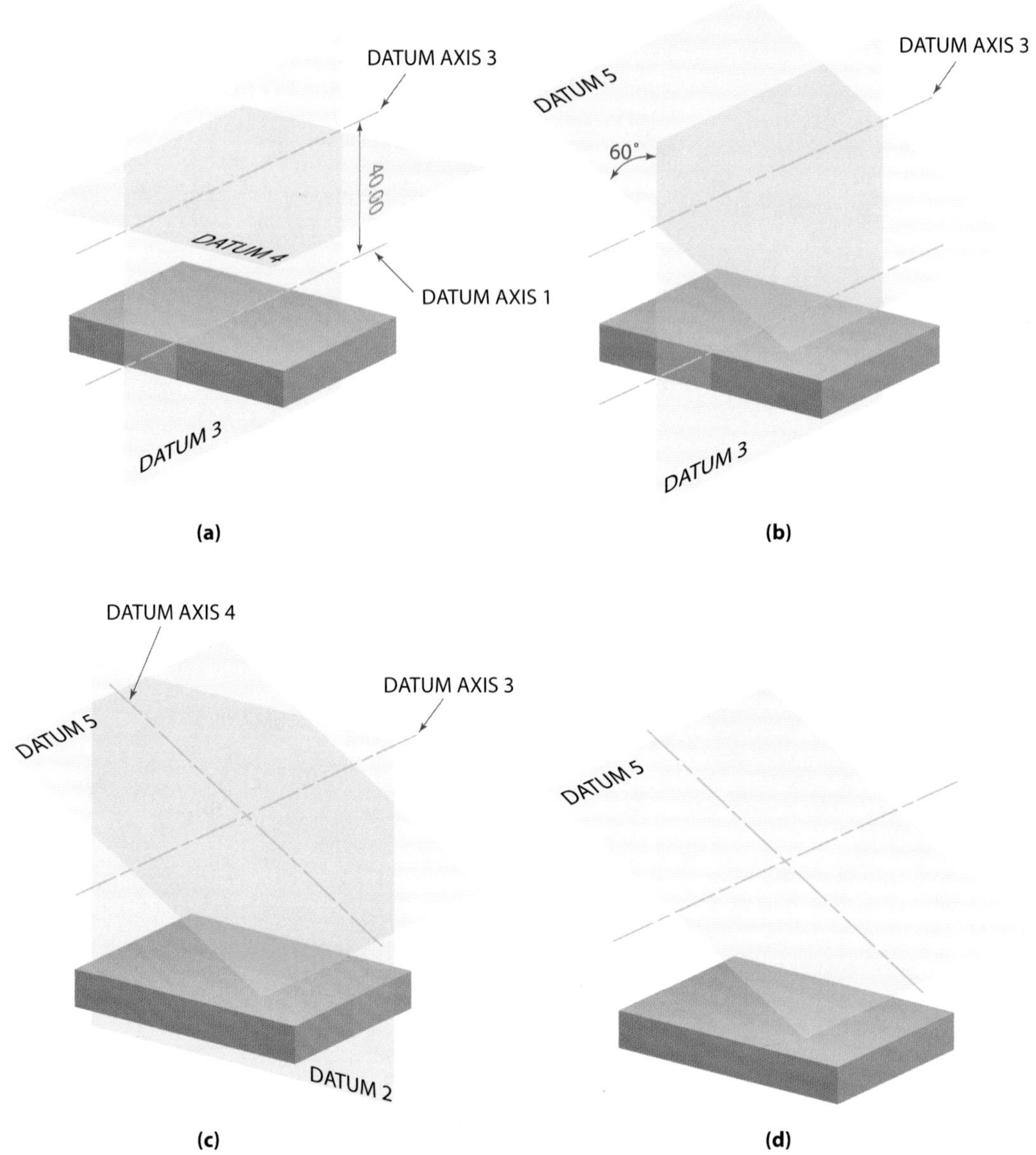

FIGURE 6.75. To create the angled feature, datum axis 3 is created in (a) above the midline of the model using intermediate datum plane 4 created above and parallel to the top of the base extrusion. Angled datum plane 5 is created through the new datum axis in (b). Datum axis 4 is created at the intersection of datum plane 5 and basic modeling plane 2 in (c) to create datum axes that can be used to locate the center of the angled feature, as shown in (d).

The next major feature to be added is the angled mounting plate. It may seem peculiar to create disconnected geometry, but you will soon see how useful this strategy can be. No sketching plane is available for creating the angled plate; so before the extrusion is created, some additional datum geometries need to be created. In addition to the datum (sketching) plane, some datum axes are needed to position the sketch at the center of the web. These datum geometries, shown in Figure 6.75, are as follows:

- Datum plane 4—located 40 mm above and parallel to datum plane 1
- Datum axis 3—at the intersection of datum plane 3 and datum plane 4

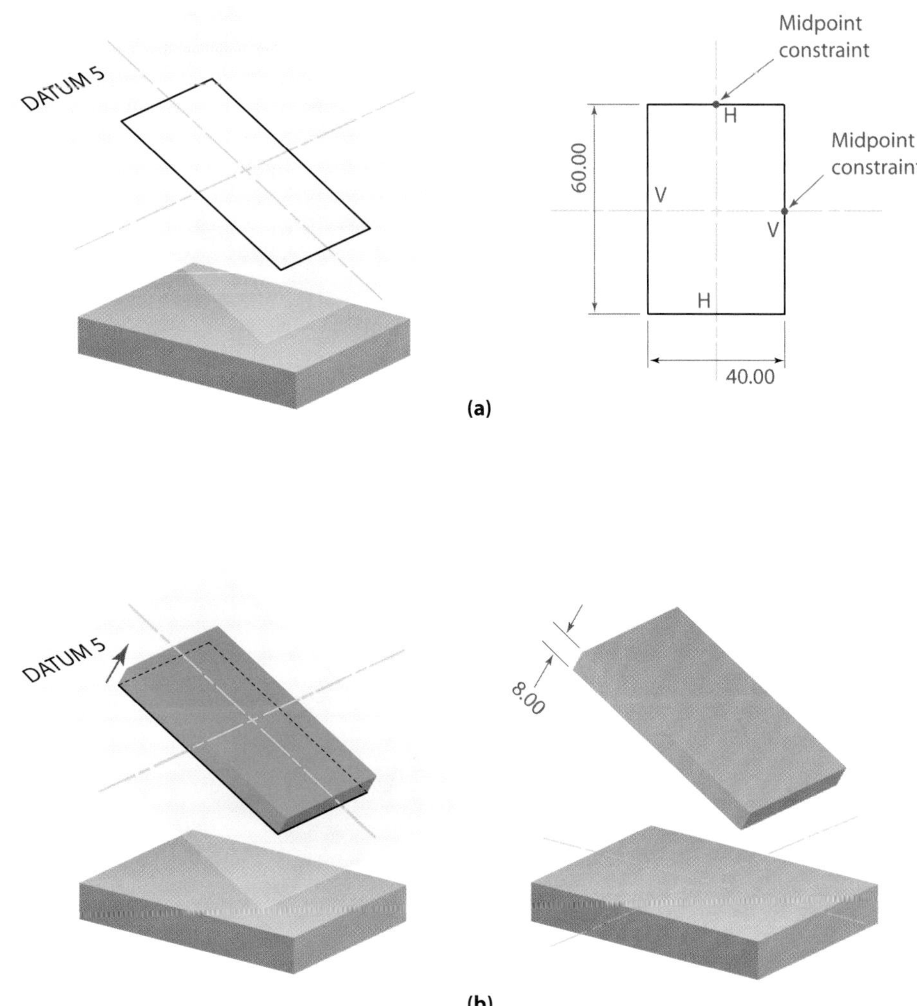

FIGURE 6.76. The profile for the angled feature is sketched and constrained on the angled datum plane in (a), constraining the midpoints of the segments to be colinear with the axes. The profile is extruded to the required depth in (b).

- Datum plane 5— through datum axis 3, 60 degrees from datum plane 3
- Datum axis 4—at the intersection of datum plane 2 and datum plane 4

Datum axes 1 and 2 mark the center of the base; and datum axes 3 and 4 mark the center of the angled mounting plate, where the web will be located.

To create the angled mounting plate, a sketch of its rectangular profile is constrained and dimensioned on datum plane 5, as shown in Figure 6.76. Note that each of the midpoints of two adjacent edges of the profile have been constrained to be colinear with datum axis 3 or 4. In this way, the rectangular profile is always centered about the axes. The sketched profile is then extruded to form the angled mounting plate.

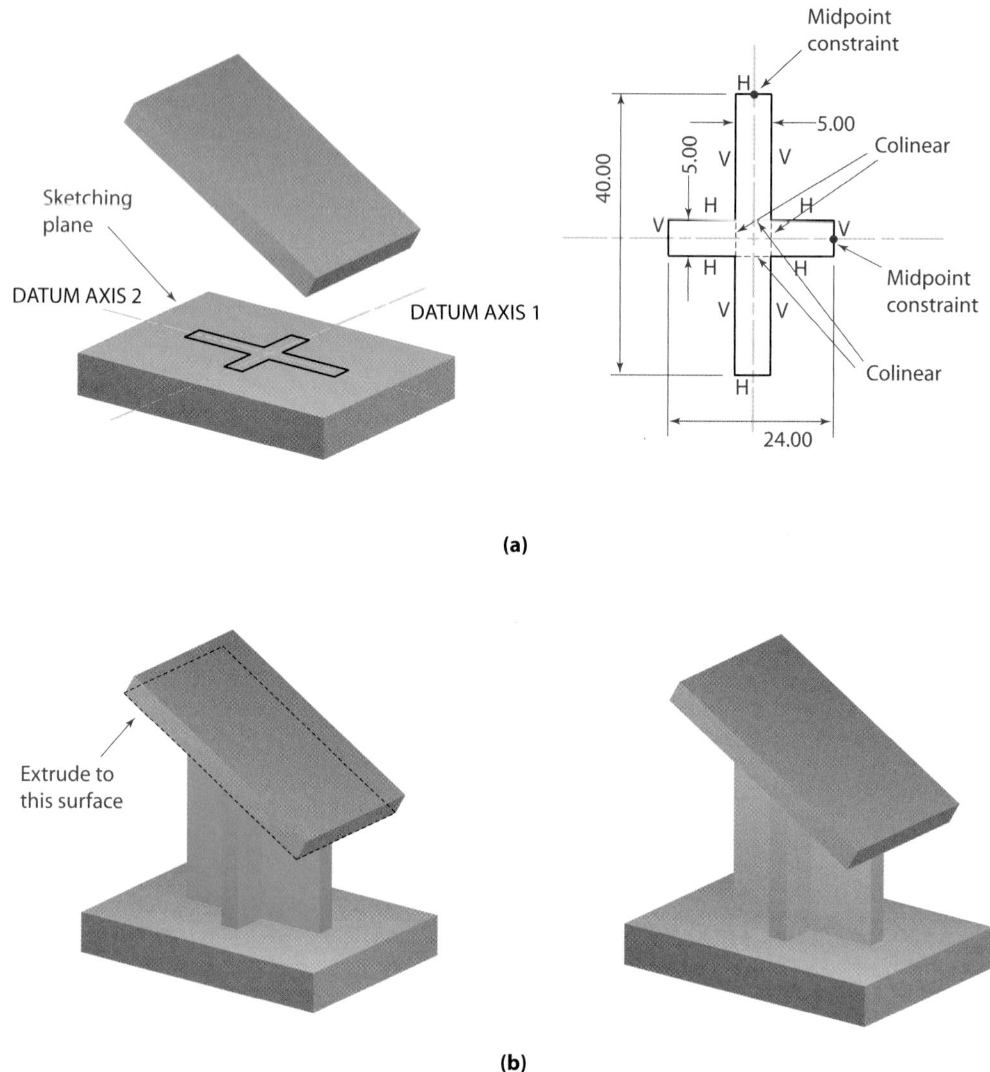

FIGURE 6.77. To create the web, the top of the base is selected as the sketching plane in (a) and the profile of the web is constrained using midpoint and colinear constraints to ensure symmetry. The profile is extruded to the underside of the angled feature in (b).

The top of the base feature is used as the sketching plane for the profile of the web. A cross-shaped profile is used, as shown in Figure 6.77. The midpoints of the line segments representing the thickness of two adjacent legs of the cross are constrained to be coincident with datum axis 1 or with datum axis 2. Note the use of colinear constraints on the edges of opposing legs of the cross. These constraints make the thickness of the legs the same on opposite sides when one of the legs is dimensioned. The profile is extruded up to the next surface that it intersects, which is the bottom of the angled mounting plate.

The rounds are added using the round and fillet feature creation tool by selecting the edges shown in Figure 6.78 and specifying the desired round radius. By grouping the rounds in one feature, their radii will be equal and will remain equal as the part is modified in the future. Adding holes to both mounting plates using a hole feature creation tool is shown in Figure 6.79. The centers of the holes are located on the upper surface of each plate, with their locations specified relative to the edges of the plates. The holes are then defined by their diameters and depth (all the way through to the next surface on each plate). Note that you would not want to use the "through all" option on these holes, as the holes from the angled plate would extend through the baseplate as well. The model is now complete.

FIGURE 6.78. The edges for rounding are selected in (a), and the desired radius of the rounds is applied in (b).

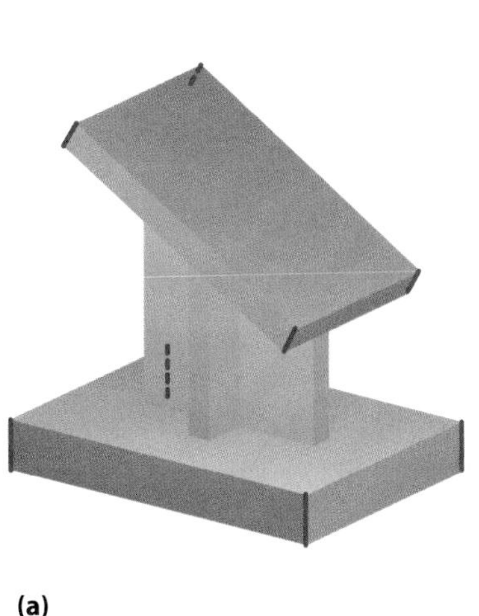

(a)

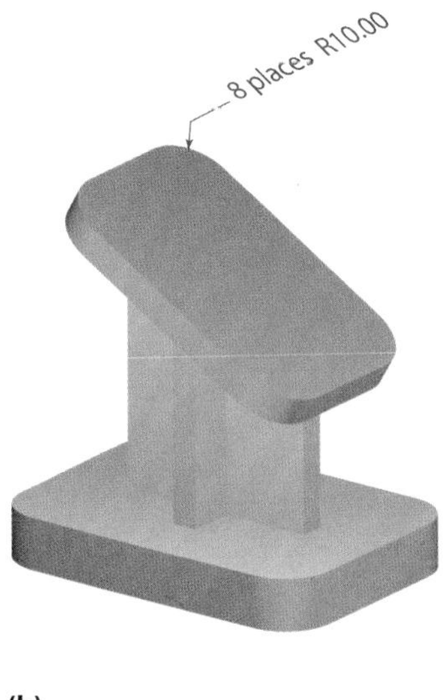

(b)

FIGURE 6.79. The holes are added to the base feature in (a) and to the angled feature in (b).

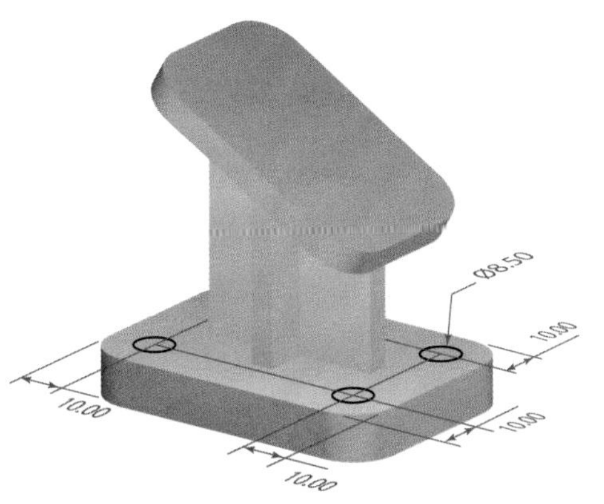

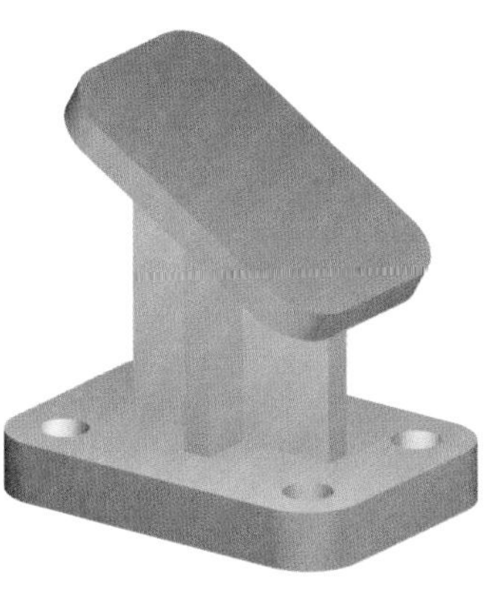

(a)

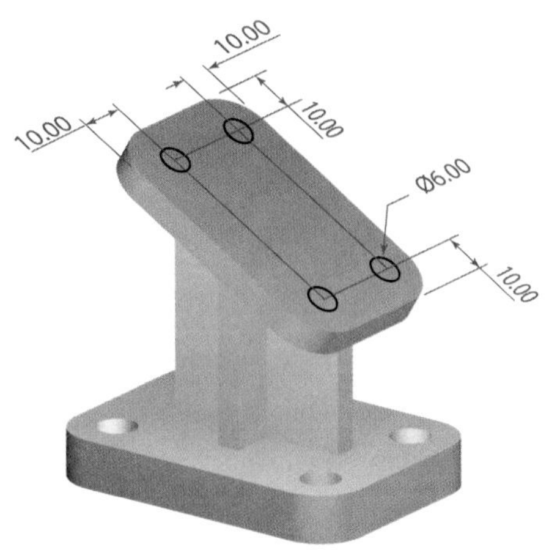

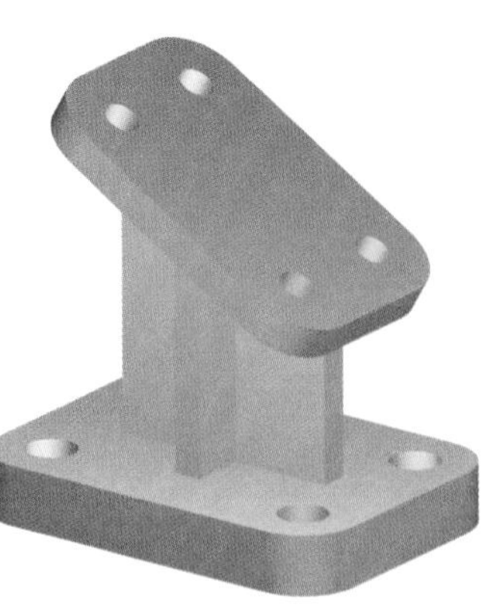

(b)

6.11.03 Step-by-Step Example 3—The Handwheel

The part shown in Figure 6.80 is a handwheel. It consists of a cylindrical hub with a D-shaped hole at its center, a circular rim with a circular cross section, and spokes that connect the hub to the rim. The handwheel is a good example of a model that should be created with a combination of rotation, sweep, and array.

A good base feature might be either the hub or the rim. Both of these features can be created by revolving a suitable profile. Some solid modelers will let you create both features in the same operation, as will be shown here. One of the basic modeling planes is selected as the sketching plane. A centerline will be required on the same sketch as the profiles, as shown in Figure 6.81. The centerline is placed coincident with the intersection of the sketching plane and another basic modeling plane. The length of the centerline is unimportant. The rectangular profile for the hub is sketched and constrained as shown in Figure 6.81(a). The circular profile of the rim is sketched and

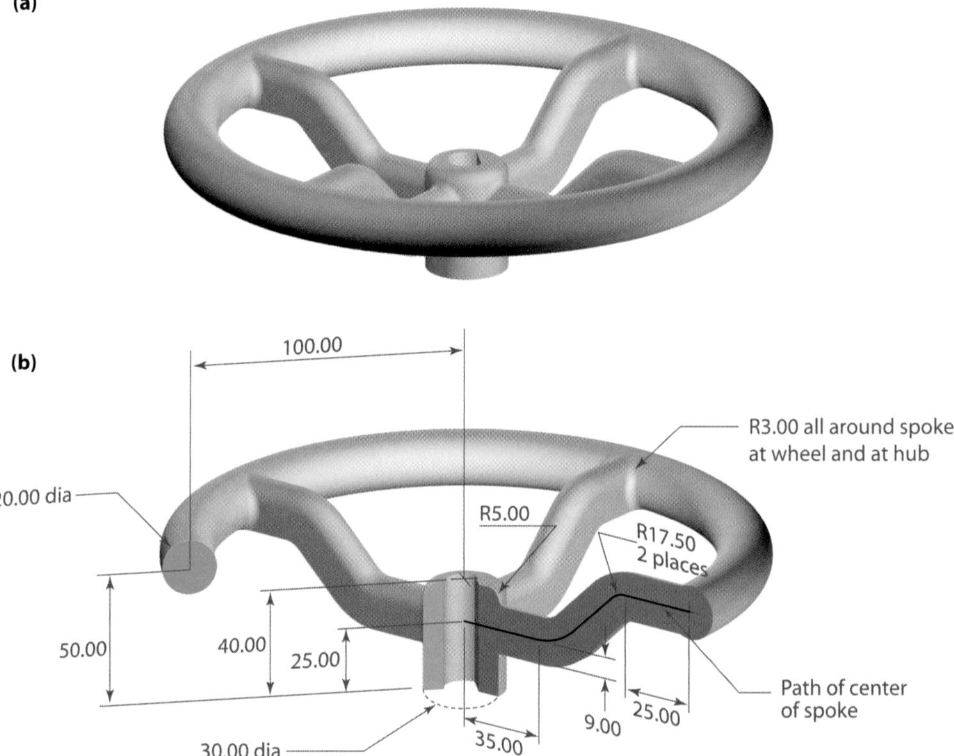

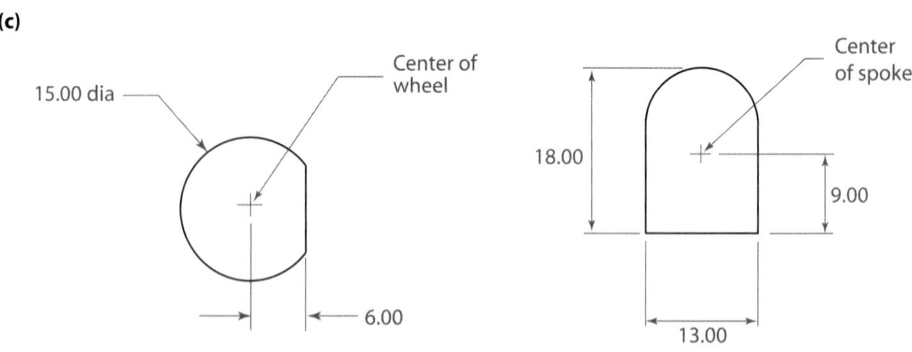

FIGURE 6.80. The handwheel to be modeled is shown in (a). It is cut away in (b) to help reveal all of its dimensions. The D-shaped center hole and a cross section of a spoke are magnified in (c) to show their dimensions.

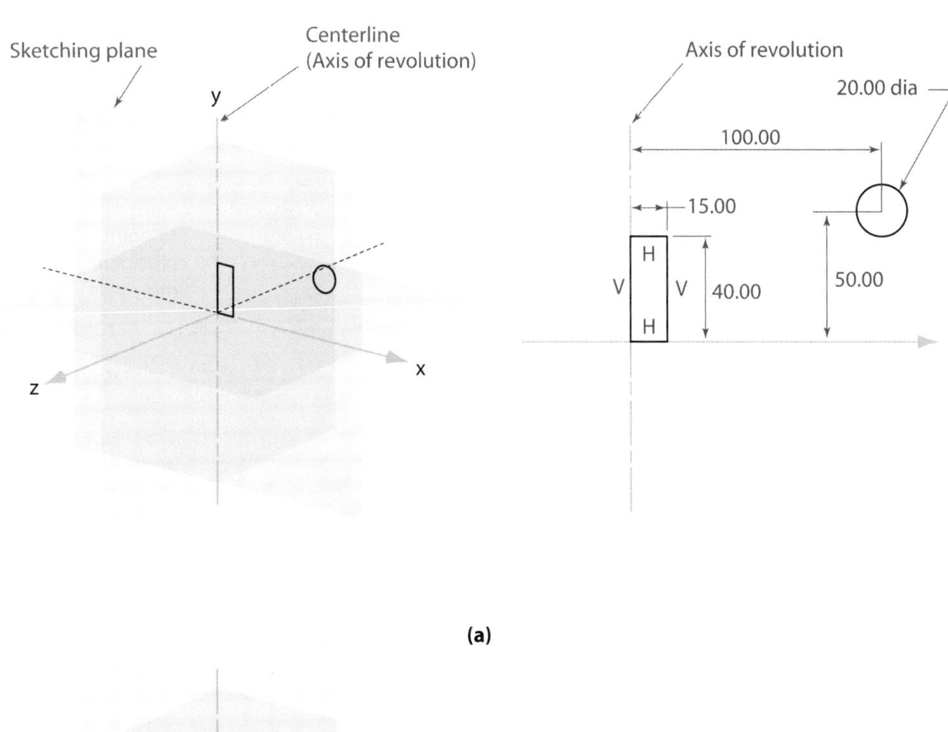

(a)

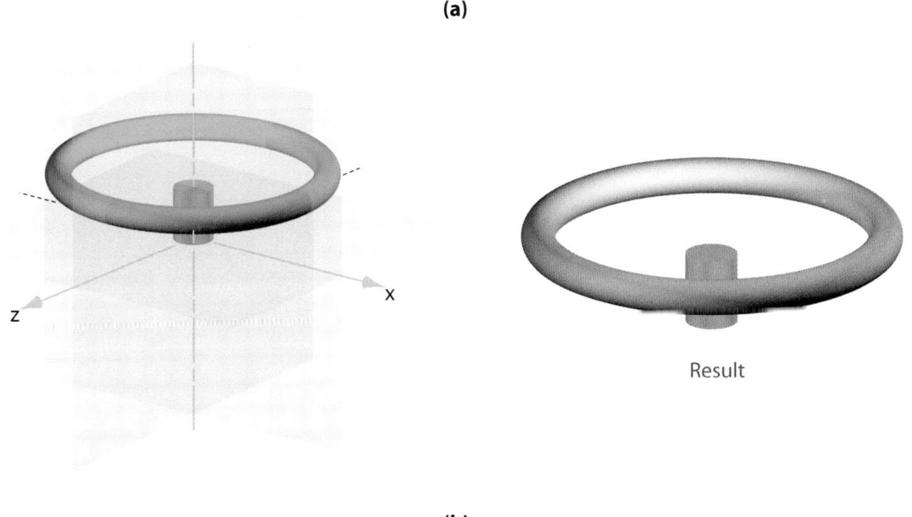

(b)

FIGURE 6.81. The profile for the base feature is sketched in (a), and the base feature is created by revolution in (b).

constrained. Notice that the two profiles are not constrained to each other, but only to the basic planes and centerline or axis of revolution. Thus, if the shape of the hub or rim is changed, the other profile in the sketch will not change. Both profiles are rotated 360 degrees about the centerline in the same operation to produce the base feature. The base feature appears as two disconnected solids, but this is fine as long as you remember to connect them later.

The next feature to be created is one of the spokes. Because it will be created using a sweep, a path will be required in addition to the cross-section profile, as shown in Figure 6.82(a). The path can be created on the same basic modeling plane that was used for sketching the base profiles. Even though the same sketching plane is used, the profiles for the base and the path are considered separate entities. The spoke cross section must be sketched on a plane perpendicular to the path. The only planes available for sketching in this example are the basic modeling planes. Therefore, you will start the swept profile on the basic modeling plane at the center of the hub, which is perpendicular to the desired path.

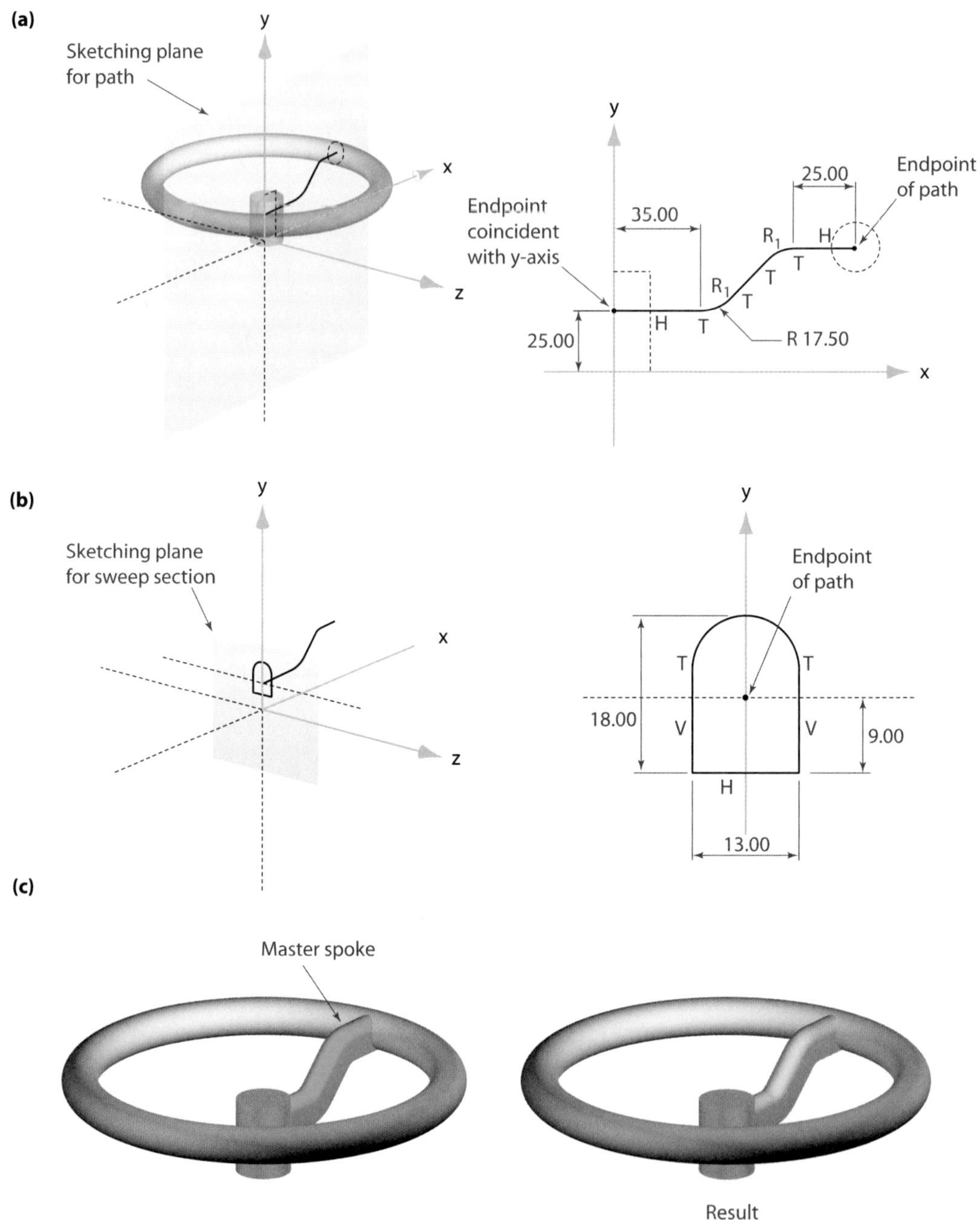

FIGURE 6.82. The profile for the path is sketched and constrained in (a). The profile for the sweep section is sketched and constrained in (b). The master spoke is created by sweeping in (c).

Some software requires the designer to create the path first, then the cross-section profile; other software reverses the order. For this example, the creation of the path will be described first. The path is comprised of three line segments and two arcs. The starting point on the hub is constrained to be coincident with the rotation axis, and the endpoint is constrained to be coincident with the center of the circle that represents the profile of the rim. Additional geometric and dimensional constraints are added, as shown in Figure 6.82(a). The profile must be created on a sketching plane that is perpendicular to the path at one of its endpoints. As previously noted, another one of the basic modeling planes satisfies this requirement, as shown in Figure 6.82; so creating a new datum plane is not necessary. The profile is sketched and constrained. Notice that the bottom of the cross-section profile does not lie on the basic modeling plane; its

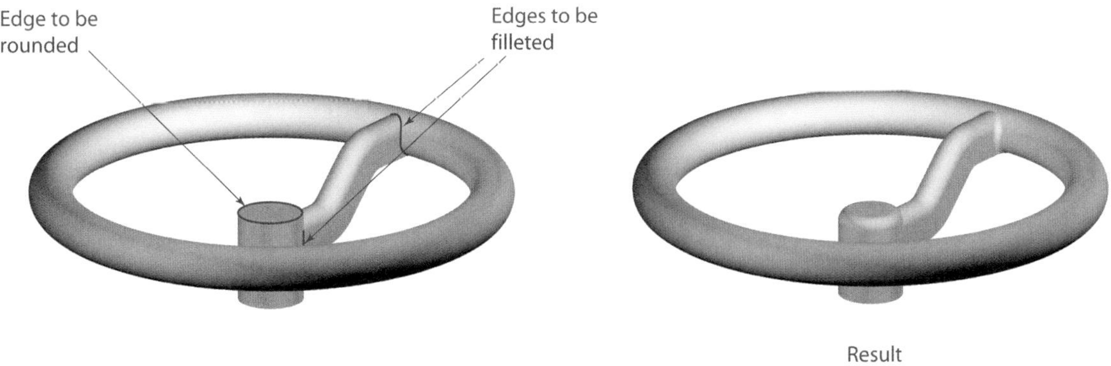

FIGURE 6.83. The edge at the top of the hub is rounded, and the intersections of the spoke are filleted to the specified dimensions.

vertical location is controlled by a dimensional constraint that is measured from the endpoint of the path curve. A sweep operation using the just-created path and profile is performed to create a solid spoke. A round is added to the top of the hub, and fillets are added at the intersections of the spoke with the hub and rim, as shown in Figure 6.83.

Multiple spokes are created with a circular array. The axis for the array is selected to be the same as the axis of rotation of the base. The features to be arrayed are selected to be the swept spoke as well as all of the fillets on it. Most solid modeling systems allow multiple features to be arrayed or patterned in a single operation. This is particularly useful when a parent feature has children that should be included in the pattern. In fact, some software systems automatically include the child features in the pattern. Some software requires a separate pattern operation for each feature. In either case, four spokes are made with equal rotational spacing, as shown in Figure 6.84.

The final feature to be added is the D-shaped hole. This feature is created as an extruded cut, as shown in Figure 6.85. The top surface of the hub is selected as the sketching plane. The profile is sketched and constrained. Note that the center of the arc for the hole is constrained to be coincident with the center axis of the hub. The profile is extruded to the next intersecting surface to create the cut and thus completes the model. Notice that the D-shaped cut for the hole was performed after the creation of the spokes. What would happen if the hole was created before the spokes were modeled? Since the path of the master spoke starts at the axis of revolution of the hub, the spoke is essentially embedded in the hub. If the hole was created first, material would be created in this region and would fill up portions of the hole when the spokes were created. A good rule of thumb to remember when modeling is to create solid geometry first (protrusions), then remove material (cuts) whenever possible.

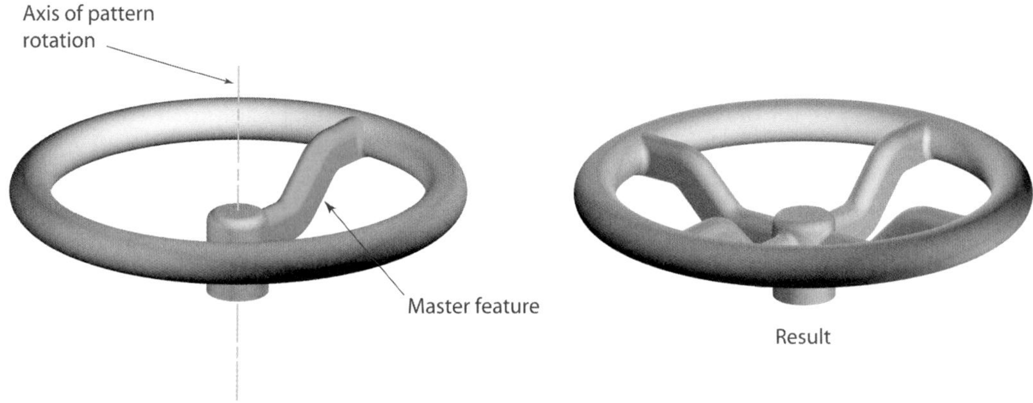

FIGURE 6.84. Multiple spokes are created by a circular pattern, or array, using the center of the wheel as the axis of rotation.

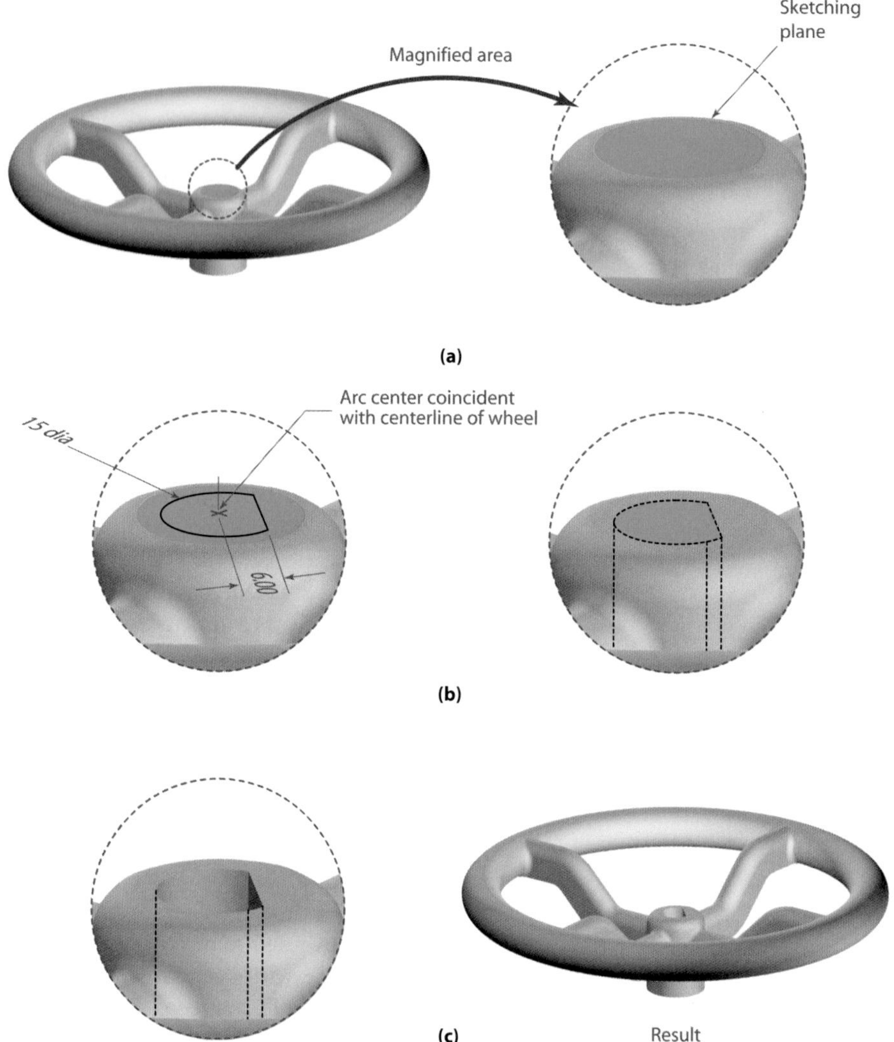

FIGURE 6.85. The D-shaped hole is created by selecting the top of the hub as the sketching plane in (a). The profile is sketched and constrained in (b). The profile cut is extruded through the hub in (c).

6.12 Extraction of 2-D Drawings

Nearly all solid modeling software packages have a facility for easily creating 2-D engineering drawings from solid models. Formal engineering drawing, which is covered in detail in later chapters, displays the part with all of its features in multiple predesignated views. It also displays the sizes and locations of the features. Solid modelers, which display a model from any viewpoint, can easily create the required views and display dimensions, thus greatly reducing the time and effort required to produce a drawing. Note that the dimensional constraints used in creating the solid model may be different from the dimension values that should be displayed on the engineering drawing. For example, the drawing is required to display all of the dimensions that are necessary for manufacturing the part. Some of these dimension values may be controlled by geometric constraints and are not included in the model as dimensional constraints. You will learn more about proper dimensioning practices in later chapters of this text. An example of a 2-D engineering drawing produced from a solid model is shown in Figure 6.86.

FIGURE 6.86. A typical solid model and a formal working drawing extracted from the model.

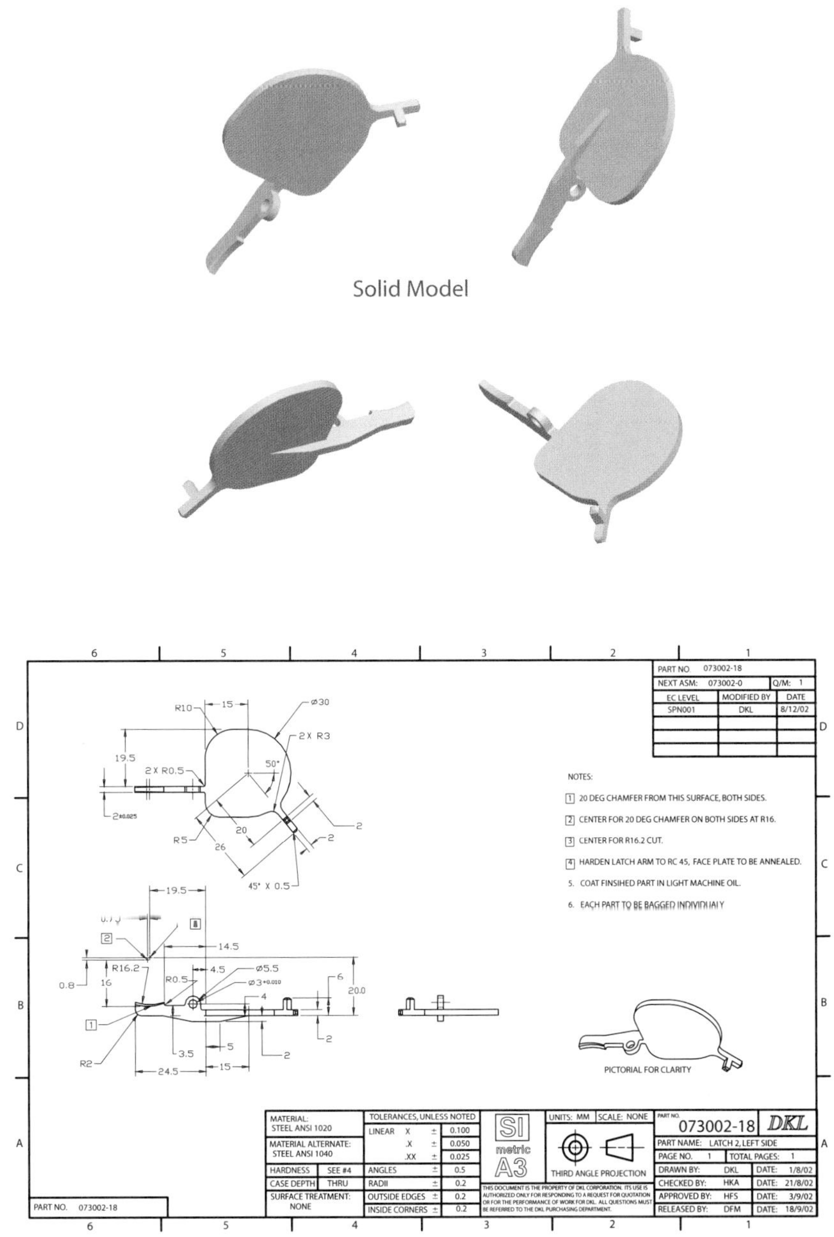

CAUTION

Inexperienced users of solid modeling software usually commit common errors in creating models. Some of these errors are merely a nuisance, such as generating extra unnecessary dimensions. Other errors do not let the user proceed with the creation process and must be resolved before the user can proceed. Still other errors let the user create a model that may appear like the one desired; however, problems manifest themselves later when the model is edited or when the part is fabricated. The following sections are a compilation of common errors made in solid modeling.

Base Profile Not Properly Positioned

A 3-D model is defined not only by its geometry but also by its location in space. When a solid model is built, the location of the model should be defined relative to the origin of the model's coordinate system. Defining this location is done by defining the location of the profile used for the base feature, most commonly by making one vertex (or point on the centerline) on the profile coincident with the origin. If this is not done, as shown in Figure 6.87, the profile will not be fully constrained and extra dimensional constraints would have to be added to define the location of the profile. These extra dimensional constraints are meaningless and add confusion to the model.

Invalid Profile

Valid profiles were discussed earlier in the chapter. Often invalid profiles are created inadvertently, usually through careless use of the computer's pointing device. Three common errors are shown in Figure 6.88. In Figure 6.88(b), the user attempted to close the sketch to make a valid profile, but missed the target endpoint and left the sketch open. Valid profiles must be closed. In Figure 6.88(c), while attempting to close the sketch, the user crossed over the first sketch element with the final sketch element. Valid profiles cannot cross, overlap, or self-intersect. A similar problem occurs when the profile contains a duplicate line segment. Overlapping lines can be very difficult to

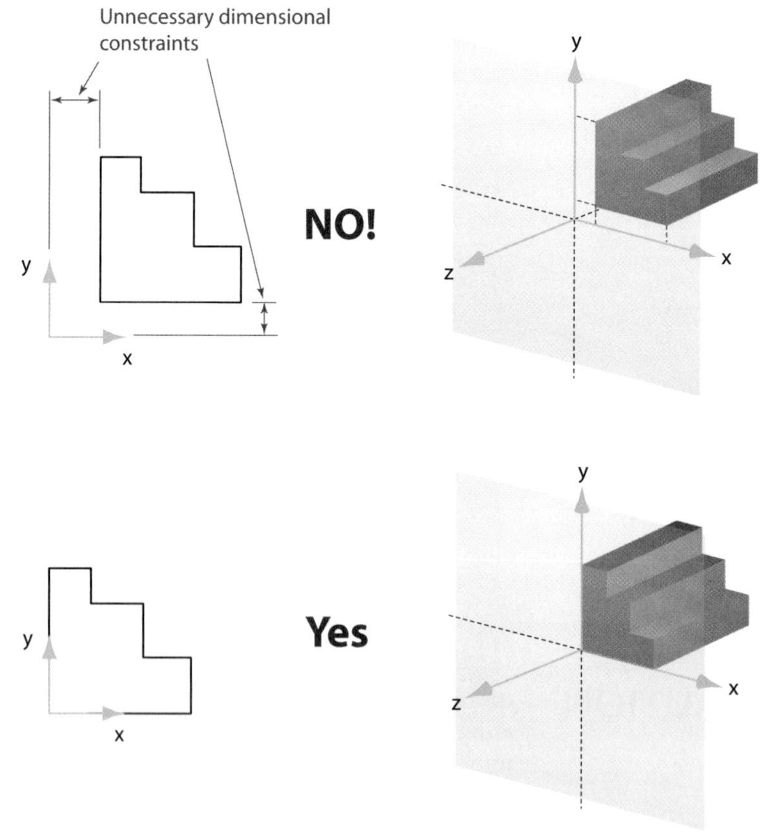

FIGURE 6.87. The base feature should be aligned with the origin whenever possible to avoid the need for extra dimensions.

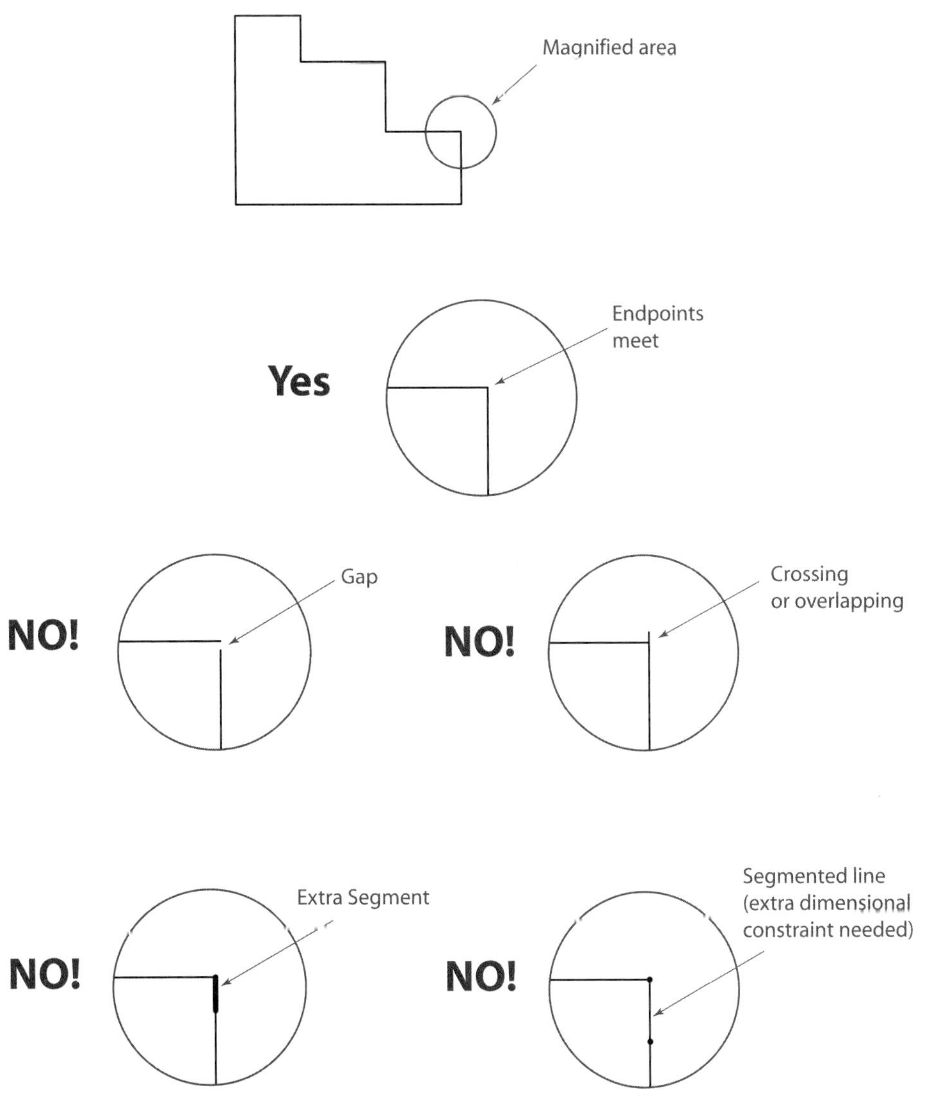

FIGURE 6.88. Careless construction leads to invalid profile errors that are sometimes difficult to find.

find; you may want to delete a line segment if you suspect that it might be duplicated, then redraw the line if no duplicate is found. In Figure 6.88(d), the user inadvertently used two line segments in place of one continuous line segment, resulting in an internal vertex along the desired edge. The profile cannot be fully constrained until the lengths of both line segments are defined. This results in an unnecessary and meaningless dimensional constraint. A profile with a segmented line like this is considered poor modeling practice. Invalid profile errors are usually difficult to see and, therefore, difficult to resolve. Some sketching editors alert the user by highlighting the location of gaps or intersections. Otherwise, you may be faced with the tedious task of searching for the source of the problem. Careful use of the computer's pointing device usually keeps these errors to a minimum.

Profile Not Constrained with Design Intent

It is sometimes tempting to use dimensional constraints instead of geometric constraints that reflect the intent of the design, because application of dimensional constraints follows traditional drafting practice and is typically easy to do in a sketching editor. The availability of geometric constraints in solid modelers offers an opportunity to include aspects of design intent that were previously unavailable in 2-D drafting. Consider the profile for the base feature of a rod clamp, shown in Figure 6.89. The profile is fully constrained, specifying the location of the center of the circular cutout. The design intent, however, is for the circular cutout to be concentric with the rounded part of the exterior of the clamp. If the overall height of the profile was changed, this design intent would not be maintained. In replacing the location dimensions of the circular cutout with a concentricity constraint (coincident arc centers), the design intent is maintained as the overall height of the profile changes. Also note that symmetry constraints on the lower edge of the part ensure that the gap is centered along the axis of the clamp and that both sides have equal thickness.

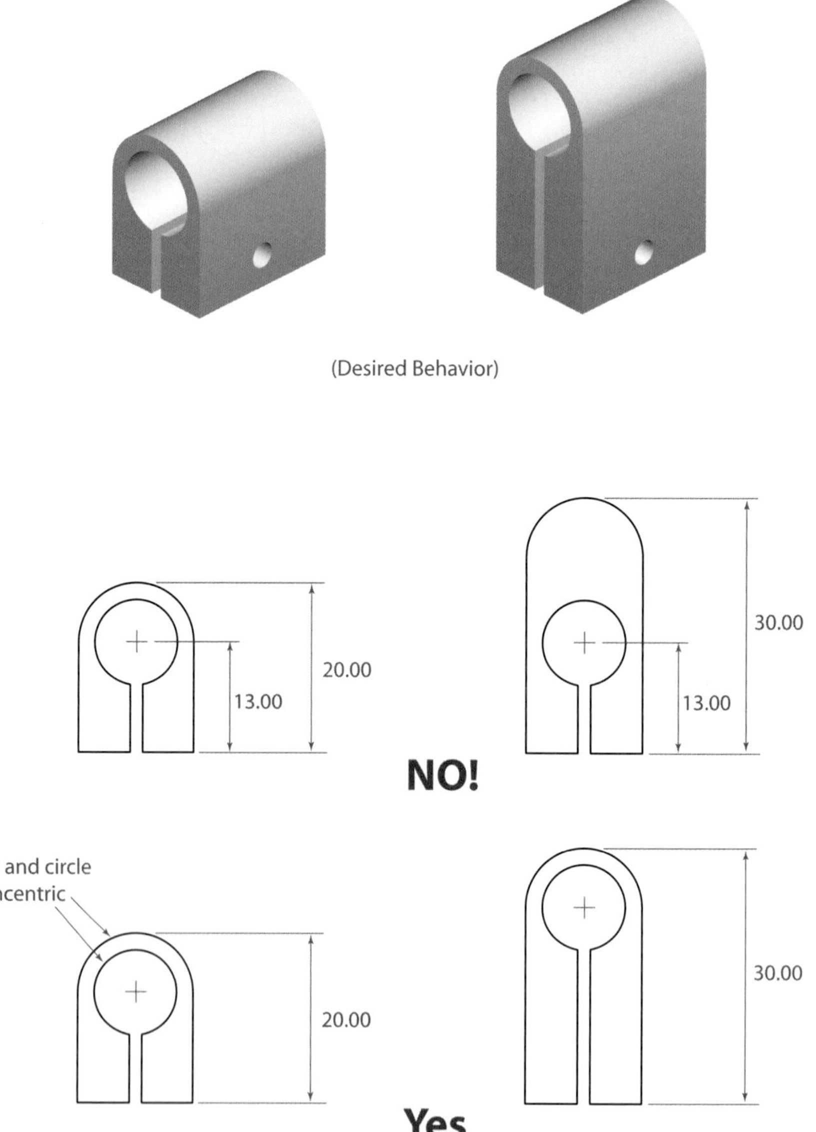

FIGURE 6.89. The desired behavior is for the outer radius of this part to be concentric with the hole. A geometric constraint rather than a dimensional constraint is preferred to guarantee this behavior, even if other dimensions are changed.

Profile Overconstrained by Automatic Constraints

For the most part, sketching editors that intelligently apply presumed geometric constraints are very useful. For example, a great deal of time is saved when lines that are sketched to be almost horizontal or almost vertical have these constraints applied automatically. However, automatic constraint application can sometimes lead to problems, particularly when the constraints are applied inadvertently. If your sketching editor has been set to search automatically for equal element sizes, as in Figure 6.90, line segments created to almost the same length in the sketch will automatically be constrained equal in length to each other. Arcs created to almost the same radius as other arcs also will be automatically constrained equal in radii to each other. If that is your intent, then fine. But if that is not your intent, then you must be careful not to create line segments or arcs that are too close in size to each other or you must delete the unintentional constraints when they appear and replace them with separate dimensional constraints on each entity. Figure 6.91 shows a case where an intelligent colinear constraint has been unintentionally applied. In Figure 6.92, an unintentional perpendicularity constraint has been applied. In all cases of unintentional constraints, the addition of the desired geometric constraints or dimensional constraints cause the profile to be overconstrained. In this case, you need to delete the unintentional constraints and then add constraints to capture your design intent.

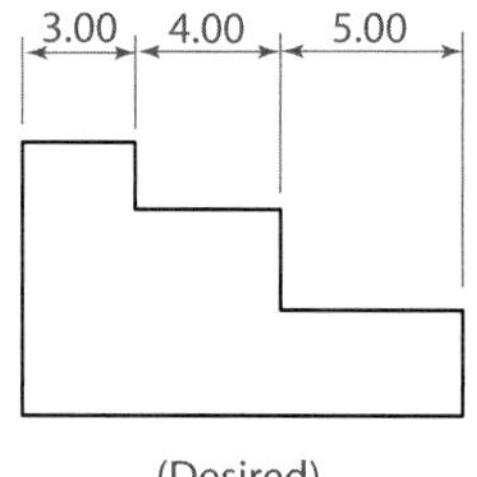

(Desired)

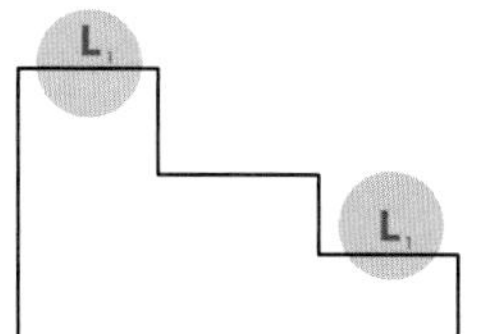

NO!

FIGURE 6.90. Automatic geometric constraints in the sketching editor sometimes cause equal length constraints to be applied inadvertently.

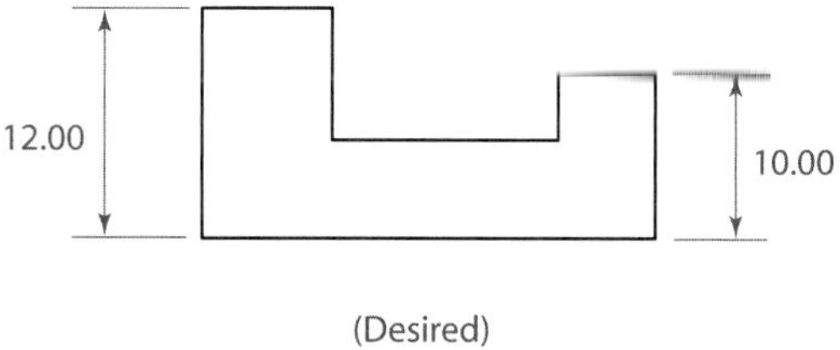

(Desired)

Colinear

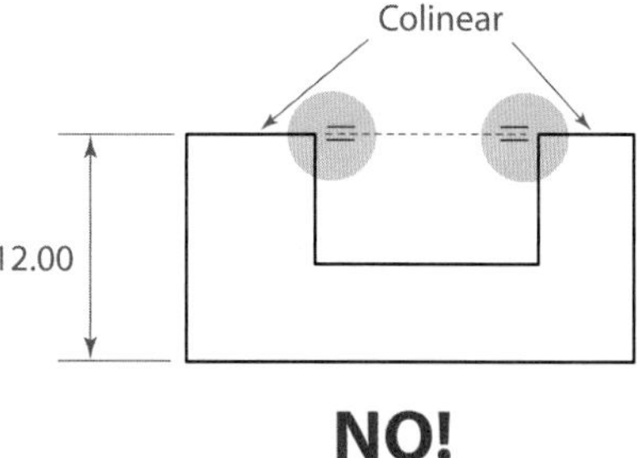

NO!

FIGURE 6.91. Automatic geometric constraints in the sketching editor sometimes cause colinear constraints to be applied inadvertently.

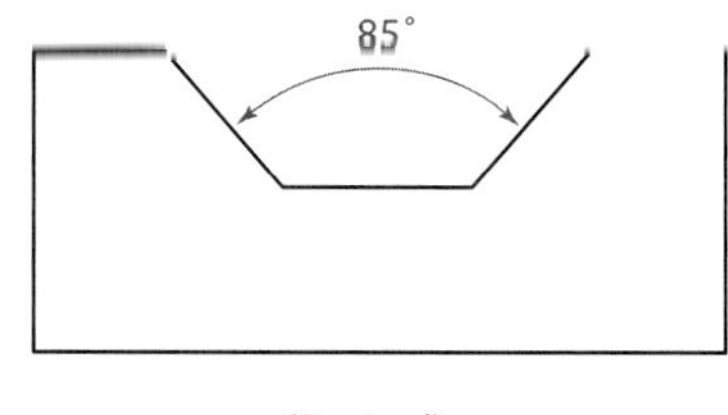

(Desired)

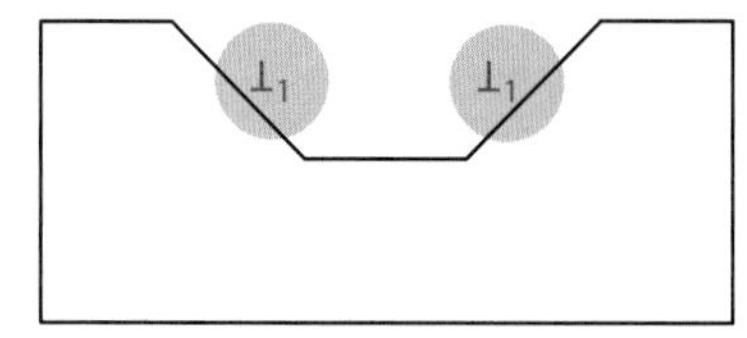

NO!

FIGURE 6.92. Automatic geometric constraints in the sketching editor sometimes cause perpendicular constraints to be applied inadvertently.

Dimensional, Instead of Geometric, Constraints

A common error in making extruded cuts and protrusions is to use a blind depth to extrude the feature a specific distance, when what is really desired is to extrude the feature to a particular surface. In the model shown in Figure 6.93, for example, the design intent is for the extruded cut to extend all the way through the base. However, the cut was created by specifying a dimensional constraint for the length of the extrusion. The specified length of the extrusion is an unnecessary, meaningless dimension, because the cut was to extend all the way through. Also, if the thickness of the base increases, as sometimes occurs when the design changes, the extruded cut may no longer extend all the way through the base. By specifying, instead, that the extruded cut is to continue until it intersects the bottom of the base, the design intent is always fulfilled.

In the model shown in Figure 6.94, the design intent is for the protrusion to extend from the angled datum plane to the top of the base. However, it was created by specifying a dimension for the length of the extrusion. The specified length of the extrusion is a meaningless dimension, because the protrusion was to extend to the top surface of the base. Also, if the thickness of the base decreases, as may occur if the design ever

FIGURE 6.93. When a cut is to extend to a specific surface or all the way through a part, cutting to a specified distance should not be done.

(Desired)

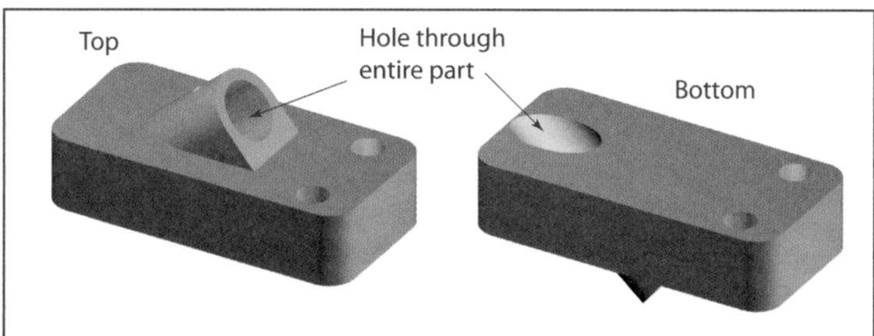

NO!

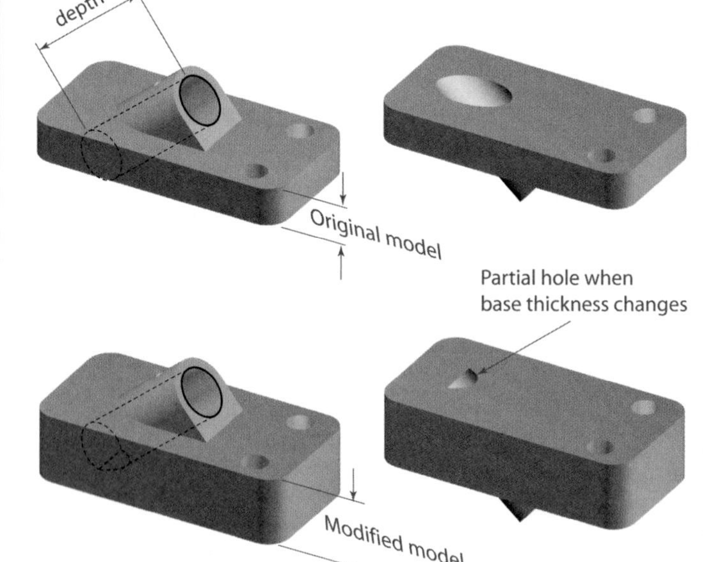

Yes

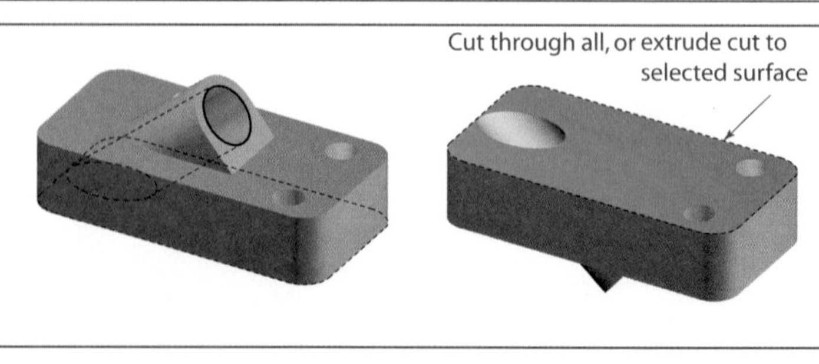

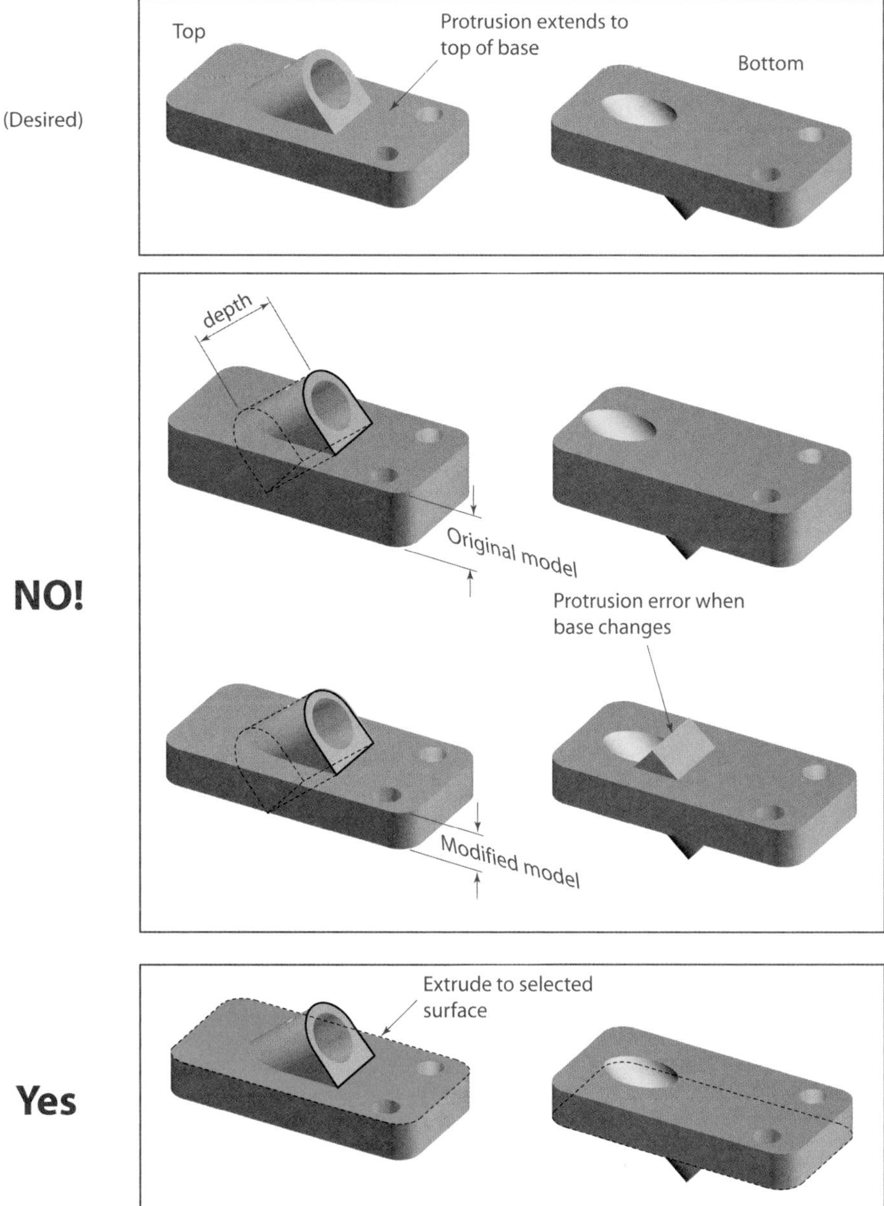

FIGURE 6.94. When a protrusion is to extend to a specific surface, extending to a specified distance should not be done.

changes, the extruded protrusion may extend beyond the bottom of the base. By specifying, instead, that the extruded protrusion is to continue until it intersects the top of the base, the design intent is always fulfilled.

Using a Cut or Protrusion Instead of a Built-in Feature

A common practice for beginning designers is to use a cut feature to create a hole, for example. Figure 6.95 shows the creation of a counterbored hole using two concentric circular cut features compared to using the built-in counterbored hole feature. Even though the geometry of the resultant holes is identical, the use of cut features can result in undesired results. What if you decreased the diameter of the counterbore to a value that was smaller than the diameter of the through hole? What if the location constraints for the two cut features had different values? The resulting geometry would not represent your intent, a counterbored hole. The built-in counterbored hole feature does not permit such changes in the geometry. Another reason to use the built-in features is to capture more design information. While it may seem easy to sketch a circle on a given surface to create

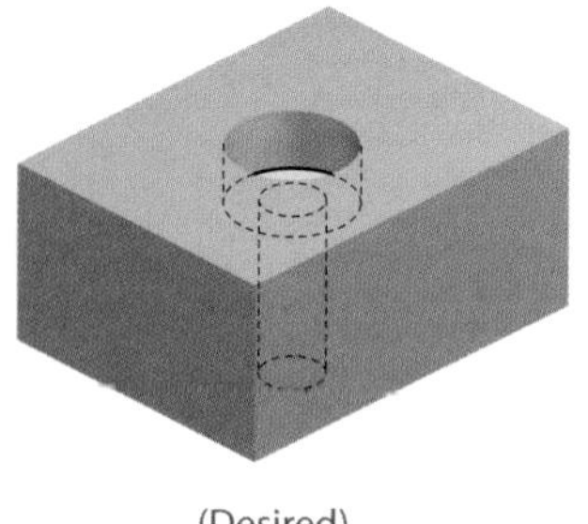

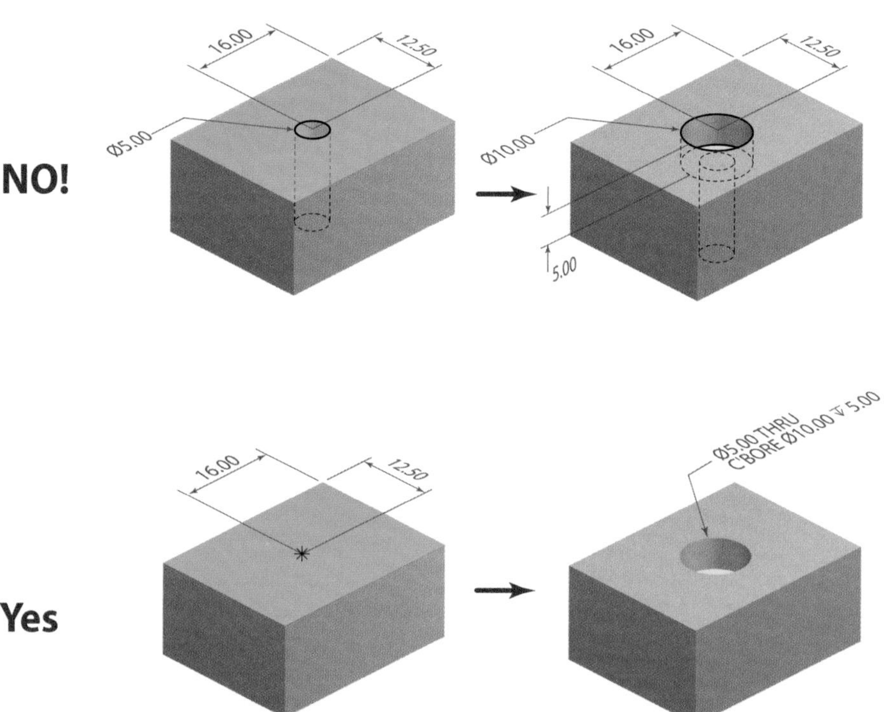

FIGURE 6.95. Functional features such as this counterbored hole should be created as features and not constructed from extruded cuts or protrusions.

a hole, the solid modeler may be integrated with a larger, more sophisticated CAD/CAM system that might not recognize the feature as a hole. If the CAD system has intelligent assembly capabilities, it may automatically align the axis of the hole with the shaft of a bolt or screw. If you are creating manufacturing plans from your solid model, the software may recognize certain types of holes and specify the appropriate fabrication sequences automatically. But when you create the hole using a cut feature, the CAD system does not know it is a hole; and you will have to specify the assembly and fabrication sequences yourself. That leaves a great deal of room for errors such as putting bolts in upside down or omitting one of the machining operations. To save time and avoid errors in later applications, it is prudent to use the appropriate feature type that imbeds the special attributes of the feature in your solid model.

6.13 Chapter Summary

In this chapter, you learned about features and parametric solid modeling. Features are distinctive shapes that compose a solid model. Parametric models have the capability to be modified by changing the sizes and other attributes of the features in the model. The history of solid modeling shows how CAD has evolved from wireframe to solid models and provides some insight regarding the strategies used to create solid models.

Part modeling can be a very complicated process, but some general strategies make it easier to create good, robust solid models. But before you go to the computer, you need to consider how the part model will be used. Later applications such as manufacturing and documentation will be easier when the part is modeled properly. Thus, you need to plan carefully and ask yourself some questions before the first feature is created: How can you decompose that complicated widget into simpler features that are available on your solid modeler? Which feature should you create first? Which features are related to each other? How is the part used? Manufactured? Can standard features such as holes be modeled to imbed design intent and/or manufacturing information in addition to simple geometric characteristics? How does the part fit into an assembly? What will the engineering drawing look like? These are just a few questions you need to consider before you begin. For now, you may not know the answers to all of these questions; but as you gain experience, you will develop an appreciation for the importance of building a robust solid model that captures your design intent.

The solid modeling process begins with identification of the features of the part, followed by selection of the base feature. Profiles for extrusion, rotation, sweeps, and blends are created with sketches and are controlled using different types of geometric, dimensional, and associative constraints. After the base feature is created, other features are added to the model; these features are dependent upon the base feature or other previously created features. Care must be taken in the creation of solid models to make flexible models that are robust and that can be used for purposes such as analysis, manufacturing, and documentation.

As modeling systems continue to develop, designers and engineers want to include more information in the model to better simulate the physical characteristics of the parts. These models, called behavioral models, might include features such as physical properties, manufacturing tolerances, surface finish, and other characteristics of the parts. Besides a person's appearance, you would need to know something about his or her education or physical abilities to determine whether the person might be able to do a particular job. Likewise, a designer or an engineer may need to know more than just the shape of an object or assembly model to determine whether it will perform its intended function. As they become more realistic and can simulate the actual behavior of the parts and assemblies, product models of the future will contain even more characteristics and features.

6.14 glossary of key terms

algebraic constraints: Constraint that define the value of a selected variable as the result of an algebraic expression containing other variables from the solid model.

associative constraints: *See* algebraic constraints.

base feature: The first feature created for a part, usually a protrusion.

blend: A solid formed by a smooth transition between two or more profiles.

blind extrusion: An extrusion made to a specified length in a selected direction.

boundary representation (b-rep): A method used to build solid models from their bounding surfaces.

chamfers: Angled cut transitions between two intersecting surfaces.

child feature: A feature that is dependent upon the existence of a previously created feature.

constraints: Geometric relationships, dimensions, or equations that control the size, shape, and/or orientation of entities in a sketch or solid model.

constructive solid geometry (CSG): A method used to build solid models from primitive shapes based on Boolean set theory.

cosmetic features: Features that modify the appearance of the surface but do not alter the size or shape of the object.

cut: A feature created by the removal of solid volume from a model.

database: A collection of information for a computer and a method for interpretation of the information from which the original model can be re-created.

6.14 glossary of key terms (continued)

datum geometries: Geometric entities such as points, axes, and planes that do not actually exist on real parts, but are used to help locate and define other features.

datum planes: The planes used to define the locations of features and entities in the construction of a solid model.

design table: A table or spreadsheet that lists all of the versions of a family model, the dimensions or features that may change, and the values in any of its versions.

design tree: *See* model tree.

dimensional constraints: Measurements used to control the size or position of entities in a sketch.

dimension name: The unique alphanumeric designation of a variable dimension.

driven dimension: A variable connected to an algebraic constraint that can be modified only by user changes to the driving dimensions.

driving dimension: A variable used in an algebraic constraint to control the values of another (driven) dimension.

double-sided extrusion: A solid formed by the extrusion of a profile in both directions from its sketching plane.

extrude through all: An extrusion that begins on the sketching plane and protrudes or cuts through all portions of the solid model that it encounters.

extrude to selected surface: An extrusion where the protrusion or cut begins on the sketching plane and stops when it intersects a selected surface.

extrusion: A solid that is bounded by the surfaces swept out in space by a planar profile as it is pulled along a path perpendicular to the plane of the profile.

family model: A collection of different versions of a part in a single model that can display any of the versions.

feature array: A method for making additional features by placing copies of a master feature on the model at a specified equal spacing.

features: Distinctive geometric shapes on solid parts; 3-D geometric entities that exist to serve some function.

feature-based solid modeling: A solid modeling system that uses features to build models.

feature pattern: *See* feature array.

feature tree: *See* model tree.

fillets: Smooth transitions of the internal edge created by two intersecting surfaces and tangent to both intersecting surfaces.

form feature: A recognizable area on a solid model that has a specific function.

geometric constraints: Definitions used to control the shape of a profile sketch through geometric relationships.

graphical user interface (GUI): The format of information on the visual display of a computer, giving its user control of the input, output, and editing of the information.

ground constraint: A constraint usually applied to a new sketch to fix the location of the sketch in space.

history tree: *See* model tree.

holes: A cut feature added to a model that will often receive a fastener for system assembly.

horizontal modeling: A strategy for creating solid models that reduces parent-child relationships within the feature tree.

master feature: A feature or collection of features that is to be copied for placement at other locations in a model.

master model: In a collection of similar parts, the model that includes all of the features that may appear in any of the other parts.

mirrored feature: A feature that is created as a mirror image of a master feature.

model tree: A list of all of the features of a solid model in the order in which they were created, providing a "history" of the sequence of feature creation.

parameters: The attributes of features, such as dimensions, that can be modified.

parametric solid modeling: A solid modeling system that allows the user to vary the dimensions and other parameters of the model.

parametric techniques: Modeling techniques where all driven dimensions in algebraic expressions must be known for the value of the dependent variables to be calculated.

parent feature: A feature used in the creation of another feature, which is called its child feature.

path: The specified curve on which a profile is placed to create a swept solid.

primary modeling planes: The planes representing the XY-, XZ-, and YZ-planes in a Cartesian coordinate system.

primitives: The set of regular shapes, such as boxes, spheres, or cylinders that are used to build solid models with constructive solid geometry methods (CSG).

principal viewing planes: The planes in space on which the top, bottom, front, back, and right and left side views are projected.

profile: A planar sketch that is used to create a solid.

6.14 glossary of key terms (continued)

protrusion: A feature created by the addition of solid volume to a model.

regeneration: The process of updating the profile or part to show its new shape after constraints are added or changed.

revolved solid: A solid formed when a profile curve is rotated about an axis.

ribs: Constant thickness protrusions that extend from the surface of a part and are used to strengthen or stiffen the part.

rounds: Smooth radius transitions of external edges created by two intersecting surfaces and tangent to both intersecting surfaces.

shelling: Removing most of the interior volume of a solid model, leaving a relatively thin wall of material that closely conforms to the outer surfaces of the original model.

sketches: Collections of 2-D entities.

sketching editor: A software tool used to create and edit sketches.

sketching plane: A plane where 2-D sketches and profiles can be created.

solid model: A mathematical representation of a physical object that includes the surfaces and the interior material, usually including a computer-based simulation that produces a visual display of an object as if it existed in three dimensions.

splines: Polynomial curves that pass through multiple data points.

suppressed: Refers to the option for not displaying a selected feature.

surface model: A CAD-generated model created to show a part as a collection of intersecting surfaces that bound a solid.

swept feature: A solid that is bound by the surfaces swept out in space as a profile is pulled along a path.

trajectory: *See* path.

unsuppressed: Refers to the option for displaying a selected feature.

variational techniques: Modeling techniques in which algebraic expressions or equations that express relationships between a number of variables and constants, any one of which can be calculated when all of the others are known.

vertex: A point that is used to define the endpoint of an entity such as a line segment or the intersection of two geometric entities.

webs: Small, thin protrusions that connect two or more thicker regions on a part.

wireframe models: CAD models created using lines, arcs, and other 2-D entities to represent the edges of the part; surfaces or solid volumes are not defined.

6.15 questions for review

1. What are some of the uses of solid models?
2. What is a feature?
3. Why are features important in solid modeling?
4. What types of features can be used as base features for your solid models?
5. Why are wireframe models inferior to solid models?
6. What are the steps in creating a solid model?
7. What are some errors that make a sketch invalid for creating a solid?
8. Why is it necessary to constrain a 2-D sketch?
9. What are the different types of geometric constraints?
10. What are associative constraints?
11. What are dimensional constraints?
12. What does it mean when a feature is a child of another feature? A parent of another feature?
13. What are some errors that constitute poor modeling practices?
14. What are some examples of good modeling-strategies?

6.16 problems

1. Create the following closed-loop profiles using the 2-D drawing capabilities of your solid modeling software. Define the geometry and sizes precisely as shown, using the necessary geometric constraints. Do not over- or underconstrain the profiles.

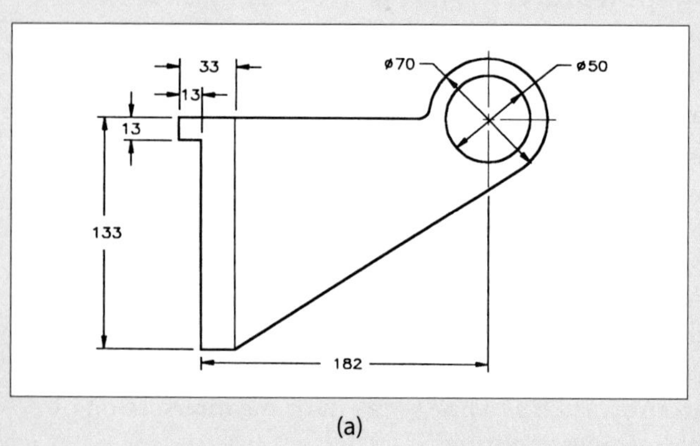

(a)

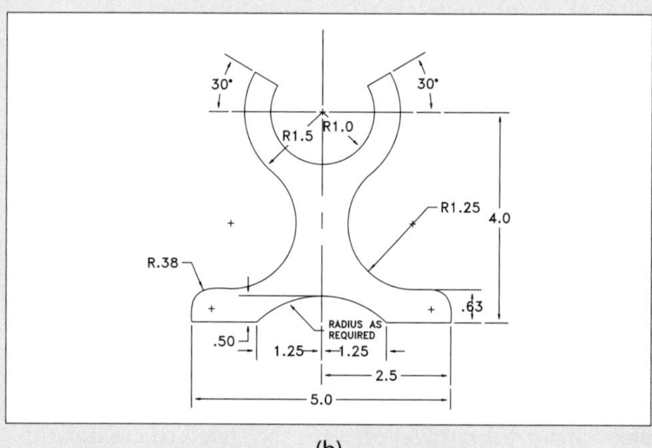

(b)

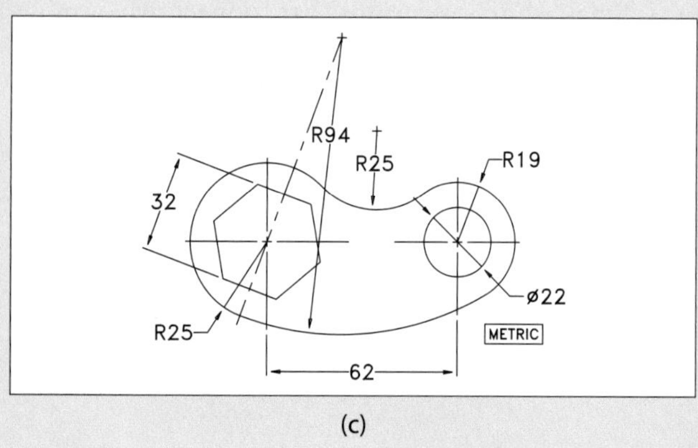

(c)

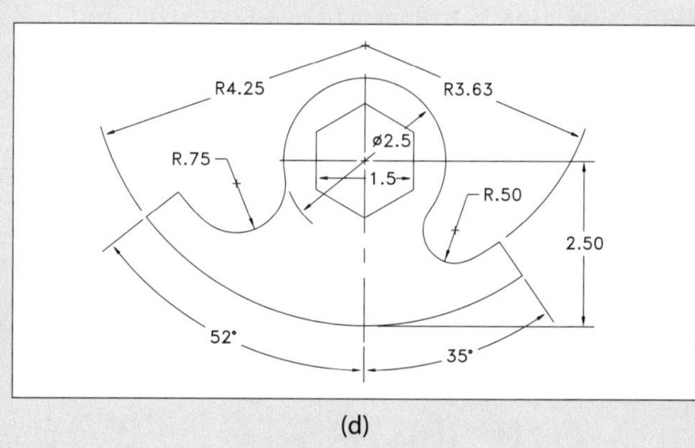

(d)

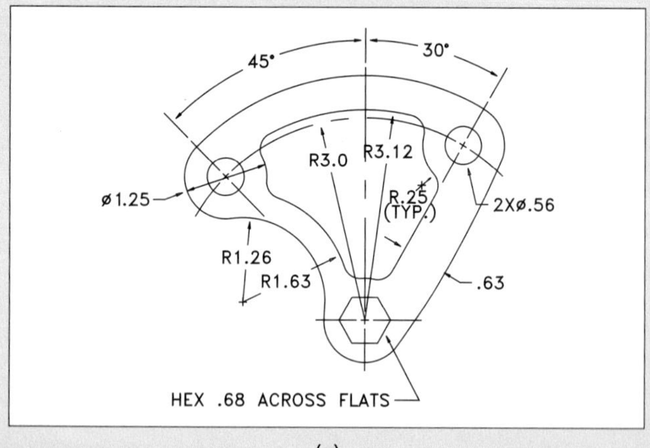

(e)

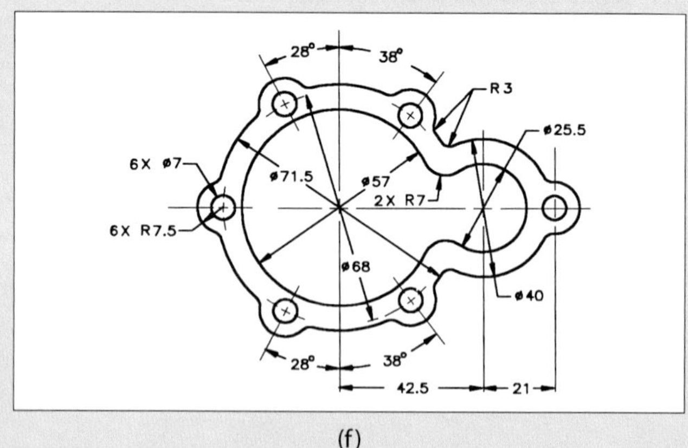

(f)

6.16 problems (continued)

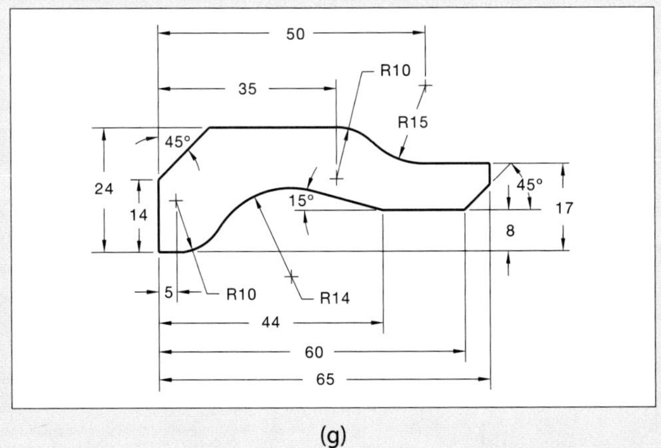

(g)

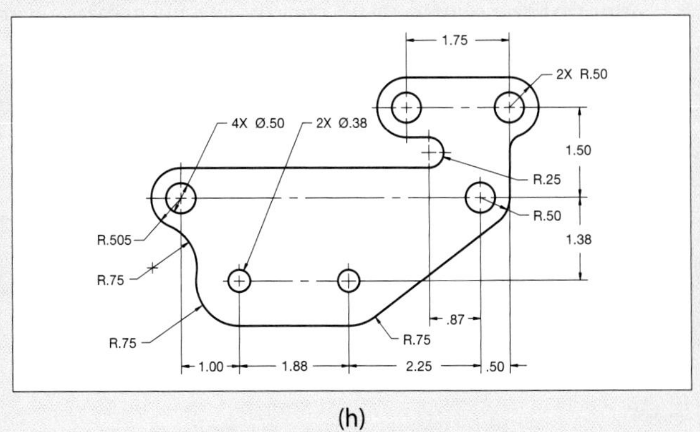

(h)

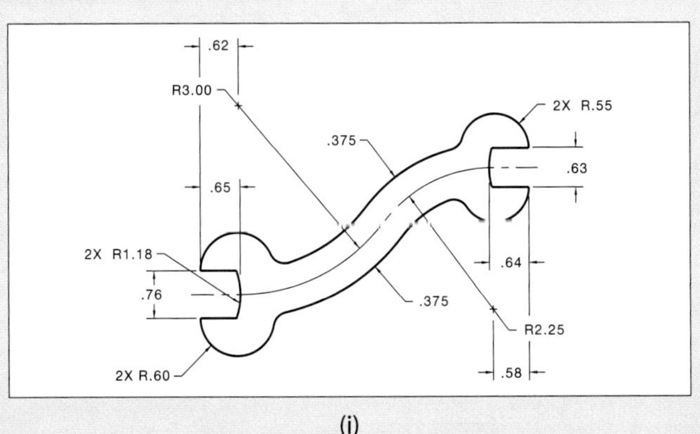

(i)

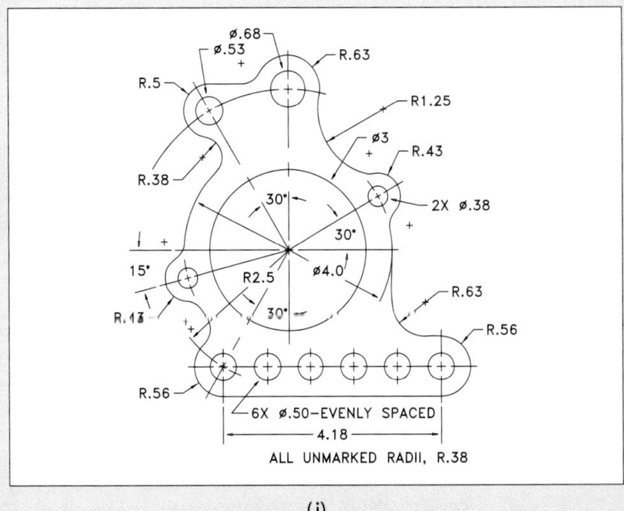

(j)

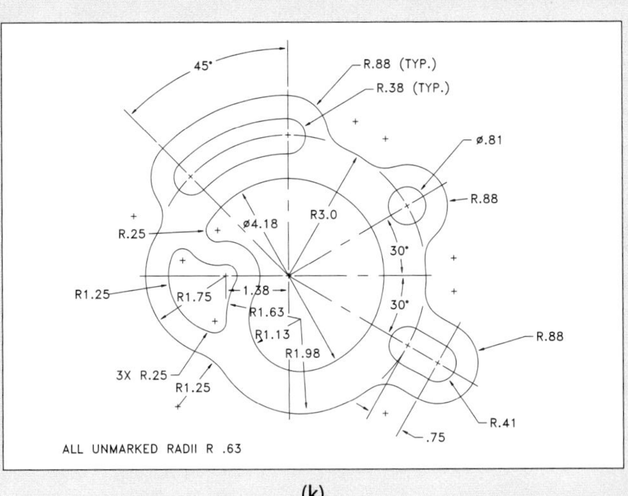

(k)

FIGURE P6.1.

6.16 problems (continued)

2. Study the following closed-loop profiles for which geometric constraints have not been added. Number each segment of the profiles and specify the necessary geometric constraints on each segment to create the final profile. Do not over- or underconstrain the profiles.

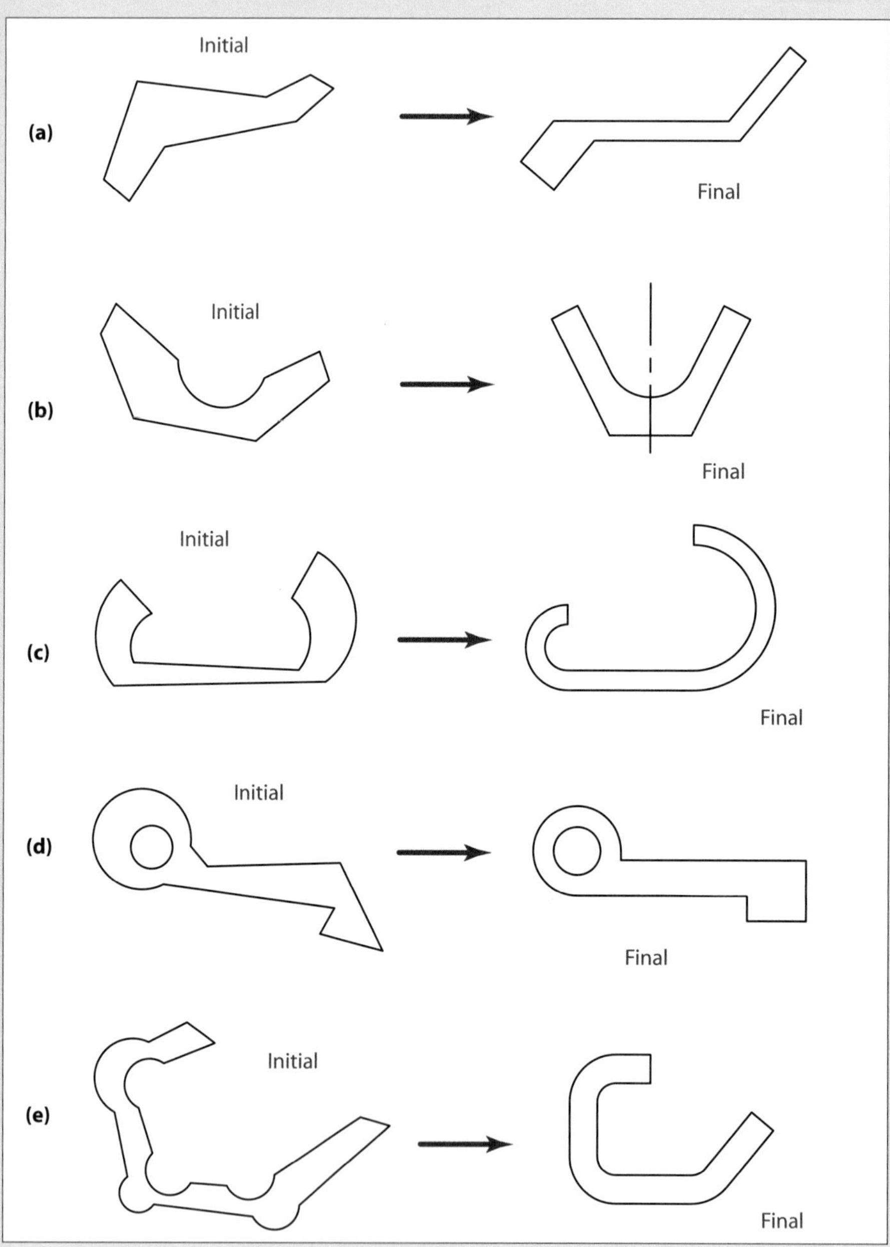

FIGURE P6.2.

6.16 problems (continued)

3. At first glance, these profiles may appear to be missing key dimensions. However, they are fully constrained by the addition of geometric constraints. Number each segment of the profiles. What were the geometric constraints used for each segment?

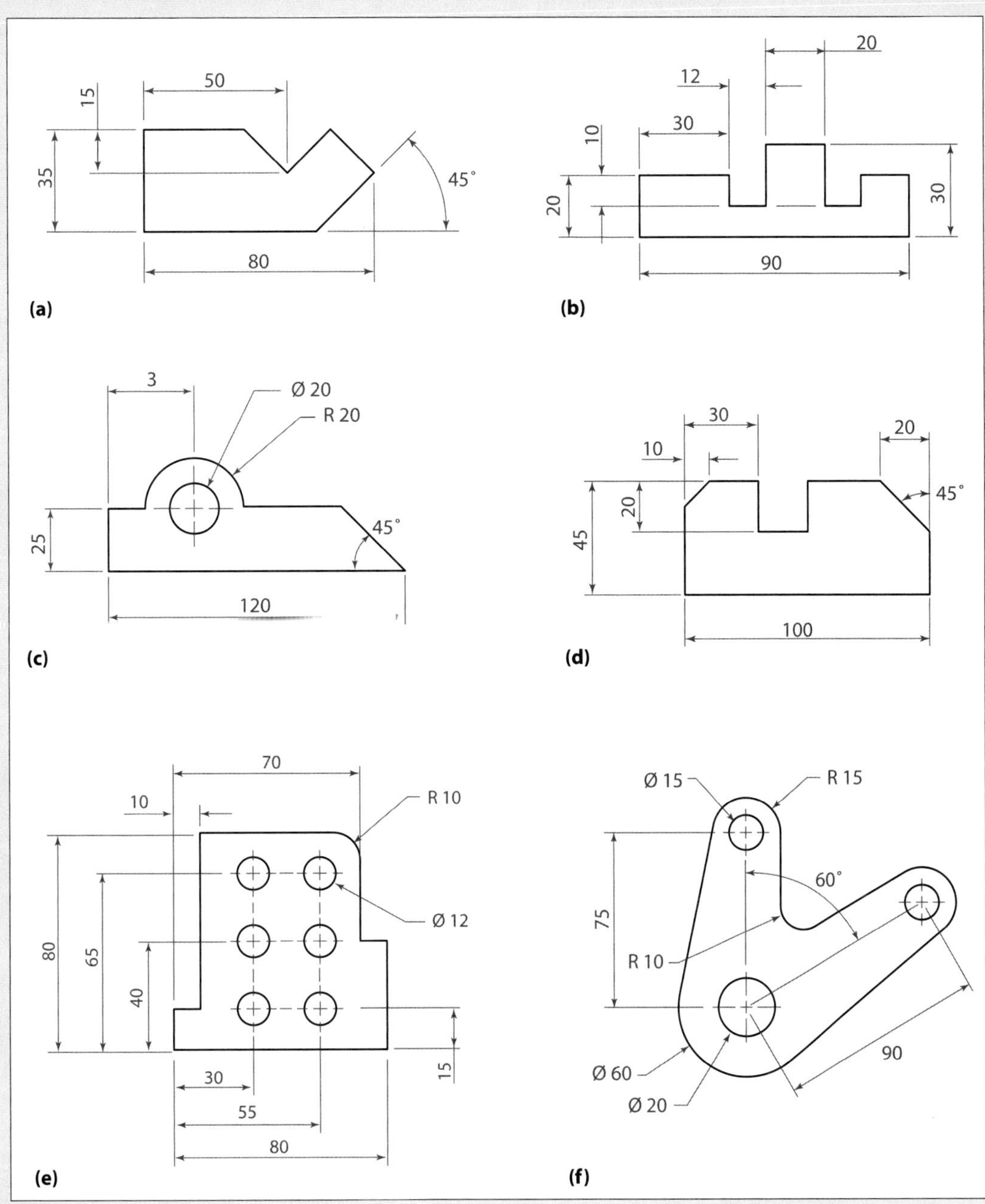

FIGURE P6.3.

6.16 problems (continued)

4. Create solid models of the following parts in your CAD system. Identify what you consider to be the base geometry for each part. Are any (child) features dependent upon the existence of other (parent) features? If so, specify the hierarchy.

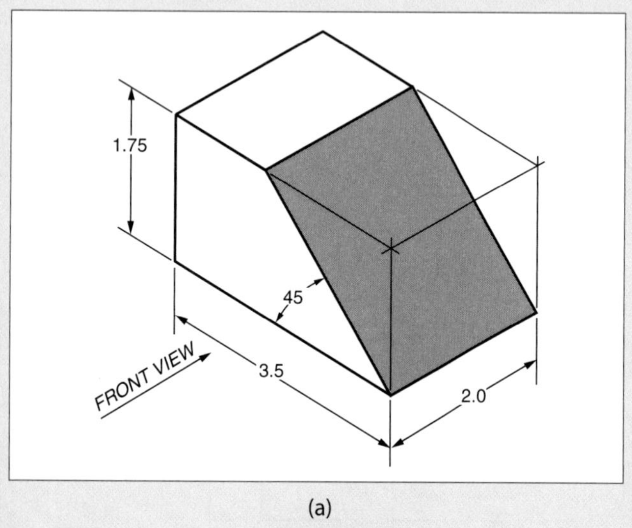

(a)

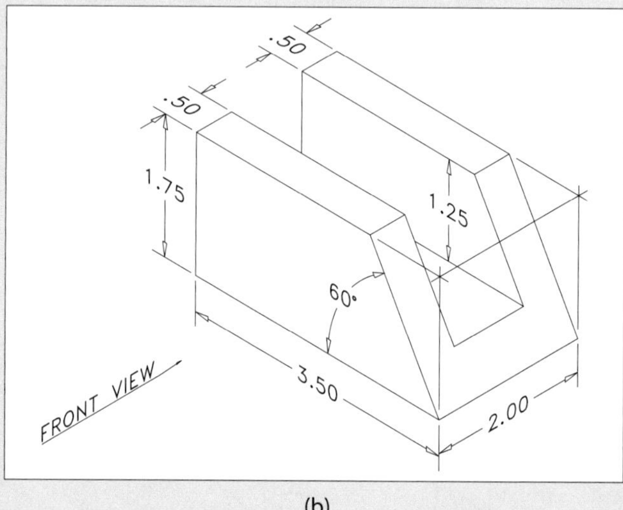

(b)

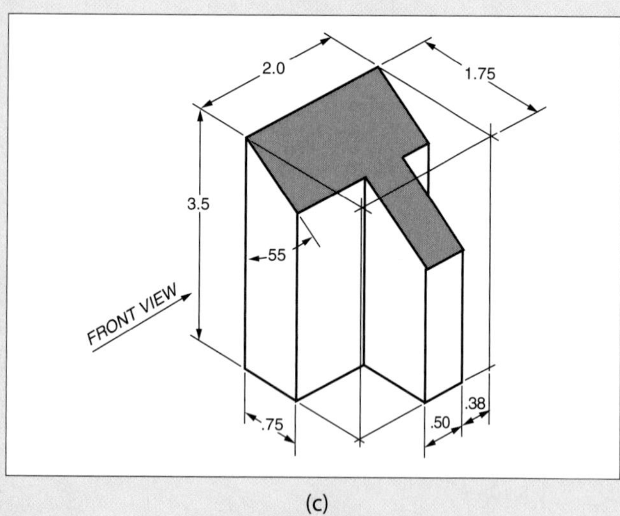

(c)

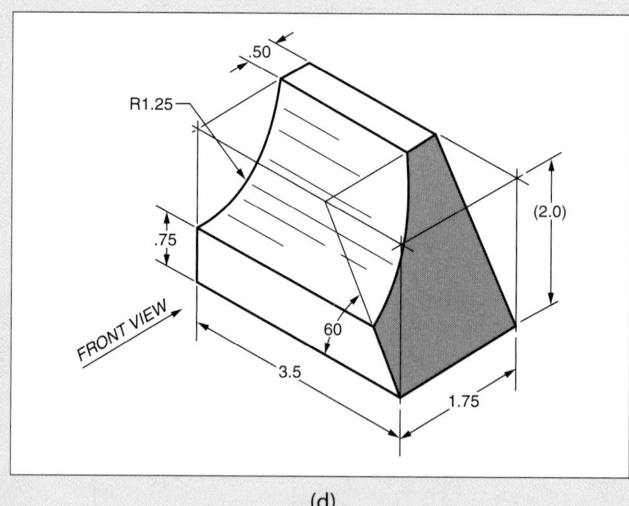

(d)

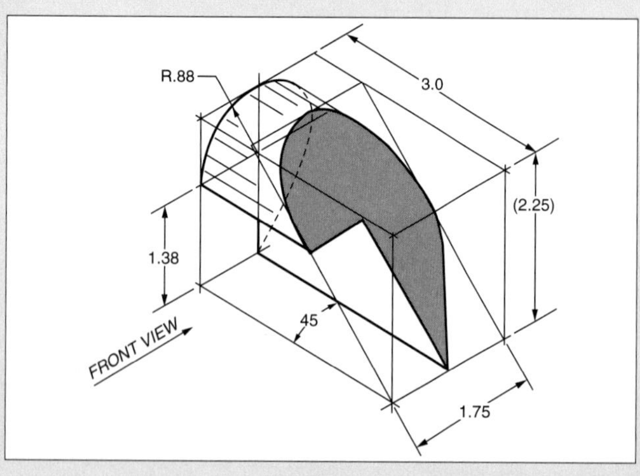

(e)

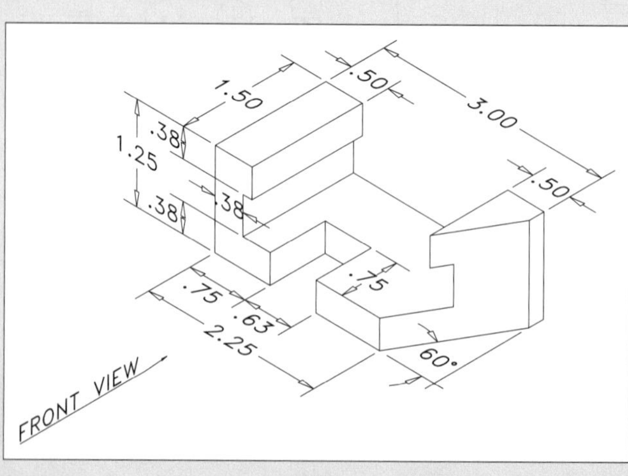

(f)

6.16 problems (continued)

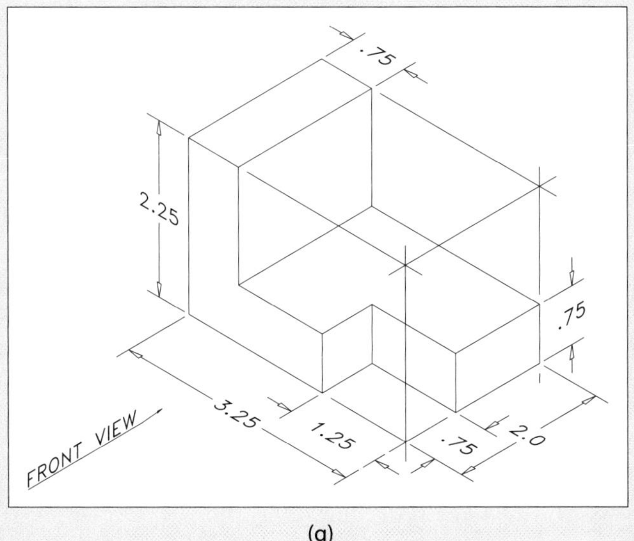

(g)

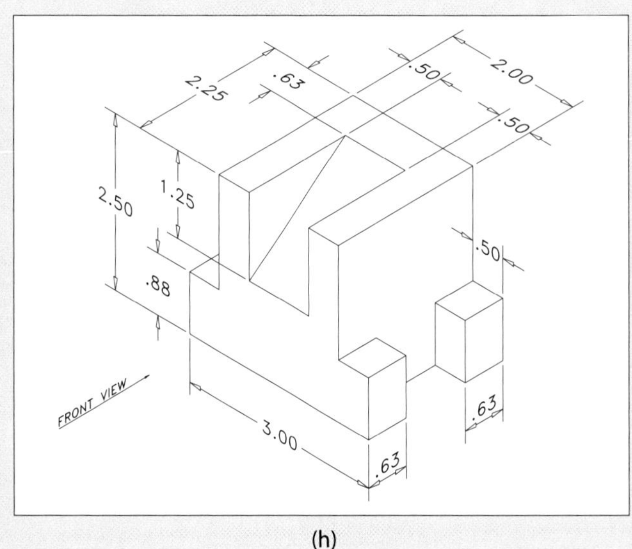

(h)

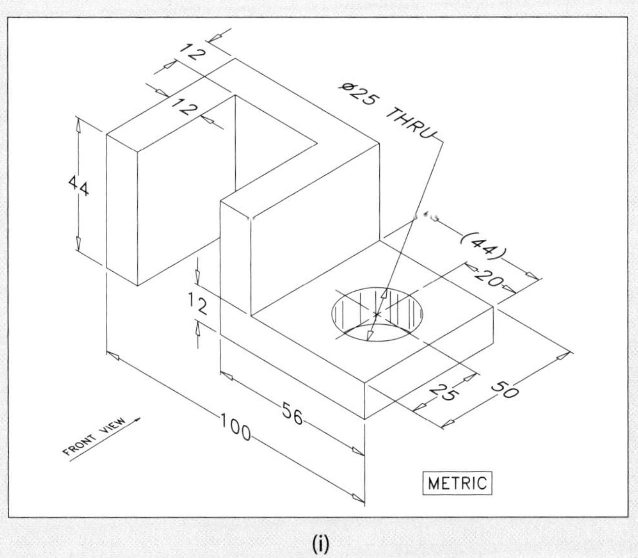

(i)

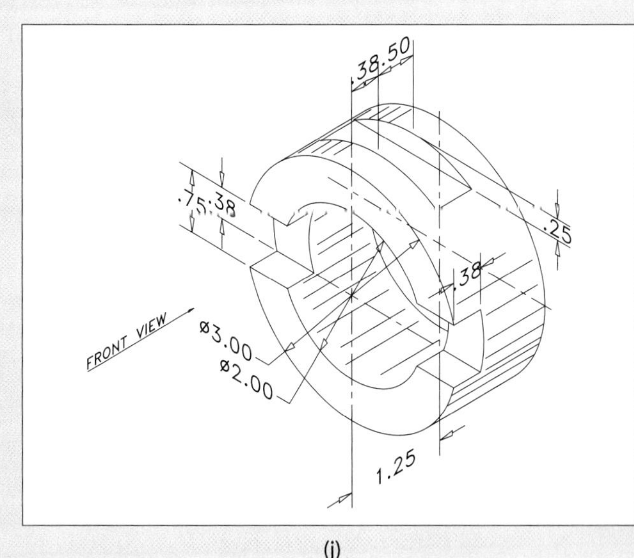

(j)

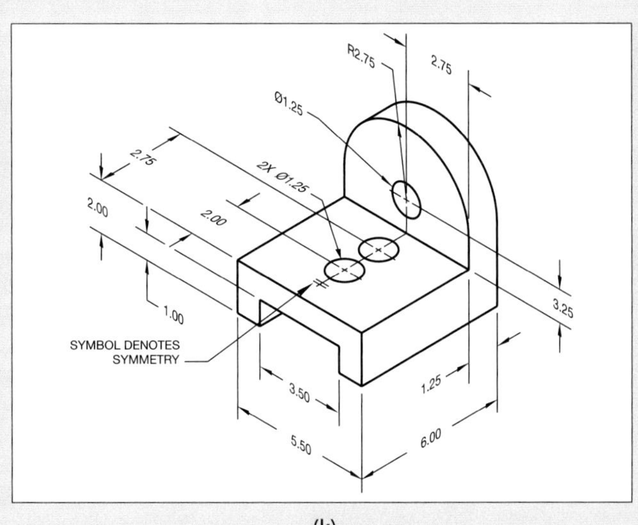

(k)

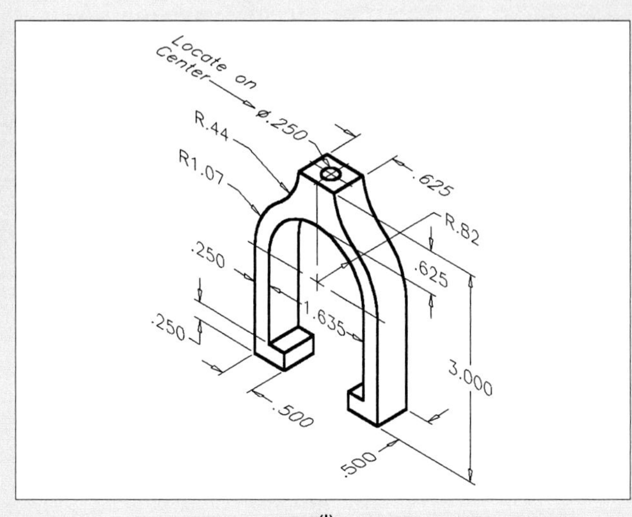

(l)

6.16 problems (continued)

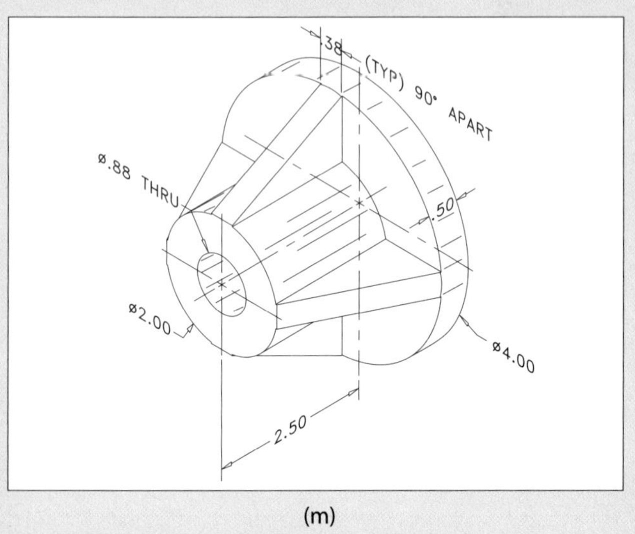

(m)

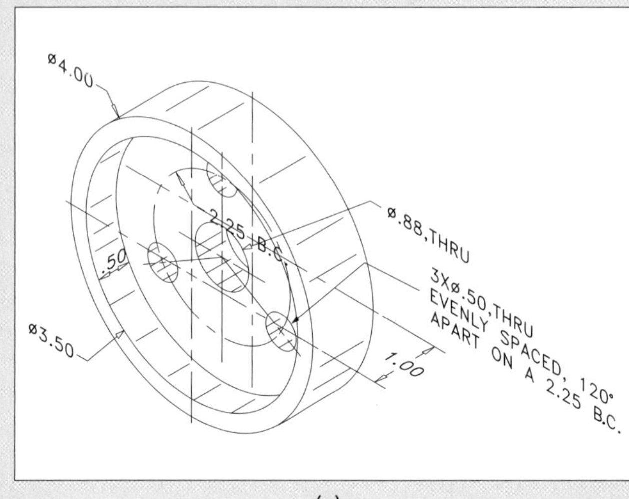

(n)

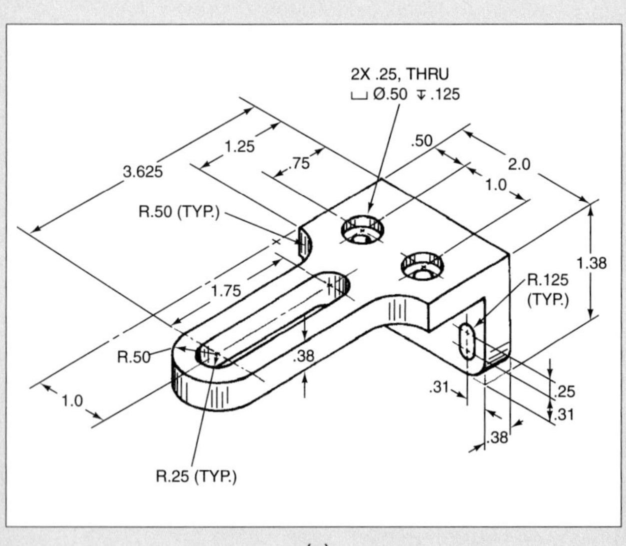

(o)

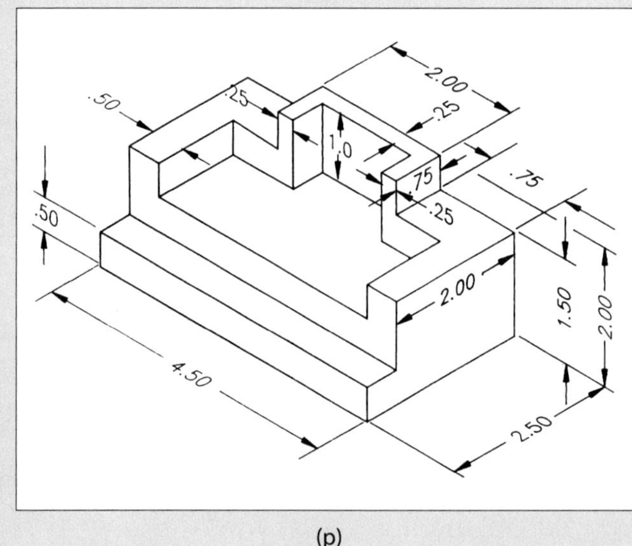

(p)

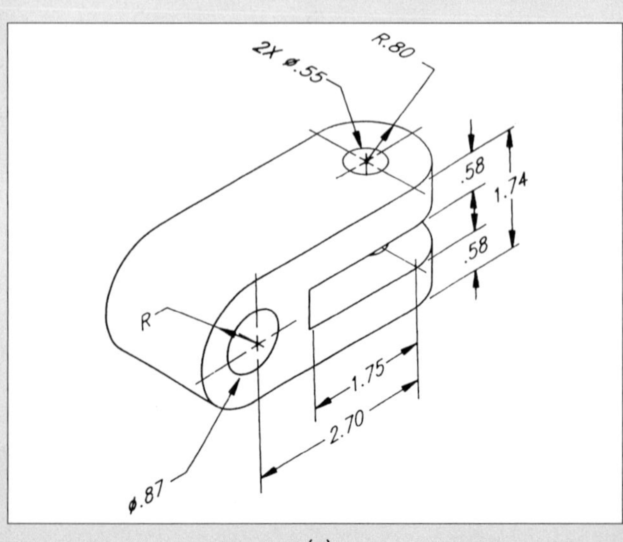

(q)

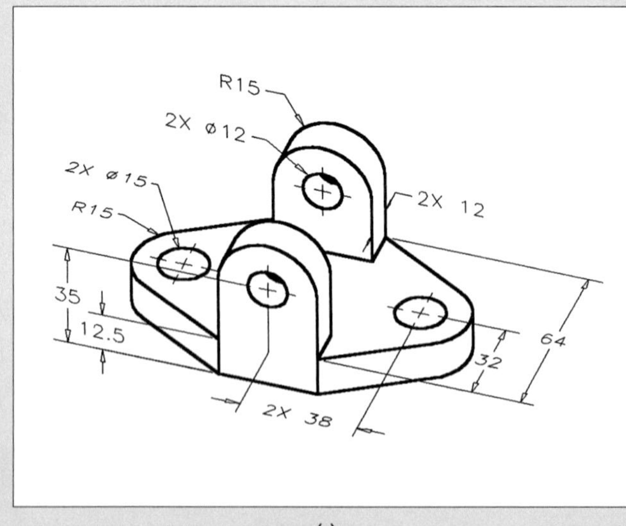

(r)

6.16 problems (continued)

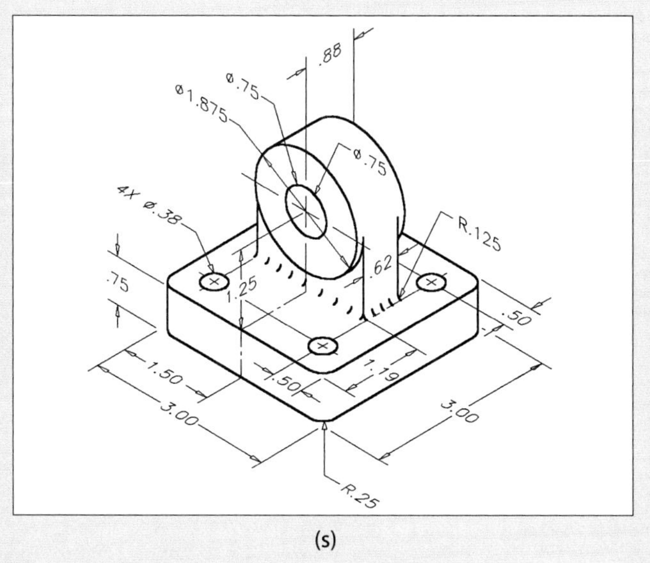

(s)

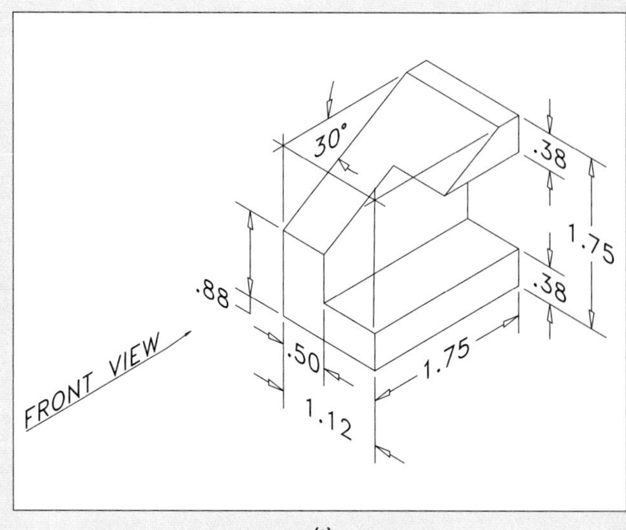

(t)

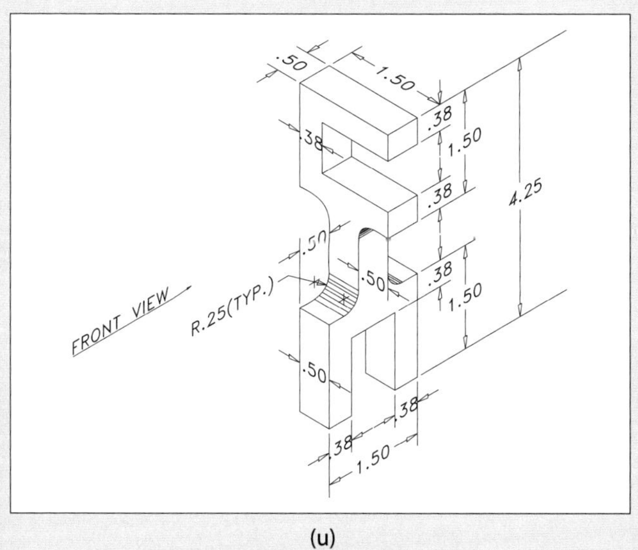

(u)

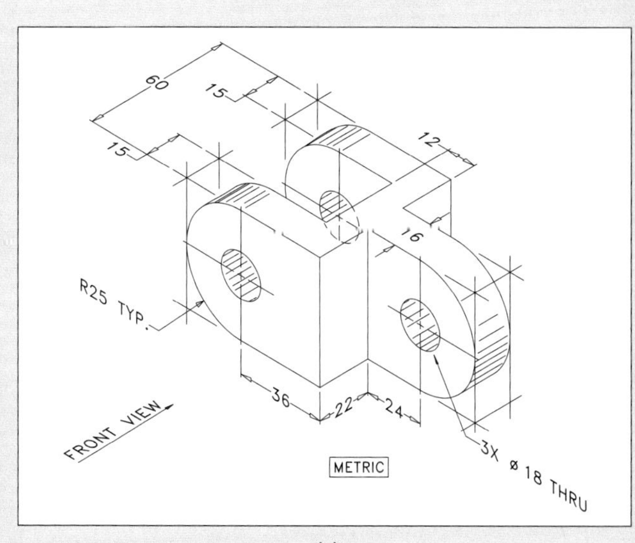

(v)

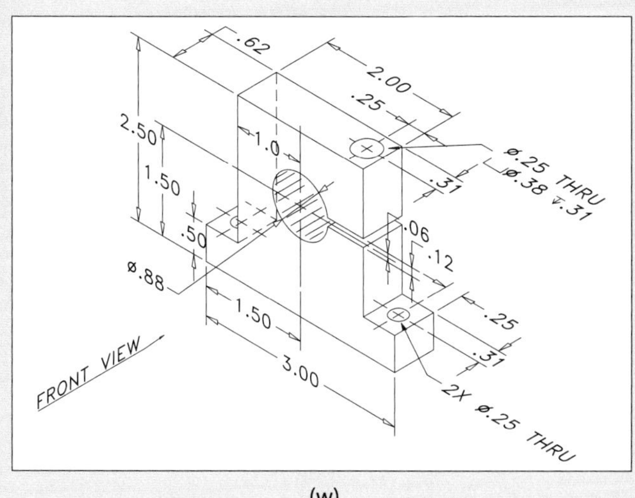

(w)

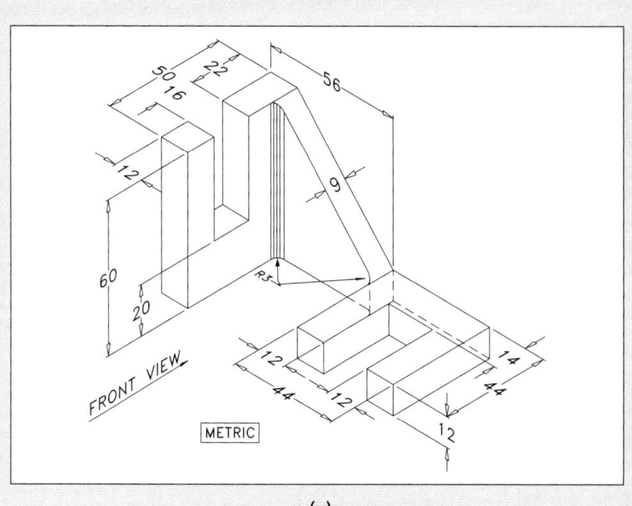

(x)

6.16 problems (continued)

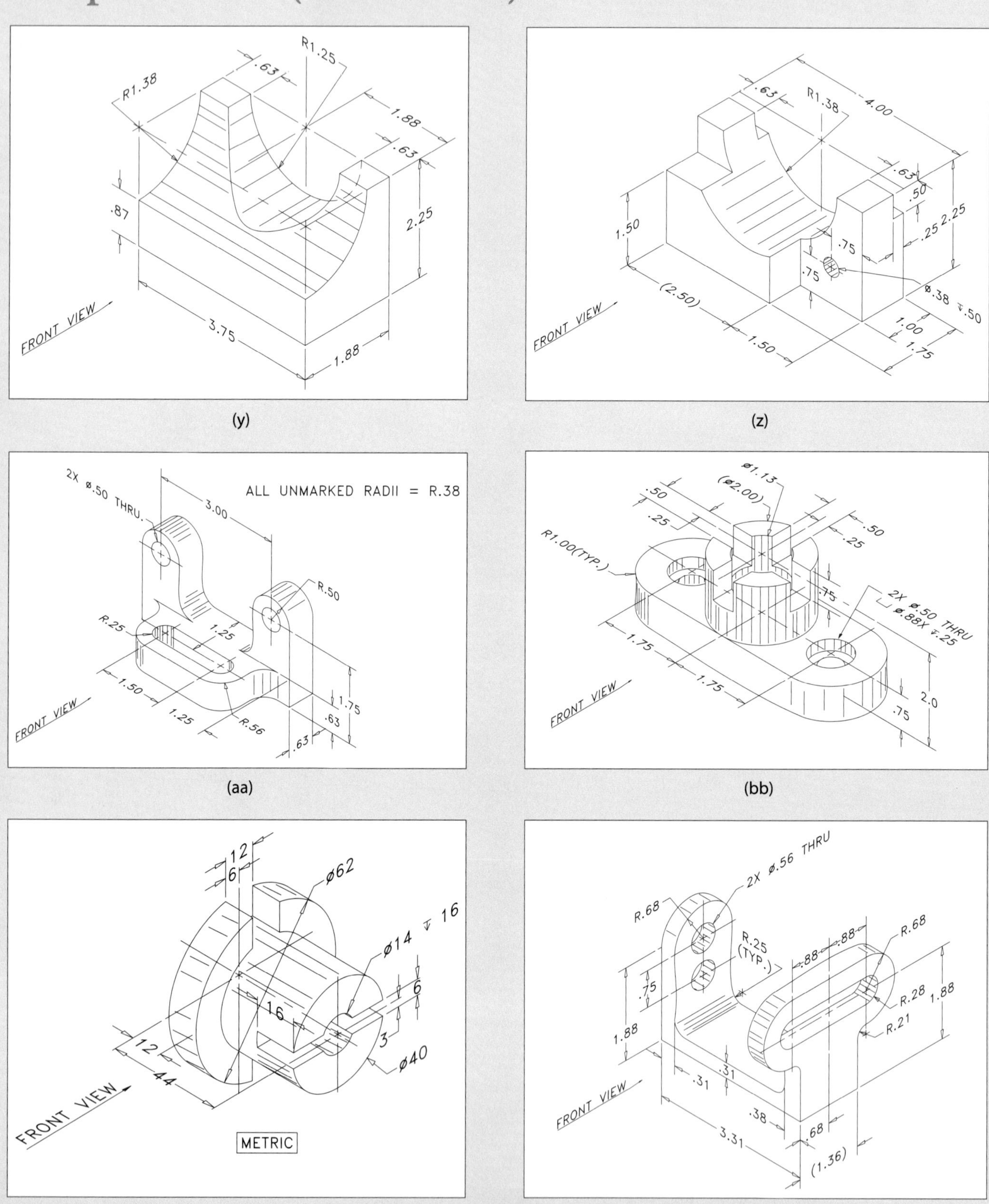

(y) (z) (aa) (bb) (cc) (dd)

6.16 problems (continued)

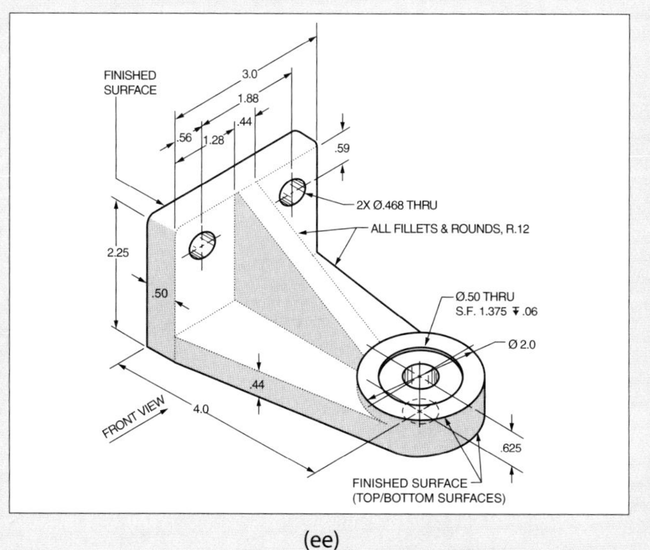

(ee)

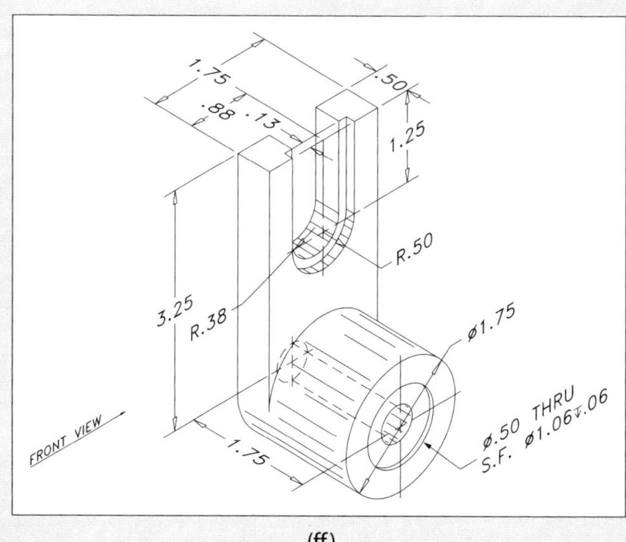

(ff)

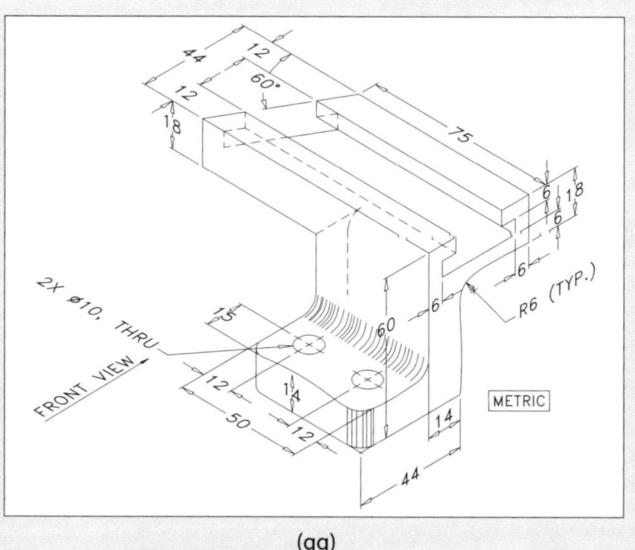

(gg)

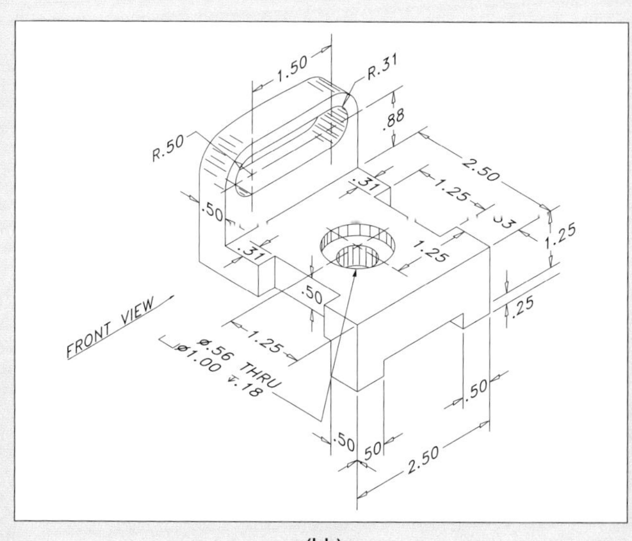

(hh)

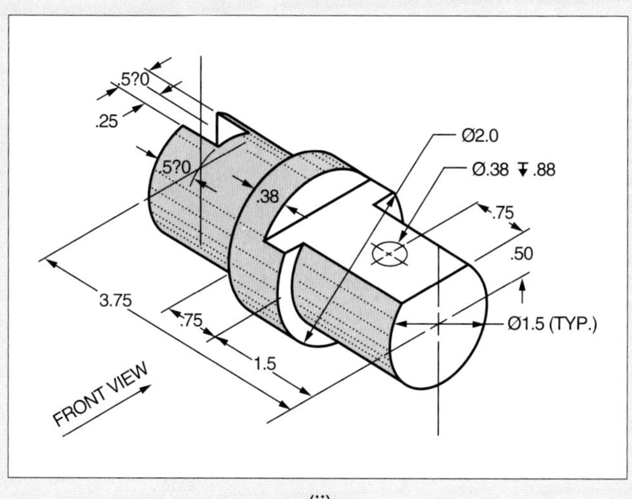

(ii)

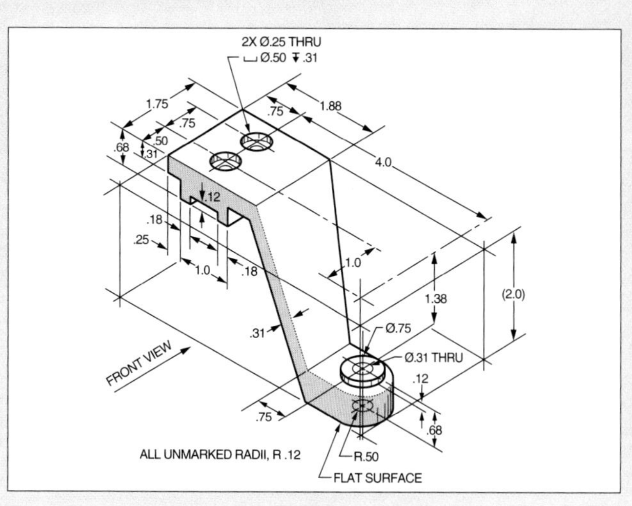

(jj)

6.16 problems (continued)

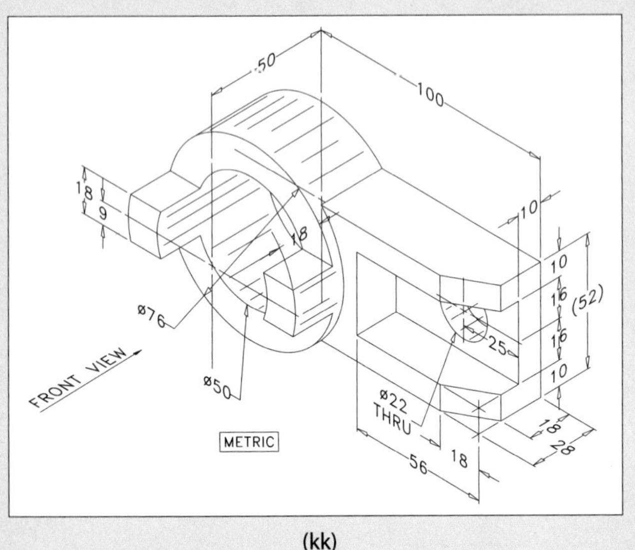

(kk)

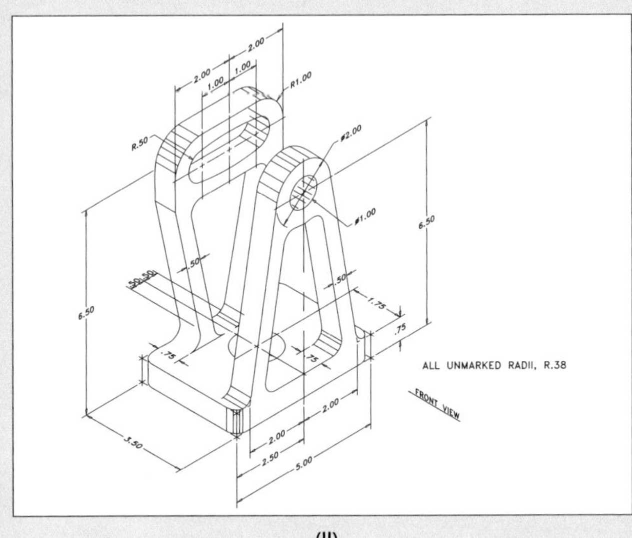

(ll)

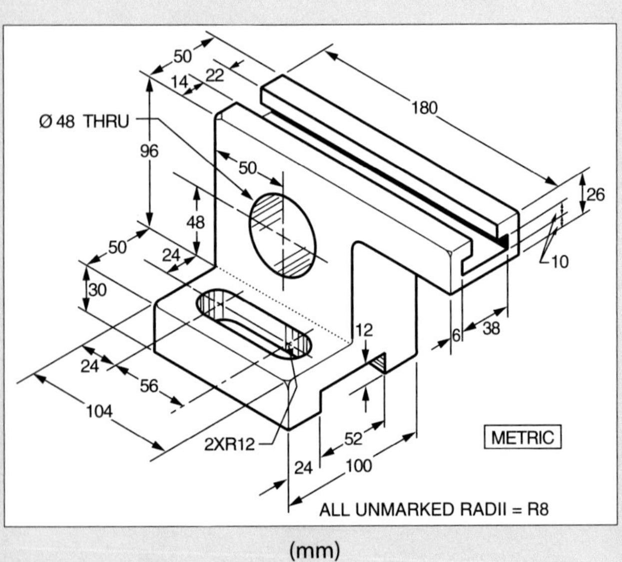

(mm)

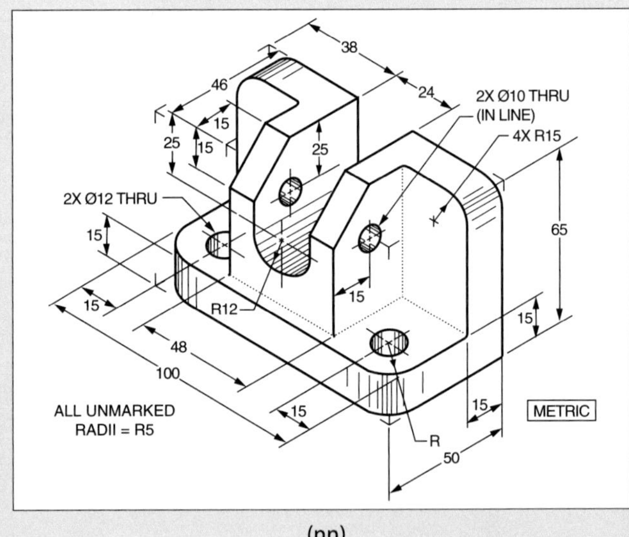

(nn)

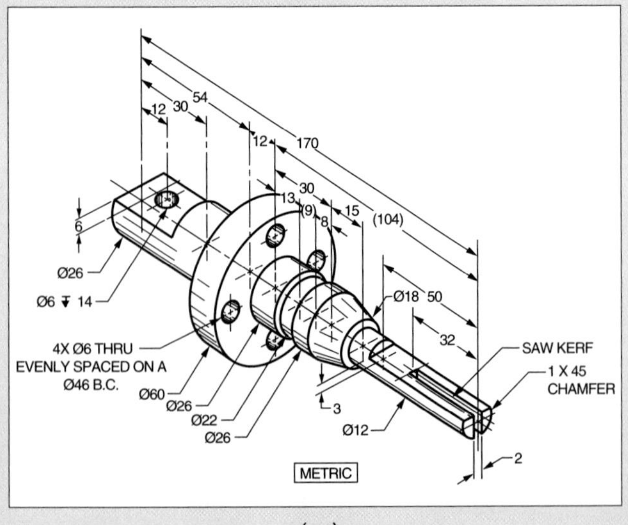

(oo)

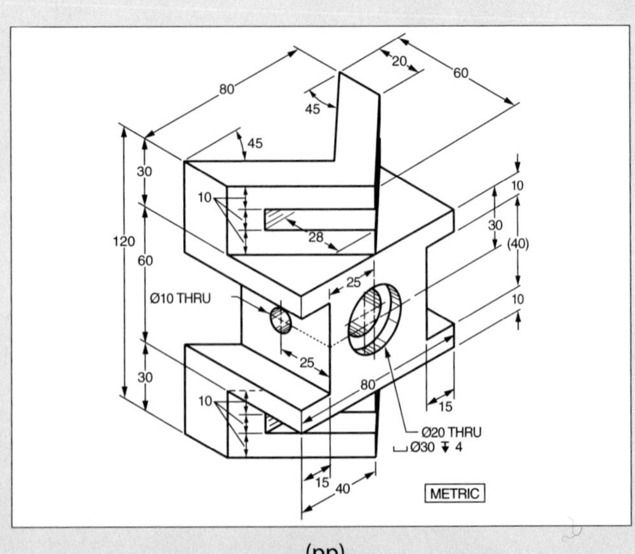

(pp)

6.16 problems (continued)

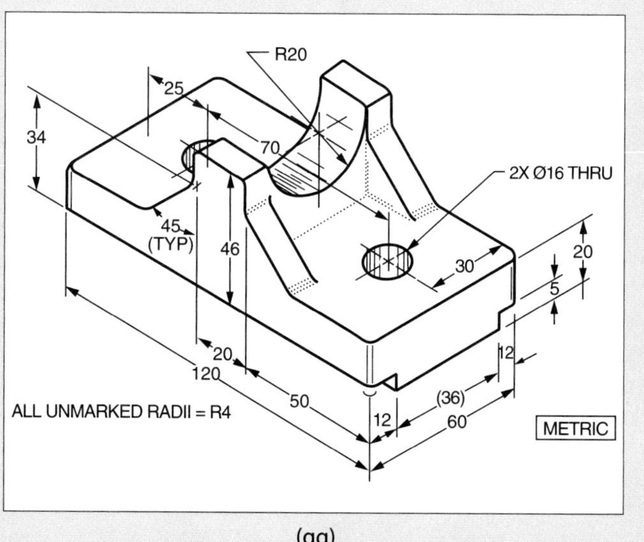

(qq)

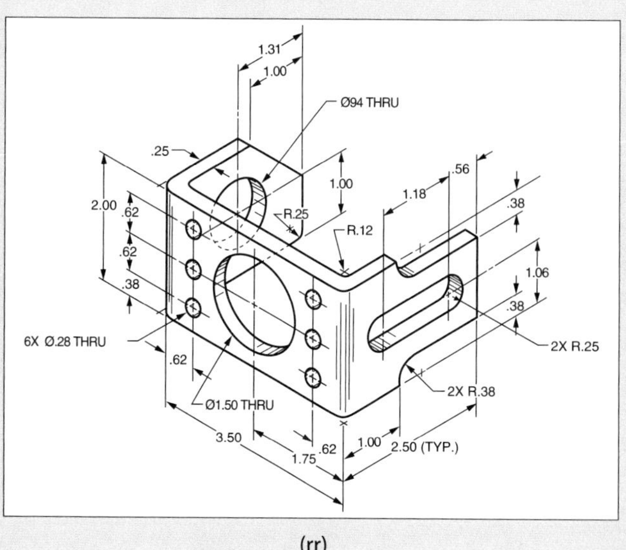

(rr)

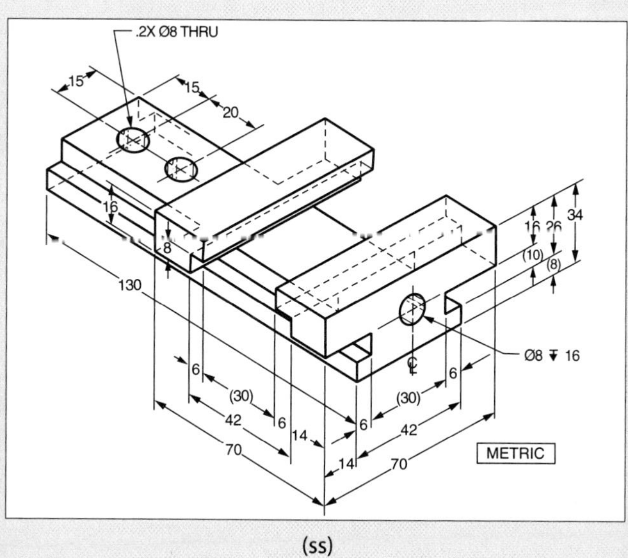

(ss)

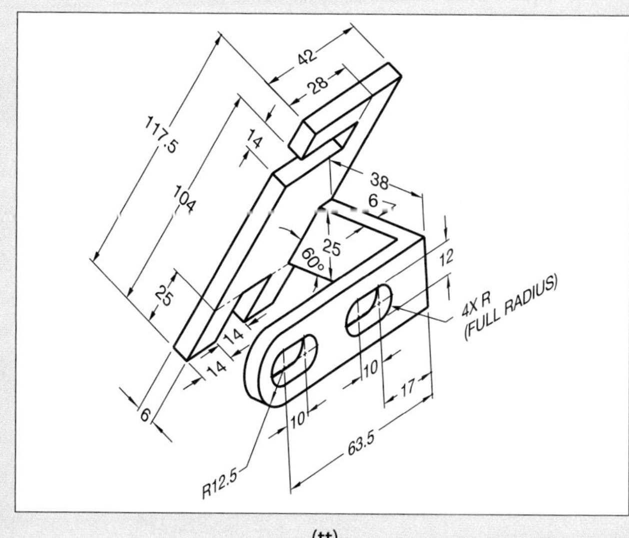

(tt)

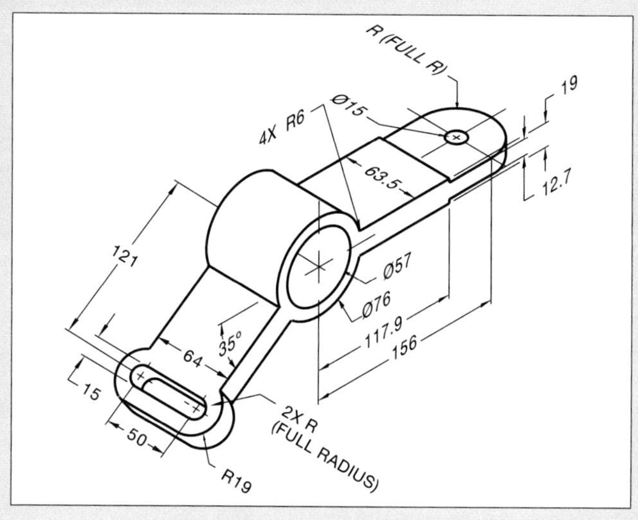

(uu)

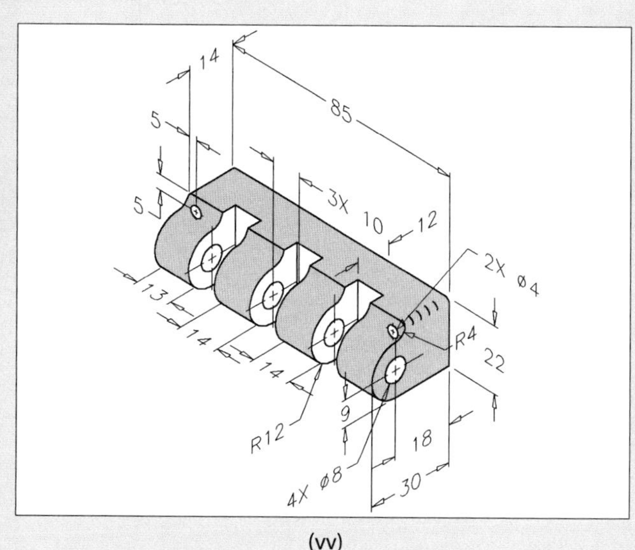

(vv)

6.16 problems (continued)

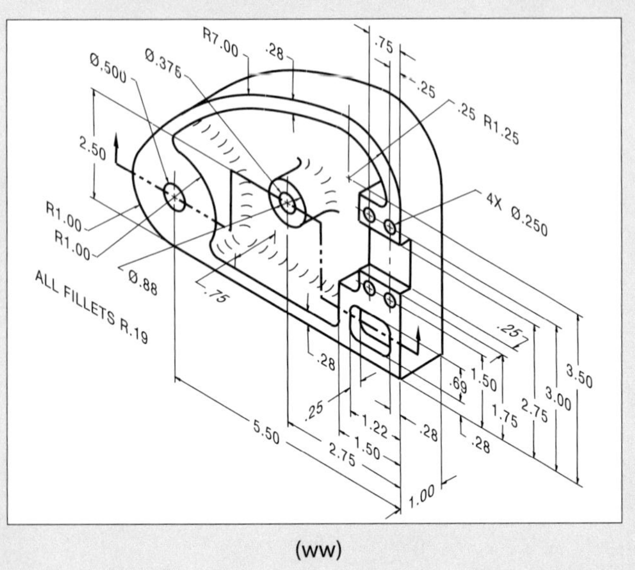

(ww)

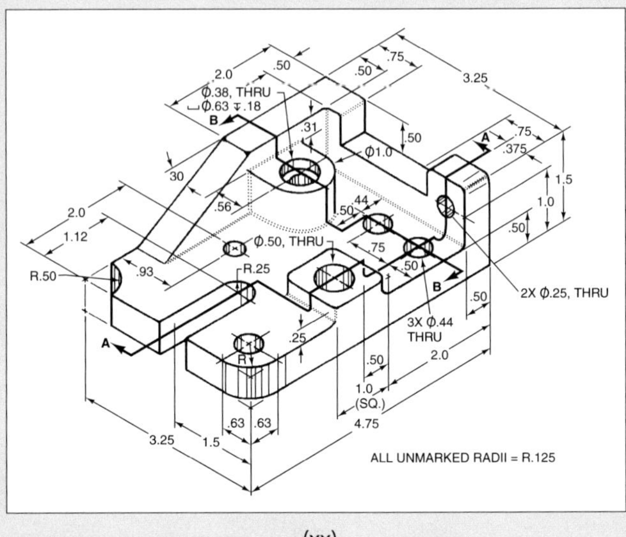

(xx)

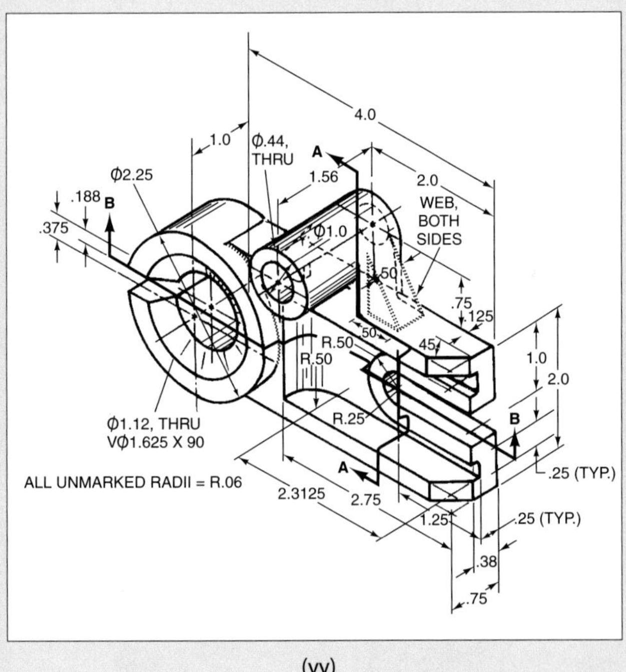

(yy)

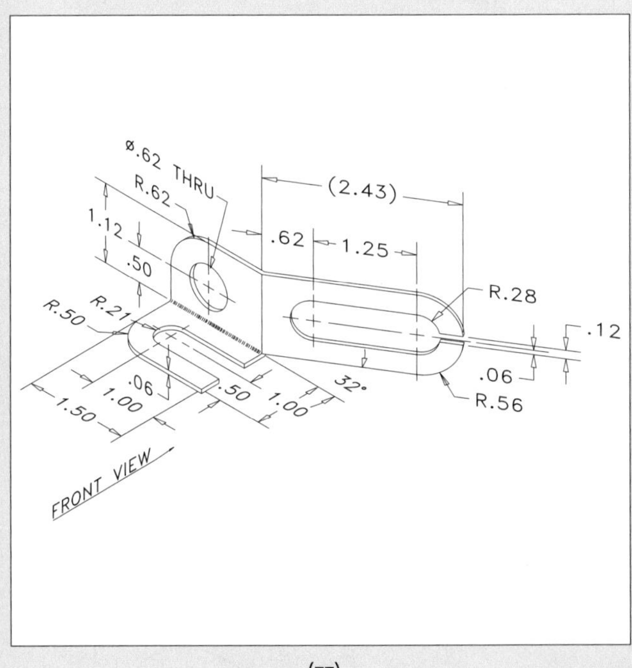

(zz)

6.16 problems (continued)

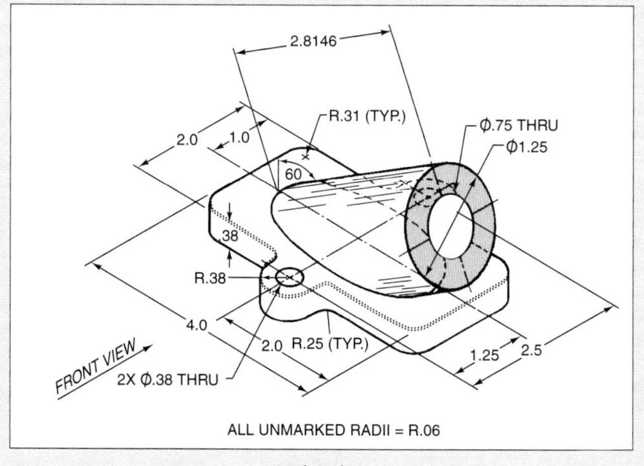

(aaa)

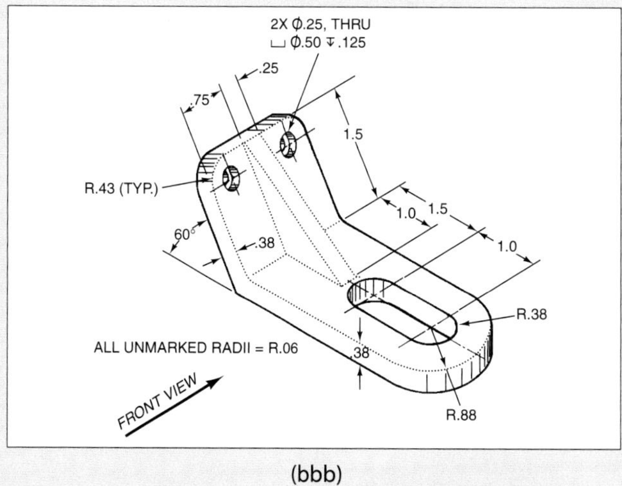

(bbb)

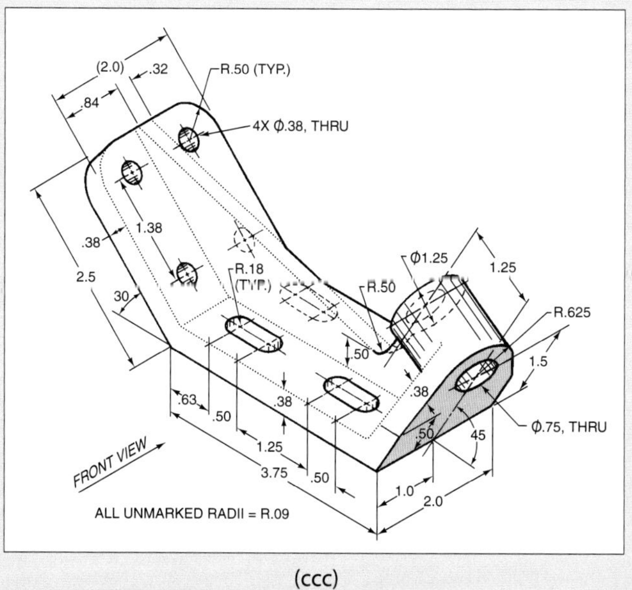

(ccc)

FIGURE P6.4.

Appendix

casestudy
TiLite Wheelchairs—"The Ultimate Ride"

The freedom to get around is important to everyone, but especially to people with disabilities. Users of manual wheelchairs must be independently mobile to enjoy work, travel, sports, and other social activities. They require comfortable, lightweight wheelchairs with features to suit their active lifestyle. A properly designed and fitted wheelchair not only is comfortable but also minimizes energy expenditure and reduces stresses on the user's body. Today people with disabilities can benefit from breakthroughs in design, materials, and manufacturing methods to obtain a unique, customized wheelchair that fits their personal abilities and active lifestyle.

Founded in 1998, TiLite's goal is to provide a twenty-first-century solution to the age-old problem of mobility. TiLite has successfully combined the unique material properties of titanium; traditional fabrication methods; and modern design tools such as parametric modeling, finite element analysis, and rapid prototyping to provide a unique line of affordable, lightweight, and custom-fit manual wheelchairs. TiLite wheelchairs are fabricated from a titanium alloy that is lightweight and has superior strength. Unlike steel, titanium does not corrode and is very durable. The unique combination of these properties is ideal for wheelchair design.

TiLite designs wheelchairs for a variety of users. Ultralightweight chairs are ideal for sports enthusiasts such as wheelchair basketball players and marathon racers. A TiLite chair has even carried the Olympic Torch. Children require chairs that are lightweight and that adapt to children's growth. Elderly users also benefit from lightweight chairs that are easy to propel. Lightweight folding models are easy to transport. TiLite wheelchairs are designed not only to be functional but also to be stylish, displaying the beautiful patina of polished titanium metal. The ability to custom-design and fabricate a unique chair to fit each individual means that there is a TiLite chair for every user. Modern CAD tools enable designers to modify their designs for custom fabrication.

VARIATIONAL DESIGN OF THE TILITE WHEELCHAIR

Over half of all manual wheelchair users suffer from repetitive motion injuries such as carpal tunnel syndrome, chronic shoulder injury, pressure sores, back pain, postural deformities, and reduced heart and lung function. Proper fit and low weight are critical to wheelchair design to reduce these injuries. Studies have shown that wheelchair users suffer fewer long-term health problems with a properly fitted wheelchair. The rear axle of the wheelchair should be positioned forward of the user's shoulder, and the center of gravity of the user and chair should be just forward of the rear axle. The seat should be as narrow as possible to keep the rear wheels close to the frame and the arms close to the body during propulsion. The wheelchair frame must be sized to fit the user's body measurements. These requirements translate into specific dimensions on the wheelchair frame.

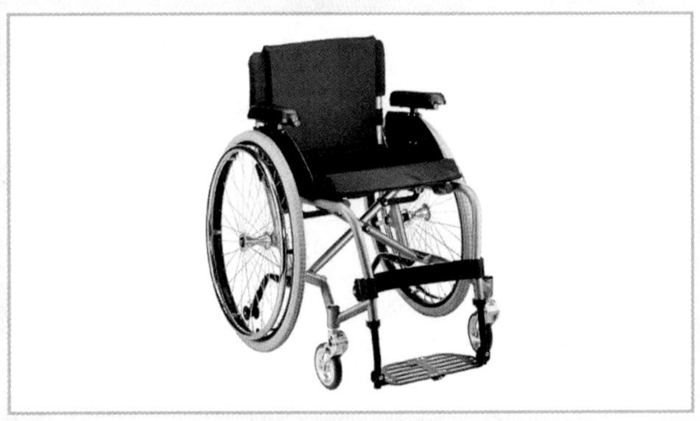

The TiLite XC Ultralightweight wheelchair.

For the TiLite TX model, the user must specify twelve measurements or parameters, such as seat width, back angle, footrest width, and wheel camber; there is a choice of between three and twenty-five possible values for each measurement. With all of these geometric variables and constraints, the designers of TiLite chairs must utilize parametric models to create solid models and drawings of the custom-fit wheelchair design. Based on the user's desired dimensions, the Design Table function in the solid modeling software allows the designer to create all of the necessary configurations of basic wheelchair frame members. When the user's desired dimensions are inserted into the design tables, all of the remaining dimensions of the parts are automatically adjusted to ensure that the parts fit together and fit the user. From these solid models, all of the necessary drawings for a custom wheelchair can be created in only a few minutes. This saves time for the designer and speeds up delivery of the wheelchair to the user.

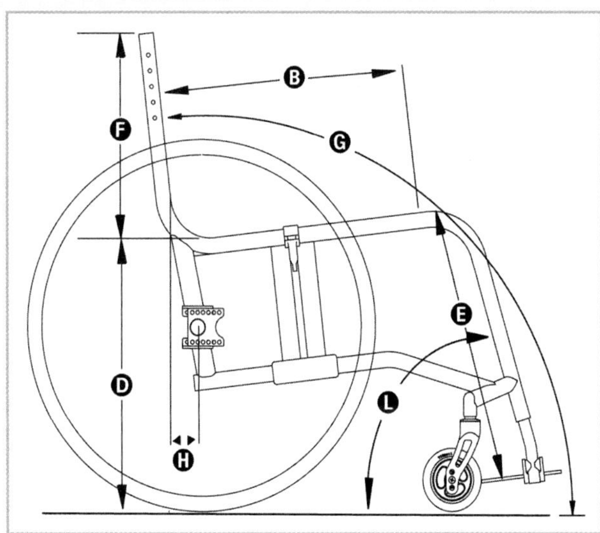

Measurements for ordering a custom wheelchair design.

ASSEMBLY MODELING OF THE TILITE WHEELCHAIR

For a custom-fit wheelchair design, a vast number of different configurations are available depending on the user's body dimensions and seating preferences. Therefore, it is impossible and impractical to build a physical prototype of each wheelchair frame configuration. Nonetheless, the wheelchair must function properly regardless of the size of the frame members. Solid modeling can be used to create virtual prototypes instead of physical models of any wheelchair design.

A folding wheelchair based on a familiar x-frame design is composed of two cross tubes (shown in gray), seat tubes (blue), and hinge members (pink). When designing a folding wheelchair, the designer must check for interference between parts in the open and folded configurations as well as all positions in between. When the sizes of the parts vary for different users, the model must be carefully checked for each configuration. This can be done most efficiently using a solid model. Stick figure models are created to represent the centerlines of each moving part, measured from their attachment points at the pin joints. These skeleton models are manipulated to make sure that the mechanism does not lock up, invert, or toggle. A trial-and-error method is used to move the locations of the hinge pins until an acceptable design is found that works for all possible sizes of the design.

Although a range of sizes is available for seat tubes and cross tubes, the goal is to have a single design for the hinge members, which may be manufactured by an outside vendor. After the positions of the hinge pins are established with the stick figure models, the hinge members are fleshed out and given a shape that will avoid interference with the cross tubes and seat tubes. A variational solid model allows the designer to check multiple configurations quickly and to ensure that all sizes of the wheelchair frame will function as desired.

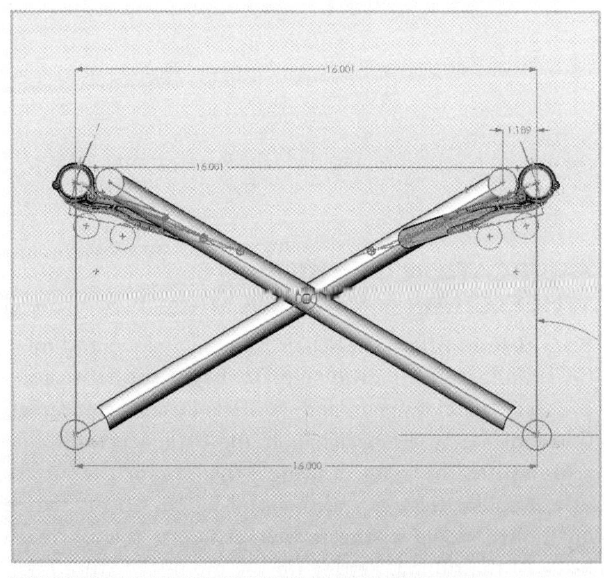

An assembly model of an open wheelchair frame.

Another important design consideration is that the wheelchair can be folded compactly. Solid models can be used to ensure that none of the parts interfere during folding. Changes in the design are easily checked for multiple sizes and configurations of the wheelchair. With the use of solid models, the designer can be assured that the wheelchair will fold to the most compact form possible and function smoothly in the folding operation.

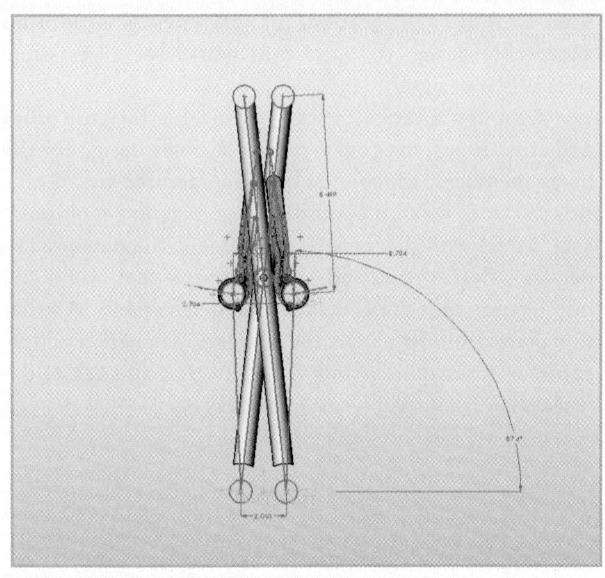

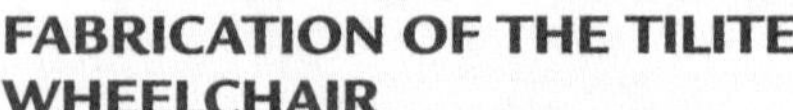

An assembly model of a folded wheelchair frame.

TiLite designer Lindy Anderlini creates full-scale drawings of wheelchair frames.

FABRICATION OF THE TILITE WHEELCHAIR

Each custom-fitted wheelchair frame is unique and must be manufactured individually. To begin the fabrication process, TiLite designer Lindy Anderlini uses the parametric models of the wheelchair frame to generate its full-scale layout. The frame drawing is then laid on the surface of a modular fixture, and the stop blocks on the fixture are bolted down in the proper positions to hold each piece of the tubing. Setting up the modular fixture takes only about ten minutes.

Each tube member is carefully bent to the proper angle using a special bending machine. The tubing is then measured and cut to the correct length based on the scale drawing, and the ends are shaped to fit snugly to the mating parts. Proper sizing and positioning of the frame members is critical to obtain sturdy weld joints. By using the full-scale drawing as a template, the manufacturer can ensure a perfect fit for every customer. After each piece of tubing has been formed and cut, they are laid in the fixture and clamped in the proper position. The parts are tack-welded together, then removed from the fixture and finish-welded. In completion of the fabrication process, the frame is drilled in the necessary locations for assembly of other parts and bead-blasted, hand-buffed, or painted according to the user's preference for surface finish.

DISCUSSION QUESTIONS/ACTIVITIES

1. Explain how a parametric solid modeler can be used to customize the design of wheelchairs to fit individual users. How is the model used in the design process? the manufacturing process?
2. Compare a custom design to a standard wheelchair with adjustable components. What are the advantages and disadvantages of each design?
3. Make a list of the important design and performance specifications for a lightweight wheelchair. How can a solid model be used to check wheelchair design to ensure that the design meets the desired specifications?

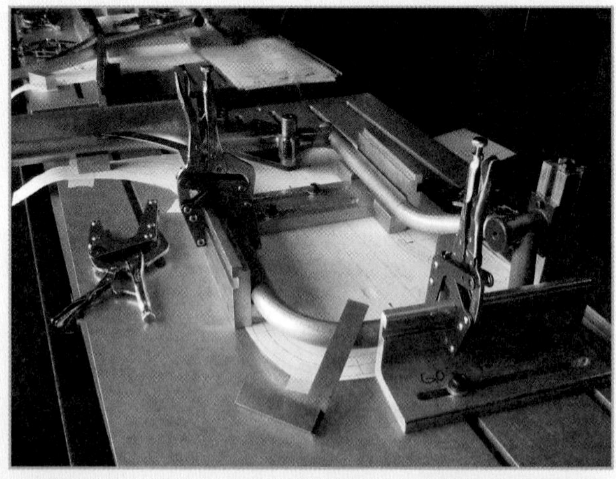

Modular.

glossary of key terms

active animation An animation in which the observer (camera) as well as objects in the scene actively move around and through the scene.

additive color model The RGB color system in which the primary colors of red, green, and blue are added together to create white.

adjacent views Orthogonal views created immediately next to each other, aligned side by side to share a common dimension, and presented on a single plane.

agenda The list of topics for discussion/action at a team meeting.

algebraic constraints Constraint that define the value of a selected variable as the result of an algebraic expression containing other variables from the solid model.

allowance The difference between the maximum material limits of mating parts. It is the minimum clearance or maximum interference between parts.

alpha channel An optional layer of image data containing an additional 8 bits of grayscale data that can be used to control transparency affecting the entire image.

ambient light Indirect light in a scene that does not come directly from a light source, but arrives at a surface by bouncing around or reflecting off other surfaces in the scene.

analysis The study of the behavior of a physical system under certain imposed conditions.

anchor edge The same edge that can be easily and confidently located on multiple views and on a pictorial for an object.

anchor point The same point, usually a vertex, that can be easily and confidently located on multiple views and on a pictorial for an object.

anchor surface The same surface that can be easily and confidently located on multiple views and on a pictorial for an object.

angle of thread The angle between the side of a thread and a line perpendicular to the axis of the thread.

ANSI Y14.5 ***(ASME Y14.5M-1994)*** Industry standard document that outlines uniform practices for displaying and interpreting dimensions and related information on drawings and other forms of engineering documentation.

approval signatures The dated signatures or initials of the people responsible for certain aspects of a formal drawing, such as the people who did the drafting or the engineer responsible for the function of the part.

arc A curved entity that represents a portion of a circle.

architects Professionals who complete conceptual designs for civil engineering projects.

Architect's scale A device used to measure or draw lines in the English system of units with a base unit of inches and fractions of an inch.

arrowhead A small triangle at the end of dimension lines and leaders to indicate the direction and extent of a dimension.

as-built drawings The marked-up drawings from a civil engineering project that show any modifications implemented in the field during construction.

as-built plans Drawings that show exactly how buildings were constructed, especially when variations exist between the final building and the plans created during the design phase.

aspect A quantitative measure of the direction of a slope face.

assembly A collection of parts and/or subassemblies that have been put together to make a device or structure that performs a specific function.

assembly constraints Used to establish relationships between instances in the development of a flexible assembly model.

assembly dimensions Dimensions that show where parts must be placed relative to other parts when the device is being put together.

associative constraints *See* algebraic constraints.

associativity The situation whereby parts can be modified and the components referenced to the parts will be modified accordingly.

attribute Spatial information that describes the characteristics of spatial features.

auxiliary views Views on any projection plane other than a primary or principal projection plane.

axis The longitudinal centerline that passes through a screw.

axonometric drawing A drawing in which all three dimensional axes on an object can be seen, with the scaling factor constant in each direction. Usually, one axis is shown as being vertical.

back light A scene light, usually located behind objects in the scene, which is used to create a defining edge that visually separates foreground objects from the background.

balloons Closed geometric shapes, usually circles, containing identification numbers and placed beside parts on a layout or assembly drawing to help identify those parts.

bar chart A chart using bars of varying heights and widths to represent quantitative data.

base feature The first feature created for a part, usually a protrusion.

base instance The one fixed instance within an assembly.

baseline dimensioning A method for specifiying the location of features on a part whereby all the locations are relative to a common feature or edge.

basic dimension A dimension that is theoretically exact. It is identified by a box around the dimension. It locates the perfect position of features from clearly identified datums.

bearing The angle that a line makes with a North-South line as seen in a plan view.

benchmarks Points established by the U.S. Geological Survey that can be used to accurately locate control points on a construction site.

bill of materials (BOM) A drawing or table in a drawing that lists all of the parts needed to build a device by (at least) the part number, part name, type of material used, and number of times the part is used in the device.

bitmap textures Texture mapping routines that are based on referencing external image files.

black box diagram A diagram that shows the major inputs and outputs from a system.

blend A solid formed by a smooth transition between two or more profiles.

blind extrusion An extrusion made to a specified length in a selected direction.

blind hole A hole that does not pass completely through a part.

blueprints The name sometimes given to construction drawings based on historical blue-on-white drawings that were produced from ink drawings.

bolt A threaded fastener that passes completely through parts and holds them together using a nut.

Boolean operations In early versions of 3-D CAD software, commands used to combine solids.

border A thick line that defines the perimeter of a drawing.

boring The general process of making a hole in a part by plunging a rotating tool bit into a part, moving a rotating part into a stationary tool bit, or moving a part into a rotating tool bit.

bottom-up modeling The process of creating individual parts and then creating an assembly from them.

boundary conditions The constraints and loads added to the boundaries of a finite element model.

boundary representation (b-rep) A method used to build solid models from their bounding surfaces.

bounding box A square box used to sketch circles or ellipses.

brainstorming The process of group creative thinking used to generate as many ideas as possible for consideration.

brainwriting A process of group creative thinking where sketching is the primary mode of communication between team members.

brazing A method for joining separate metal parts by heating the parts, flowing a molten metal (solder) between them, and allowing the unit to cool and harden.

brief A small graphic using word content alone.

broach A long, shaped cutting tool that moves along the length of a part when placed against it. It is used to create uniquely shaped holes and slots.

broaching A process of creating uniquely shaped holes and slots using a long, shaped cutting tool that moves along the length of a part in a single stroke when placed against the part.

broken-out section The section view produced when the cutting plane is partially imbedded into the object, requiring an irregular portion of the object to be removed before the hypothetically cut surface can be seen.

buffer Measured in units of distance or time, a zone around a map feature. A buffer is useful for proximity analysis.

bump mapping A technique used to create the illusion of rough or bumpy surface detail through surface normal perturbation.

business diagram A diagram used in an organization to show organizational hierarchy, task planning or analysis, or relationships between different groups or sets of information.

butt joint A joint between two parts wherein the parts are butted, or placed next to each other.

CAD Computer-aided drawing. The use of computer hardware and software for the purpose of creating, modifying, and storing engineering drawings in an electronic format.

CAD designers Designers who create 3-D computer models for analysis and detailing.

cabinet oblique drawing An oblique drawing where one half the true length of the depth dimension is measured along the receding axes.

caliper A handheld device used to measure objects with a fair degree of accuracy.

cap screw A small threaded fastener that mates with a threaded hole.

casting A method of creating a part by pouring or injecting molten material into the cavity of a mold, allowing it to harden, and then removing it from the mold.

cavalier oblique drawing An oblique drawing where the true length of the depth dimension is measured along the receding axes.

center-of-mass (centroid) The origin of the coordinate axes for which the first moments are zero.

centerline A series of alternating long and short dashed lines used to identify an axis of rotational symmetry.

centermark A small right-angle cross that is used to identify the end view of an axis of rotational symmetry.

Central Meridian The line of longitude that defines the center and is often the x-origin of a projected coordinate system.

Central Parallel The line of latitude that divides a map into north and south halves and is often the y-origin of a projected coordinate system.

chain dimensioning A method for specifiying the location of features on a part whereby the location of each feature is successively specified relative to the location of the previous feature.

chamfers Angled cut transitions between two intersecting surfaces.

charts Charts, graphs, tables, and diagrams of ideas and quantitative data.

child feature A feature that is dependent upon the existence of a previously created feature.

chief designer The individual who oversees other members of the design team and manages the overall project.

circle A closed curved figure where all points on it are equidistant from its center point.

Clarke Ellipsoid of 1866 A reference ellipsoid having a semimajor axis of approximately 6,378,206.4 meters and a flattening of 1/294.9786982. It is the basis for the North American Datum of 1927 (NAD27) and other datums.

clearance A type of fit where space exists between two mating parts.

clearances The minimum distances between two instances in an assembly.

clip A geoprocessing command that extracts the features from a coverage that reside entirely within a boundary defined by features in another coverage.

clipping plane A 3-D virtual camera technique that allows you to selectively exclude, and not view or render, unnecessary objects in a scene that are either too close or too far away.

CODEC Video compression-decompression algorithm.

collision detection A built-in software capability for calculating and graphically animating the results of collisions between multiple objects based on object properties of speed, mass, and gravity.

color mapping Sometimes called diffuse mapping, color mapping replaces the main surface color of a model with an external image map or texture.

combining solids The process of cutting, joining, or intersecting two objects to form a third object.

components References of object geometry used in assembly models.

compositing The technique and art of rendering in layers or passes, editing the image on each layer as needed, and compiling the edited layers or images into a single unified final image.

computer-aided design (CAD) The process by which computers are used to model and analyze designed products.

computer-aided manufacturing (CAM) The process by which parts are manufactured directly from 3-D computer models.

concept mapping The creative process by which the central idea is placed in the middle of a page and related concepts radiate out from that central idea.

conceptual design The initial idea for a design before analysis has been performed.

concurrent engineering The process by which designers, analysts, and manufacturers work together from the start to design a product.

consensus A process of decision making where an option is chosen that everyone supports.

constraint A boundary condition applied to a finite element model to prevent it from moving through space.

constraints Geometric relationships, dimensions, or equations that control the size, shape, and/or orientation of entities in a sketch or solid model.

constructive solid geometry (CSG) A method used to build solid models from primitive shapes based on Boolean set theory.

construction drawings Working drawings, often created by civil engineers, that are used to build large-scale, one-of-a-kind structures.

construction line A faint line used in sketching to align items and define shapes.

continuation blocks Header blocks used on the second and subsequent pages of multipage drawings.

contour dimensioning Placing each dimension in the view where the contour or shape of the feature shows up best.

contour interval The vertical distance between contours.

contour rule A drawing practice where each dimension should be placed in the view where the contour shape is best shown.

contours Lines or curves that represent the same elevation across the landscape.

control points Points at a construction site that are referenced to an origin by north, south, east, or west coordinates.

coordinate measuring machine A computer-based tool used to digitize object geometry for direct input to a 3-D CAD system.

corner views An isometric view of an object created from the perspective at a given corner of the object.

cosmetic features Features that modify the appearance of the surface but do not alter the size or shape of the object.

cover sheet The first page in a set of construction drawings showing a map of the location of the project and possibly an index.

crest The top surface or point joining the sides of a thread.

critical path The sequence of activities in a project that have the longest duration.

critical path method (CPM) A tool for determining the least amount of time in which a project can be completed.

cross-section The intersection between a cutting plane and a 3-D object.

curved surface Any nonflat surface on an object.

cut (noun) A feature created by the removal of solid volume from a model.

cut (verb) To remove the volume of interference between two objects from one of the objects.

cutaway diagram A diagram that allows the reader to see a slice of an object.

cutting-plane An imaginary plane that intersects with an object to form a cross section.

cutting plane A theoretical plane used to hypothetically cut and remove a portion of an object to reveal its interior details.

cutting plane line On an orthographic view of an object, the presentation of the edge view of a cutting plane used to hypothetically cut and remove a portion of that object for viewing.

cutting segment On a stepped cutting plane for an offset section view, that portion of the plane that hypothetically cuts and reveals the interior detail of a feature of interest.

data structure The organization of data within a specific computer system that allows the information to be stored and manipulated effectively.

database A collection of information for a computer and a method for interpretation of the information from which the original model can be re-created.

datum A theoretical plane or axis established by real features on an object for the purpose of defining the datum reference frame.

datum geometries Geometric entities such as points, axes, and planes that do not actually exist on real parts, but are used to help locate and define other features.

datum planes The planes used to define the locations of features and entities in the construction of a solid model.

datum reference frame A system of three mutually perpendicular planes used as the coordinate system for geometric dimensioning.

decimal degrees (DD) A measuring system in which values of latitude and longitude are expressed in decimal format rather than in degrees, minutes, and seconds, such as 87.5°.

deep drawing Creating a thin-shelled part by pressing sheet metal into a deep cavity mold.

default tolerances Usually appearing in the drawing header, the tolerances to be assumed for any dimension show on a part when that dimension does not specify any tolerances.

degrees, minutes, seconds (DMS) A measuring system for longitude and latitude values, such as 87° 30' 00", in which 60 seconds equals 1 minute and 60 minutes equals 1 degree.

density The mass per unit volume for a given material.

depiction An illustration describing and simplifying factual information on a real-world system.

depth of thread The distance between the crest and the root of a thread, measured normal to the axis.

depth-mapped shadows Also called shadow-mapped shadows, depth-mapped shadows use a precalculated depth map to determine the location, density, and edge sharpness of shadows.

descriptive geometry A two-dimensional graphical construction technique used for geometric analysis of three-dimensional objects.

design (noun) An original manifestation of a device or method created for performing one or more useful functions.

design (verb) The process of creating a design (noun).

design analysts Individuals who analyze design concepts by computer methods to determine their structural, thermal, or vibration characteristics.

design documentation The set of drawings and specifications that illustrate and thoroughly describe a designed product.

design process The multistep, iterative process by which products are conceived and produced.

design table A table or spreadsheet that lists all of the versions of a family model, the dimensions or features that may change, and the values in any of its versions.

design tree *See* model tree.

detail designers The individuals who create engineering drawings, complete with annotation, from 3-D computer models or from engineering sketches.

detail drawing A formal drawing that shows the geometry, dimensions, tolerances, materials, and any processes needed to fabricate a part.

detail sections Drawings included in a set of construction plans that show how the various components are assembled.

devil's advocate The team member who challenges ideas to ensure that all options are considered by the group.

diagram An illustration that explains information, represents a process, or shows how pieces are put together.

diametric drawing An axonometric drawing in which the scaling factor is the same for two of the axes.

die A special tool made specifically to deform raw or stock material into a desired outline of a part or feature in a single operation.

die casting A method of casting where the mold is formed by cutting a cavity into steel or another hard material. *See* casting.

digital elevation model (DEM) The representation of continuous elevation values over a topographic surface by a regular array of z-values referenced to a common datum. It is typically used to represent terrain relief.

dimension A numerical value expressed in appropriate units of measure and used to define the size, location, geometric characteristic, or surface texture of a part or part feature.

dimension line A thin, dark, solid line that terminates at each end with arrowheads. The value of a dimension typically is shown in the center of the dimension line.

dimension name The unique alphanumeric designation of a variable dimension.

dimensional constraints Measurements used to control the size or position of entities in a sketch.

direct dimensioning Dimensioning between two key points to minimize tolerance accumulation.

directional light A computer-generated light source designed to simulate the effect of light sources, such as the sun, that are so far away from objects in the scene that lighting and shadow patterns in the scene appear to be parallel.

displacement A change in the location of points on an object after it has been subjected to external loads.

dissolve A geoprocessing command that removes boundaries between adjacent polygons that have the same value for a specified attribute.

draft The slight angling of the walls of a cast, forged, drawn, or stamped part to enable the part to be removed from the mold more easily.

drawing A collection of images and other detailed graphical specifications intended to represent physical objects or processes for the purpose of accurately re-creating those objects or processes.

drill bit A long, rotating cutting tool with a sharpened tip used to make holes.

drilling A process of making a hole by plunging a rotating tool bit into a part.

drill press A machine that holds, spins, and plunges a rotating tool bit into a part to make holes.

driven dimension A variable connected to an algebraic constraint that can be modified only by user changes to the driving dimensions.

driving dimension A variable used in an algebraic constraint to control the values of another (driven) dimension.

double-sided extrusion A solid formed by the extrusion of a profile in both directions from its sketching plane.

EC Level A number included in the title block of a drawing indicating that the part has undergone a revision.

edge tracking A procedure by which successive edges on an object are simultaneously located on a pictorial image and on a multiview image of that object.

edge view (of a plane) A view in which the given plane appears as a straight line.

EDM Electric discharge machining; a process by which material is eroded from a part by passing an electric current between the part and an electrode (or a wire) through an electrolytic fluid.

electrical plan A plan view showing the layout of electrical devices on a floor in a building.

elevation view In the construction of a perspective view, the object as viewed from the front, as if created by orthogonal projection.

elevation views Views of a structure that show changes in elevation (side or front views).

ellipse A closed curve figure where the sum of the distance between any point on the figure and its two foci is constant.

end mill A rotating cutting tool that, when placed in the spindle of a milling machine, can remove material in directions parallel or perpendicular to its rotation axis.

engineer (noun) A person who engages in the art of engineering.

engineer (verb) To plan and build a device that does not occur naturally within the environment.

Engineer's scale A device used to measure or draw lines in the English system of units with a base unit of inches and tenths of an inch.

engineering The profession in which knowledge of mathematical and natural sciences gained by study, experience, and practice is applied with judgment to develop and utilize economically the materials and forces of nature for the benefit of humanity.

engineering animation A dynamic virtual 3-D prototype of a mechanism or system that can be assembled (usually from preexisting or newly created 3-D CAD part models) and/or shown in operation over a period of time.

engineering change (EC) number A dated number that defines the degree to which the specifications of a part have been updated.

engineering design The process by which many competing factors of a product are weighed to select the best alternative in terms of cost, sustainability, and function.

engineering scale A device used to make measurements in much the same way a ruler is used.

explanation diagram An illustration explaining the way something works; a basic process; or the deconstruction of an object, a plan, or a drawing.

exploded assembly drawing A formal drawing, usually in pictorial form, that shows the orientation and sequence in which parts are put together to make a device.

exploded configuration A configuration of an assembly that shows instances separated from one another. An exploded configuration is used as the basis for an assembly drawing.

extension line A thin, dark, solid line extending from a point on an object, perpendicular to a dimension line used to indicate the extension of a surface or point to a location preferably outside the part outline.

external thread Threads that are formed on the outside of a cylindrical feature, such as on a bolt or stud.

extrude through all An extrusion that begins on the sketching plane and protrudes or cuts through all portions of the solid model that it encounters.

extrude to selected surface An extrusion where the protrusion or cut begins on the sketching plane and stops when it intersects a selected surface.

extrusion (in fabrication) A process for making long, solid shapes with a constant cross section by squeezing raw material under elevated temperatures and pressure through an orifice shaped with that cross section.

extrusion (in 3D modeling) A solid that is bounded by the surfaces swept out in space by a planar profile as it is pulled along a path perpendicular to the plane of the profile.

fabricate To make something from existing materials.

family model A collection of different versions of a part in a single model that can display any of the versions.

false easting A value applied to the origin of a coordinate system to modify the x-coordinate readings, usually to make all of the coordinate values positive.

false northing A value applied to the origin of a coordinate system to modify the y-coordinate readings, usually to make all of the coordinate values positive.

fastener A manufactured part whose primary function is to join two or more parts.

feature array A method for making additional features by placing copies of a master feature on the model at a specified equal spacing.

feature generalization The process of going from the specific to the general in analyzing data.

features Distinctive geometric shapes on solid parts; 3-D geometric entities that exist to serve some function.

feature-based solid modeling A solid modeling system that uses features to build models.

feature control frame The main alphabet of the language of geometric dimensioning and tolerancing. These boxes contain the geometric characteristic symbol, the geometric tolerances, and the relative datums.

feature pattern *See* feature array.

feature tree *See* model tree.

feature with size A cylindrical or spherical surface or a set of two opposed elements or opposed parallel surfaces associated with a size dimension. Typical features with size are holes, cylinders, spheres, and opposite sides of a rectangular block.

feature without size A planar surface or a feature where the normal vectors point in the same direction.

field A column in a table that stores the values for a single attribute.

fill light A light that softens and extends the illumination of the objects provided by the key light.

fillets Smooth transitions of the internal edge created by two intersecting surfaces and tangent to both intersecting surfaces.

Finite Element Analysis An advanced computer-based design analysis technique that involves subdividing an object into several small elements to determine stresses, displacements, pressure fields, thermal distributions, or electromagnetic fields.

first-angle projection The process of creating a view of an object by imprinting its image, using orthogonal projection, on an opaque surface behind that object.

fishbone diagram A diagram that shows the various subsystems in a device and the parts that make up each subsystem.

fixture A mechanical device, such as a clamps or bracket, used for holding a workpiece in place while it is being modified.

flash Bits of material that are left on a part from a casting or molding operation and found along the seams where the mold pieces separate to allow removal of the part.

floor plan A plan view of a single floor in a building that shows the layout of the rooms.

flowchart A quality improvement tools used to document, plan, or analyze a process or series of tasks.

foreshortened (line or plane) Appearing shorter than its actual length in one of the primary views.

forging A process of deforming metal with a common shape at room temperature into a new but similar shape by pressing it into a mold under elevated pressure.

form The shape of the thread cross section when cut through the axis of the thread cylinder.

form feature A recognizable area on a solid model that has a specific function.

forward kinematics In a hierarchical link, total motion in which the motion of the parent is transferred to the motion of the child.

foundation plan A plan view of the foundation of a building showing footings and other support structures.

foundation space The rectilinear volume that represents the limits of the volume occupied by an object.

foundation space The rectilinear volume that represents the limits of the volume occupied by an object.

frame rate The rate of speed, usually in frames per second, in which individual images or frames are played when an animation is viewed.

frontal surface A surface on an object being viewed that is parallel to the front viewing plane.

full section The section view produced when a single cutting plane is used to hypothetically cut an object completely into two pieces.

function curve A graphical method of displaying and controlling an object's transformations.

functional gage An inspection tool built uniquely for the purpose of quickly checking a specific dimension or geometric condition on a part to determine whether or not it fall within tolerance limits.

fused deposition A process where parts are gradually built up by bits of molten plastic that are deposited by a heated tip at selected locations and then solidified by cooling.

Gantt chart A tool for scheduling a project timeline.

general sections Sections through entire structures that show the layout of rooms but provide little detail.

geographic coordinate system A spatial reference system using a grid network of parallels and meridians to locate spatial features on the earth's surface.

geographically referenced data Information that is referenced to a specific geographic location, usually on the earth's surface.

geometric constraints Definitions used to control the shape of a profile sketch through geometric relationships.

geometric dimensioning and tolerancing (GD&T) A 3-D mathematical system that allows a designer to describe the form, orientation, and location of features on a part within precise tolerance zones.

geometric transformation Transformations used to alter the position, size, or orientation of a part, camera, or light over a specified period of time.

georeferenced data *See* geographically referenced data.

glass box A visualization aid for understanding the locations and orientations of images of an object produced by third-angle projection on a drawing. The images of an object are projected, using orthogonal projection, on the sides of a hypothetical transparent box that is then unfolded into a single plane.

global positioning systems (GPS) A system of geosynchronous, radio-emitting and receiving satellites used for determining positions on the earth.

graphical user interface (GUI) The format of information on the visual display of a computer, giving its user control of the input, output, and editing of the information.

green engineering The process by which environmental and life cycle considerations are examined from the outset in design.

grinding A method of removing small amounts of material from a part using a rotation abrasive wheel, thus creating surfaces of very accurate planar or cylindrical geometries.

ground constraint A constraint usually applied to a new sketch to fix the location of the sketch in space.

ground line (GL) In the construction of a perspective view, a line on the elevation view that represents the height of the ground.

GRS80 spheroid The satellite-based spheroid for the Geodetic Reference System 1980.

half section The section view produced when a single cutting plane is used to hypothetically cut an object up to a plane or axis of symmetry, leaving that portion beyond the plane or axis intact.

header A premade outline on which working drawings are created to ensure that all information required for fabrication and record keeping is entered.

heating and ventilation plan A plan view of the ventilation systems on a specific floor of a building, including ductwork and devices such as air conditioning units.

hidden lines The representation, using dashed lines, on a drawing of an object of the edges that cannot be seen because the object is opaque.

hierarchy The parent-child relationships between instances in an assembly.

hierarchical link A series of user-defined or linked objects that have a parent-child-grandchild relationship.

history tree *See* model tree.

holes A cut feature added to a model that will often receive a fastener for system assembly.

horizon line (HL) In the construction of a perspective view, the line that represent the horizon, which is the separation between the earth and the sky at a long distance. The left and right vanishing points are located on the HL. The PP and the HL are usually parallel to each other.

horizontal modeling A strategy for creating solid models that reduces parent-child relationships within the feature tree.

horizontal surface A surface on an object being viewed that is parallel to the top viewing plane.

identity A geometric integration of spatial datasets that preserves only the geographic features from the first input layer; the second layer merely adds more information to the dataset.

image A collection of printed, displayed, or imagined patterns intended to represent real objects, data, or processes.

inclined surface A flat surface on an object being viewed that is perpendicular to one primary view and angled with respect to the other two views.

index A list of all sheets of drawings contained in a set of construction plans.

industrial designers The individuals who use their creative abilities to develop conceptual designs of potential products.

infographic A shortened form of *informational graphic* or *information graphic*.

information graphics Often referred to as *infographic*, visual explanations.

injection molding A process for creating a plastic part by injecting molten plastic into a mold under pressure, allowing the material to solidify, and removing the part from the mold.

instances Copies of components that are included within an assembly model.

instructional diagram A diagram showing how a specific action within an object occurs.

instruments In engineering drawing, mechanical devices used to aid in creating accurate and precise images.

interchangeable manufacturing A process by which parts are made at different locations and brought together for assembly. For many industries, this process opens the door for third-party companies to produce replacement parts or custom parts.

interference A fit where two mating parts have intersecting nominal volumes, requiring the deformation of the parts . For example, the diameter of the shaft is larger than the diameter of the hole. When assembled, the intent is that the shaft will not spin in the hole.

interference The amount of overlap between two instances in an assembly.

internal thread Threads that are formed on the inside of a hole.

international sheet sizes The internationally accepted paper dimensions used when drawings are created or printed to their full intended size.

intersect To create a new object that consists of the volume of interference between two objects.

intersection A geometric integration of spatial datasets that preserves features or portions of features that fall within areas common to all input datasets.

inverse kinematics A bidirectional set of constraints that allows motion of a set of linked objects by moving the very end of the hierarchically linked chain and having the rest of the links move in response.

investment casting A method of casting where the mold is formed by successive dipping of a master form into progressively coarser slurries, allowing each layer to harden between each dipping. *See* casting.

isometric axes A set of three coordinate axes that are portrayed on the paper at 120 degrees relative to one another.

isometric dot paper Paper used for sketching purposes that includes dots located along lines that meet at 120 degrees.

isometric drawing An axonometric drawing in which the scaling factor is the same for all three axes.

isometric grid paper Paper used for sketching purposes that includes grid lines at 120 degrees relative to one another.

isometric lines Lines on an isometric drawing that are parallel or perpendicular to the front, top, or profile viewing planes.

isometric pictorial A sketch of an object that shows its three dimensions where isometric axes were used as the basis for defining the edges of the object.

item number A number used to identify a part on a layout or assembly drawing.

join To absorb the volume of interference between two objects to form a third object.

key A small removable part similar to a wedge that provides a positive means of transferring torque between a shaft and a hub.

key light A light that creates an object's main illumination, defines the dominant angle of the lighting, and is responsible for major highlights on objects in a scene.

keyframe A specific frame located at a specified time within an animation where an object's location, orientation, and scale are defined perfectly.

keyseat A rectangular groove cut in a shaft to position a key.

keyway A rectangular groove cut in a hub to position a key.

landscape The drawing orientation in which the horizontal size is larger than the vertical size.

lap joint A joint between two parts wherein the parts are overlapped.

laser scanning (three-dimensional) A process where cameras and lasers are used to digitize an object based on the principle of triangulation.

lathe A machine used to make axially symmetric parts or features using a material removal process known as turning.

latitude An imaginary line around the Earth's surface in which all of the points on the line are equidistant from the Equator.

layout drawing A formal drawing that shows a device in its assembled state with all of its parts identified.

lead The distance a screw thread advances axially in one full turn.

leader A thin, dark, solid line terminating with an arrowhead at one end and a dimension, note, or symbol at the other end.

left-handed system Any 3-D coordinate system that is defined by the left-hand rule.

level of detail The number of polygon mesh triangles used to define the surface shape of a 3-D model. For rendering speed, as a general case, objects close to the camera in a scene require a higher number of polygons to more accurately define their surfaces while more distant objects can be effectively rendered with fewer polygons.

life cycle The amount of time a product will be used before it is no longer effective.

line A spatial feature that has location and length but no area and is represented by a series of nodes, points, and arcs.

line chart A graph showing the relationship between two sets of data, where line segments are used to link the data to show trends in their changes.

list A boxed series of components, definitions, tips, etc.

location A dimension associated with the position of a feature on a part.

location grid An imaginary alphanumeric grid, similar to that of a street map, on a drawing that is used to specify area locations on the drawing.

longitude An imaginary north-south line on the Earth's surface that extends from the North Pole to the South Pole.

machine screw A threaded fastener wherein the threads are cut along the entire length of the cylindrical shaft. Machine screws can mate with a threaded hole or nut.

major diameter The largest diameter on an internal or external thread.

main assembly A completed device usually composed of multiple smaller parts and/or subassemblies.

main title block A bordered area of a drawing (and part of the drawing header) that contains important information about the identification, fabrication, history, and ownership of the item shown on the drawing.

manufacturing drawings Working drawings, often created by mechanical engineers, that are used to mass-produce products for consumers.

map A diagram of the location of events with geography.

map projection A systematic arrangement of parallels and meridians on a plane surface representing the geographic coordinate system.

mapping coordinates Also called UVW coordinates, mapping coordinates are special coordinate systems designed to correctly place and control the shape of external and procedurally generated images on the surfaces of 3-D models.

mass A property of an object's ability to resist a change in acceleration.

mass properties analysis A computer-generated document that gives the mechanical properties of a 3-D solid model.

master feature A feature or collection of features that is to be copied for placement at other locations in a model.

master model In a collection of similar parts, the model that includes all of the features that may appear in any of the other parts.

matt object An object with a combined material and alpha channel map.

maximum material condition The condition in which a feature of size contains the maximum amount of material within the stated limits of size.

measuring line (ML) In the construction of a perspective view, a vertical line used in conjunction with the elevation view to locate vertical points on the perspective drawing.

measuring wall In the construction of a perspective view, a line that extends from the object to the vanishing point to help establish the location of horizontal points on the drawing.

mechanical dissection The process of taking apart a device to determine the function of each part.

mechanical stress Developed force applied per unit area that tries to deform an object.

mental rotations The ability to mentally turn an object in space.

meridian A line of longitude through the North and South Poles that measure either E or W in a geographic coordinate system.

mesh The series of elements and nodal points on a finite element model.

Metric scale A device used to measure or draw lines in the metric system of units with drawings scales reported as ratios.

metrology The practice of measuring parts.

milling A process of removing material from a part using a rotating tool bit that can remove material in directions parallel or perpendicular to the tool bit's rotation axis.

milling machine A machine used to make parts through a material removal process known as milling.

minor diameter The smallest diameter on an internal or external thread.

mirrored feature A feature that is created as a mirror image of a master feature.

model A mathematical representation of an object or a device from which information about its function, appearance, or physical properties can be extracted.

model builders Engineers who make physical mockups of designs using modern rapid prototyping and CAM equipment.

model tree A list of all of the features of a solid model in the order in which they were created, providing a "history" of the sequence of feature creation.

mold A supported cavity shaped like a desired part into which molten material is poured or injected.

moment-of-inertia The measure of an object's ability to resist rotational acceleration about an axis.

morphological chart A chart used to generate ideas about the desirable qualities of a product and all of the possible options for achieving them.

motion blur The amount of movement of a high-speed object recorded as it moves through a single frame.

motion path Spline curves that serve as a trajectory for the motion of objects in animation.

multiple thread A thread made up of two or more continuous ridges side by side.

multiple views The presentation of an object using more than one image on the same drawing, each image representing a different orientation of the object.

multiview Refers to a drawing that contains more than one image of an object and whose adjacent images are generated from orthogonal viewing planes.

node A point at the beginning and end of a line feature or a point that defines a polygon feature.

normal surface A surface on an object being viewed that is parallel to one of the primary viewing planes.

note taker The person who records the actions discussed and taken at team meetings and then prepares the formal written notes for the meeting.

notes Additional information or instructions placed on a drawing that are not contained on the dimensions, tolerances, or header.

nut The threaded mate to a bolt used to hold two or more pieces of material together.

oblate ellipsoid An ellipsoid created by rotating an ellipse around its minor axis.

oblique axes A set of three coordinate axes that are portrayed on the paper as two perpendicular lines, with the third axis meeting them at an angle, typically 45 degrees.

oblique pictorial A sketch of an object that shows one face in the plane of the paper and the third dimension receding off at an angle relative to the face.

oblique surface A flat surface on an object being viewed that is neither parallel nor perpendicular to any of the primary views.

offset section The section view produced by a stepped cutting plane that is used to hypothetically cut an object completely into two pieces. Different portions of the plane are used to reveal the interior details of different features of interest.

one-off A one-of-a-kind engineering project for which no physical prototypes are created.

optimization Modification of shapes, sizes, and other variables to achieve the best performance based on predefined criteria.

organizational chart A chart representing the relationships of entities of an organization in terms of responsibility or authority.

orthogonal projection The process by which the image of an object is created on a viewing plane by rays from the object that are perpendicular to that plane.

outline assembly drawing *See* layout drawing.

parallel An imaginary line parallel to the equator that corresponds to a measurement of latitude either N or S in a geographic coordinate system.

parameters The attributes of features, such as dimensions, that can be modified.

parametric solid modeling A solid modeling system that allows the user to vary the dimensions and other parameters of the model.

parametric techniques Modeling techniques where all driven dimensions in algebraic expressions must be known for the value of the dependent variables to be calculated.

parent feature A feature used in the creation of another feature, which is called its child feature.

part An object expected to be delivered from a fabricator as a single unit with only its external dimensions and functional requirements specified.

part name A very short descriptive title given to a part, subassembly, or device.

part number Within a company, a string of alphanumeric characters used to identify a part, a subassembly, an assembly, or a device.

particle system Specialized software modules used to generate, control, and animate very large numbers of small objects involved in complex events.

parts list *See* bill of materials.

path The specified curve on which a profile is placed to create a swept solid.

passive animation An animation in which the observer remains still while the action occurs around him or her.

patents A formal way to protect intellectual property rights for a new product.

pattern A master part from which molds can be made for casting final parts.

perspective drawing A drawing in which all three-dimensional axes on an object can be seen, with the scaling factor linearly increasing or decreasing in each direction. Usually one axis is shown as being vertical. This type of drawing generally offers the most realistic presentation of an object.

pictorial A drawing that shows the 3-D aspects and features of an object.

picture plane (PP) In the construction of a perspective view, the viewing plane through which the object is seen. The PP appears as a line (edge view of the viewing plane) in the plan view.

pie chart A circular chart that is divided into wedges like a pie, representing a piece of the whole.

pin A cylindrical (or slightly tapered) fastener typically used to maintain a desired position or orientation between parts.

pitch The distance from one point on a thread to the corresponding point on the adjacent thread as measured parallel to its axis.

pitch diameter The diameter of an imaginary cylinder that is halfway between the major and minor diameters of the screw thread.

pivot point An independent, movable coordinate system on an object that can be used for location, orientation, and scale transformations.

pixel The contraction for "picture element"; the smallest unit of information within a grid or raster data set.

plan and profile drawings Construction drawings typically used for roads or other linear entities that show the road from above as well as from the side, with the profile view usually drawn with an exaggerated vertical scale.

plan view In the construction of a perspective view, the object as viewed from the top, as if created by orthogonal projection.

plan views Drawings created from a viewpoint above the structure (top view).

planar coordinate system A 2-D measurement system that locates features on a plane based on their distance from an origin (0,0) along two perpendicular axes.

point A spatial feature that has only location, has neither length nor area, and is represented by a pair of xy coordinates.

point light A computer-generated light source, also called an omni light, that emits light rays and casts shadows uniformly in all directions. Also called an omnidirectional light.

point tracking A procedure by which successive vertices on an object are simultaneously located on a pictorial image and a multiview image of that object.

polygon A spatial feature that has location, area, and perimeter and is represented by a series of nodes, points, and arcs that must form a closed boundary.

portrait The drawing orientation in which the vertical size is larger than the horizontal size.

preferred configuration The drawing presentation of an object using its top, front, and right-side views.

primary modeling planes The planes representing the XY-, XZ-, and YZ-planes in a Cartesian coordinate system.

primitives The set of regular shapes, such as boxes, spheres, or cylinders that are used to build solid models with constructive solid geometry methods (CSG).

principal viewing planes The planes in space on which the top, bottom, front, back, and right and left side views are projected.

problem identification The first stage in the design process where the need for a product or a product modification is clearly defined.

procedural textures Texture mapping routines based on algorithms written into the rendering software that can generate a specialized colored pattern such as wood, water, a checker pattern, a tile pattern, stucco, and many others without reference to external image files.

process check A method for resolving differences and making adjustments in team performance.

process diagram An illustration that explains how system elements work and how interactions occur.

professional engineer (PE) An individual who has received an engineering degree, who has worked under the supervision of a PE for a number of years, and who has passed two examinations certifying knowledge of engineering practice.

profile A planar sketch that is used to create a solid.

profile surface A surface on an object being viewed that is parallel to a side viewing plane.

profile views Views of a structure that show horizontal surfaces in edge view (side or front views).

project In engineering, a collection of tasks that must be performed to create, operate, or retire a system or device.

projection ray A line perpendicular to the projection plane. It transfers the 2-D shape from the object to an adjacent view. Projection rays are drawn lightly or are not shown at all on a finished drawing.

prototype The initial creation of a product for testing and analysis before it is mass-produced.

protrusion A feature created by the addition of solid volume to a model.

qualitative data Information collected using words and ideas.

quantitative data Numeric information.

quantity per machine (Q/M) The number of times a part is required to build its next highest assembly.

radius-of-gyration The distance from an axis where all of the mass can be concentrated and still produce the same moment-of-inertia.

rapid prototyping Various methods for creating a part quickly by selective hardening of a powder or liquid raw material at room temperature.

raster data model A representation of the geographic location as a surface divided into a regular grid of cells or pixels.

raytraced shadows Shadows calculated by a process called raytracing, which traces the path that a ray of light would take from the light source to illuminate or shade each point on an object.

raytracing A method of rendering that builds an image by tracing rays from the observer, bouncing them off the surfaces of objects in the scene, and tracing them back to the light sources that illuminate the scene.

reaming A process for creating a hole with a very accurate final diameter using an accurately made cylindrical cutting tool similar to a drill bit to remove final bits of material after a smaller initial hole is created.

rebars Steel bars added to concrete for reinforcement or for temperature control.

receding dimension The portion of the object that appears to go back from the plane of the paper in an oblique pictorial.

record A set of related data fields, often a row in a database, containing the attribute values for a single feature.

reference dimensions Unneeded dimensions shown for the convenience of the reader used to show overall dimensions that could be extracted from other dimensions on the part or from other drawings.

reference line Edges of the glass box or the intersection of the perpendicular planes. The reference line is drawn only when needed to aid in constructing additional views. The reference line should be labeled in constructing auxiliary views to show its association between the planes it is representing; for example, H/F for the hinged line between the frontal and horizontal planes. A reference line is also referred to as a fold line or a hinged line.

reflection The process of obtaining a mirror image of an object from a plane of reflection.

reflection mapping Mapping that allows the use of grayscale values in an image file to create the illusion of a reflection on the surface of a part. White creates reflective highlights, while black is transparent to the underlying color of the surface.

regeneration The process of updating the profile or part to show its new shape after constraints are added or changed.

related views Views adjacent to the same view that share a common dimension that must be transferred in creating auxiliary views.

removed section The section view produced when a cutting plane is used to hypothetically remove an infinitesimally thin slice of an object for viewing.

rendering The process where a software program uses all of the 3-D geometric object and lighting data to calculate and display a finished image of a 3-D scene in a 2-D viewport.

retaining rings Precision-engineered fasteners that provide removable shoulders for positioning or limiting movement in an assembly.

reverse engineering A systematic methodology for analyzing the design of an existing device.

revolved section The section view produced when a cutting plane is used to hypothetically create an infinitesimally thin slice, which is rotated 90 degrees for viewing, on an object.

revolved solid A solid formed when a profile curve is rotated about an axis.

ribs Constant thickness protrusions that extend from the surface of a part and are used to strengthen or stiffen the part.

right-hand rule Used to define a 3-D coordinate system whereby by pointing the fingers of the right hand down the x-axis and curling them in the direction of the y-axis, the thumb will point down the z-axis.

right-handed system Any 3-D coordinate system that is defined by the right-hand-rule.

rivet A cylindrical pin with heads at both ends, one head being formed during the assembly process, forming a permanent fastener often used to hold sheet metal together.

rolling A process for creating long bars with flat, round, or rectangular cross sections by squeezing solid raw material between large rollers. This can be done when the material is in a hot, soft state (hot rolling) or when the material is near room temperature (cold rolling).

root The bottom surface or point of a screw thread.

rounds Smooth radius transitions of external edges created by two intersecting surfaces and tangent to both intersecting surfaces.

sand casting A casting process where the mold is made of sand and binder material hardened around a master pattern that is subsequently removed to form the cavity. *See* casting.

sawing A cutting process that uses a multitoothed blade that moves rapidly across and then through the part.

scatter plot A graph using a pattern of dots showing the relationship between two sets of data.

schedule of materials A list of the materials, such as doors and windows, necessary for a construction project.

schematic diagram A diagram explaining how components work together, what the measurements are, how components are set up, or how pieces are connected.

screw thread A helix or conical spiral formed on the external surface of a shaft or on the internal surface of a cylindrical hole.

scripting A programming capability that allows a user to access and write code at or near the source code level of the software.

secondary title block An additional bordered area of a drawing (and part of the drawing header) that contains important information about the identification, fabrication, and history of the item shown on the drawing.

section lines Shading used to indicate newly formed or cut surfaces that result when an object is hypothetically cut.

section view A general term for any view that presents an object that has been hypothetically cut to reveal the interior details of its features, with the cut surfaces perpendicular to the viewing direction and filled with section lines for improved presentation.

sectioned assembly drawing A formal drawing, usually in pictorial form, that shows the device in its assembled form but with sections removed from obscuring parts to reveal formerly hidden parts.

selective laser sintering A process where a high-powered laser is used to selectively melt together the particles on a bed of powdered metal to form the shape of a desired part.

self-tapping screw A fastener that creates its own mating thread.

sequence diagram A group of diagrams that includes process diagrams, timelines, and step-by-step diagrams.

set of construction plans A collection of drawings, not necessarily all of them plan views, needed to construct a building or infrastructure project.

set screw A small screw used to prevent parts from moving due to vibration or rotation, such as to hold a hub on a shaft.

shading Marks added to surfaces and features of a sketch to highlight 3-D effects.

shading algorithms Algorithms designed to deal with the diffuse and specular light transmission on the surface of an object.

shelling Removing most of the interior volume of a solid model, leaving a relatively thin wall of material that closely conforms to the outer surfaces of the original model.

sidebar Small infographics used within a body of text that are subdivided into briefs, lists, and bio profiles.

single thread A thread that is formed as one continuous ridge.

sintering A process where a part is formed by placing powdered metal into a mold and then applying heat and pressure to fuse the powder into a single solid shape.

site plan A plan view showing the construction site for an infrastructure project.

site survey Data regarding the existing topography and structures gathered during the preliminary design stages by trained surveying crews.

six standard views (or six principal views) The drawing presentation of an object using the views produced by the glass box (i.e., the top, front, bottom, rear, left-side, and right-side views).

size The general term for the size of a feature, such as a hole, cylinder, or set of opposed parallel surfaces.

sketches Collections of 2-D entities.

sketching editor A software tool used to create and edit sketches.

sketching plane A plane where 2-D sketches and profiles can be created.

slope The rate of change of elevation (rise) over a specified distance (run). Measured in percent or degrees.

solid model A mathematical representation of a physical object that includes the surfaces and the interior material, usually including a computer-based simulation that produces a visual display of an object as if it existed in three dimensions.

solid modeling Three-dimensional modeling of parts and assemblies originally developed for mechanical engineering use but presently used in all engineering disciplines.

spatial data A formalized schema for representing data that has both geographic location and descriptive information.

spatial orientation The ability of a person to mentally determine his own location and orientation within a given environment.

spatial perception The ability to identify horizontal and vertical directions.

spatial relations The ability to visualize the relationship between two objects in space, i.e., overlapping or nonoverlapping.

spatial visualization The ability to mentally transform (rotate, translate, or mirror) or to mentally alter (twist, fold, or invert) 2-D figures and/or 3-D objects.

specifications (specs) The written instructions that accompany a set of construction plans used to build an infrastructure project.

spheroid *See* oblate ellipsoid.

spindle That part of a production cutting machine that spins rapidly, usually holding a cutting tool or a workpiece.

splines Polynomial curves that pass through multiple data points.

split line The location where a mold can be disassembled for removal of a part once the molten raw material inside has solidified.

spotlight A computer-generated light that simulates light being emitted from a point in space through a cone or beam, with the angle and direction of light controlled by the user.

spring pin A hollow pin that is manufactured by cold-forming strip metal in a progressive roll-forming operation. Spring pins are slightly larger in diameter than the hole into which they are inserted and must be radially compressed for assembly.

sprue Bits of material that are left on the part from a casting or molding operation and found at the ports where the molten material is injected into the mold or at the ports where air is allowed to escape.

stage That part of a machine that secures and slowly moves a cutting tool or workpiece in one or more directions.

stamping A process for cutting and shaping sheet metal by shearing and bending it inside forms with closely fitting cutouts and protrusions.

standard commercial shape A common shape for raw material as would be delivered from a material manufacturer.

State Plane Coordinate System The planar coordinate system developed in the 1930s for each state to permanently record the locations of the original land survey monuments in the United States.

station point (SP) In the construction of a perspective view, the theoretical location of the observer who looks at the object through the picture plane.

statistical tolerancing A way to assign tolerances based on sound statistical practices rather than conventional tolerancing practices.

step segment On a stepped cutting plane for an offset section view, that portion of the plane that connects the cutting segments and is usually perpendicular to them but does not intersect any interior features.

step-by-step diagram An illustration that visually explains a complex process; it is a type of a sequence diagram.

stereolithography A process for creating solid parts from a liquid resin by selectively focusing heat or ultraviolet light into a pool of the resin, causing it to harden and cure in the selected areas.

storyboard A sequential set of keyframe sketches or drawings, including brief descriptions, indications of object and camera movement, lighting, proposed frame numbers, and timelines sufficient to produce a complete animation project.

stud A fastener that is a steel rod with threads at both ends.

subassembly A logical grouping of assembly instances that is treated as a single entity within the overall assembly model.

subassemblies Collections of parts that have been put together for the purpose of installing the collections as single units into larger assemblies.

successive cuts A method of forming an object with a complex shape by starting with a basic shape and removing parts of it through subtraction of other basic shapes.

suppressed Refers to the option for not displaying a selected feature.

surface area The total area of the surfaces that bound an object.

surface model A CAD-generated model created to show a part as a collection of intersecting surfaces that bound a solid.

surface modeling The technique of creating a 3-D computer model to show a part or an object as a collection of intersecting surfaces that bound the part's solid shape.

surface normal A vector that is perpendicular to each polygon contained in a polygon mesh model.

surface tracking A procedure by which successive surfaces on an object are simultaneously located on a pictorial image and a multiview image of that object.

sustainable design A paradigm for making design decisions based on environmental considerations and life cycle analysis.

swept feature A solid that is bound by the surfaces swept out in space as a profile is pulled along a path.

symmetry The characteristic of an object in which one half of the object is a mirror image of the other half.

system A collection of parts, assemblies, structures, and processes that work together to perform one or more prescribed functions.

table Data organized in columns and rows.

tangent edge The intersection line between two surfaces that are tangent to each other.

tap The machine tool used to form an interior thread. Tapping is the process of making an internal thread.

tap drill A drill used to make a hole in material before the internal threads are cut.

tapped hole A hole that has screw threads inside it.

task credit matrix A table that lists all team members and their efforts on project tasks.

team contract The rules under which a team agrees to operate (also known as a code of conduct, an agreement to cooperate, or rules of engagement).

team leader The person who calls the meetings, sets the agenda, and maintains the focus of team meetings.

team roles The roles that team members fill to ensure maximum effectiveness for a team.

technical diagram A diagram depicting a technical illustration's measurements, movement, dissection, or relationship of parts.

telephoto As seen through a camera lens with a focal length longer than 80 degrees, creating a narrow field of view and resulting in a flattened perspective.

texture mapping The technique of adding variation and detail to a surface that goes beyond the level of detail modeled in the geometry of an object.

thematic layer Features of one type that are generally placed together in a single georeferenced data layer.

thematic layer overlay The process of combining spatial information from two thematic layers.

third-angle projection The process of creating a view of an object by imprinting its image, using orthogonal projection, on translucent surface in front of that object.

thread note Information on a drawing that clearly and completely identifies a thread.

thread series The number of threads per inch on a standard thread.

3-D coordinate system A set of three mutually perpendicular axes used to define 3-D space.

3-D printing A process for creating solid objects from a powder material by spraying a controlled stream of a binding fluid into a bed of that powder, thus fusing the powder in the selected areas.

three-axis mill A milling machine whose spindle, which holds the rotating cutting tool, can be oriented along any one of three Cartesian axes.

three-dimensional (3-D) modeling Mathematical modeling where the appearance, volumetric, and inertial properties of parts, assemblies, or structures are created with the assistance of computers and display devices.

through hole A hole that extends all the way through a part.

tick mark A short dash used in sketching to locate points on the paper.

timekeeper The person who keeps track of the meeting agenda, keeping the team on track to complete all necessary items within the alotted time frame.

timeline A specific type of sequence diagram used to highlight significant moments in history.

title block Usually the main title block, which is a bordered area of the drawing (and part of the drawing header) that contains important information about the identification, fabrication, history, and ownership of the item shown on the drawing.

tolerance The total amount a specific dimension is permitted to vary. It is the difference between the upper and lower limits of the dimension.

tool bit A fixed or moving replaceable cutting implement with one or more sharpened edges used to remove material from a part.

tool runout The distance a tool may go beyond the required full thread length.

tooling Tools and fixtures used to hold, align, create, or transport a part during its production.

top-down modeling The process of establishing the assembly and hierarchy before individual components are created.

trail Dashed lines on an assembly drawing that show how various parts or subassemblies are inserted to create a larger assembly.

trajectory *See* path.

transparency/opacity mapping A technique used to create areas of differing transparency on a surface or an object.

trimetric drawing An axonometric drawing in which the scaling factor is different for all three axes.

true shape (of a plane) The actual shape and size of a plane surface as seen in a view that is parallel to the surface in question.

two-dimensional (2-D) drawing Mathematical modeling or drawing where the appearance of parts, assemblies, or structures are represented by a collection of two-dimensional geometric shapes.

tumbling A process for removing sharp external edges and extraneous bits of material from a part by surrounding it in a pool of fine abrasive pellets and then shaking the combination.

turning A process for making axially symmetric parts or features by rotating the part on a spindle and applying a cutting to the part.

undercut feature A concave feature in which the removed material expands outward anywhere along its depth.

union A topological overlay of two or more polygon spatial datasets that preserves the features that fall within the spatial extent of either input dataset; that is, all features from both datasets are retained and extracted into a new polygon dataset.

Universal Transverse Mercator (UTM) The planar coordinate system that divides the earth's surface between 84° N and 80° S into 60 zones, each 6° longitude wide.

unsuppressed Refers to the option for displaying a selected feature.

US sheet sizes The accepted paper dimensions used in the United States when drawings are created or printed to their intended size.

vanishing point (VP) In the construction of a perspective view, the point on the horizon where all parallel lines in a single direction converge.

variational techniques Modeling techniques in which algebraic expressions or equations that express relationships between a number of variables and constants, any one of which can be calculated when all of the others are known.

vector data model A data model that uses nodes and their associated geographic coordinates to construct and define spatial features.

Venn diagram A type of business diagram that shows the mathematical or logical relationships and overlapping connections between different groups or sets of information.

vertex A point that is used to define the endpoint of an entity such as a line segment or the intersection of two geometric entities.

video compression One of a number of algorithms designed to reduce the size and storage requirements of video content.

viewing direction The direction indicated by arrows on the cutting plane line from the eye to the object of interest that corresponds to the tail and point of the arrow, respectively.

viewing plane A hypothetical plane between an object and its viewer onto which the image of the object, as seen by the viewer, is imprinted.

visual storytelling diagram An illustration that displays empirical data or clarification of ideas.

visual thinking A method for creative thinking, usually through sketching, where visual feedback assists in the development of creative ideas.

visualization The ability to create and manipulate mental images of devices or processes.

volume The quantity of space enclosed within an object's boundary surfaces.

volume of interference The volume that is common between two overlapping objects.

wall sections Sectional views of walls from foundation to roof for a construction project.

washer A flat disk with a center hole to allow a fastener to pass through it.

webs Small, thin protrusions that connect two or more thicker regions on a part.

weighted decision table A matrix used to weigh design options to determine the best possible design characteristics.

welding A method for joining two or more separate parts by applying heat to the edges where they meet and melting the edges together along with a filler of essentially the same material composition as the parts.

wide-angle As seen through a camera lens with a focal length shorter than 30 degrees, creating a wide field of view and resulting in a distorted and exaggerated perspective.

wire drawing The process of reducing the diameter of a solid wire by pulling it through a nozzle with a reducing aperture.

wireframe models CAD models created using lines, arcs, and other 2-D entities to represent the edges of the part; surfaces or solid volumes are not defined.

working drawings A collection of all drawings needed to fabricate and put together a device or structure.

workpiece A common name for a part while it is still in the fabrication process, that is, before it is a finished part.

z-buffer rendering A scene-rendering technique that uses visible-surface determination in which each pixel records (in addition to color) its distance from the camera, its angle, light source orientation, and other information defining the visible structure of the scene.

z-value The value for a given surface location that represents an attribute other than position. In an elevation or terrain model, the z-value represents elevation.

index

G

H

I

J

K

L

M

N

O

P

R

S

T

U

V

W

X

Y

Z